T0138799

Pocket Guide to Bacterial Infections

POCKET GUIDES TO
BIOMEDICAL SCIENCES

Series Editor
Lijuan Yuan
Virginia Polytechnic Institute and State University

https://www.crcpress.com/Pocket-Guides-to-Biomedical-Sciences/bookseries/CRCPOCGUITOB

The *Pocket Guides to Biomedical Sciences* series is designed to provide a concise, state-of-the-art, and authoritative coverage on topics that are of interest to undergraduate and graduate students of biomedical majors, health professionals with limited time to conduct their own searches, and the general public who are seeking for reliable, trustworthy information in biomedical fields.

Pocket Guide to Bacterial Infections

Edited by
K. Balamurugan

Co-editor
U. Prithika

CRC Press is an imprint of the
Taylor & Francis Group, an **informa** business

CRC Press
Taylor & Francis Group
6000 Broken Sound Parkway NW, Suite 300
Boca Raton, FL 33487-2742

Printed on acid-free paper

International Standard Book Number-13: 978-1-138-05489-9 (Paperback)
International Standard Book Number-13: 978-1-138-05491-2 (Hardback)

Library of Congress Cataloging-in-Publication Data

Names: Balamurugan, K. (Krishnaswamy), editor.
Title: Pocket guide to bacterial infections / editor(s), K. Balamurugan.
Other titles: Pocket guides to biomedical sciences.
Description: Boca Raton : Taylor & Francis, 2019. | Series: Pocket guides to biomedical sciences | Includes bibliographical references.
Identifiers: LCCN 2018043418| ISBN 9781138054899 (paperback : alk. paper) | ISBN 9781138054912 (hardback : alk. paper) | ISBN 9781315166377 (general) | ISBN 9781351679817 (pdf) | ISBN 9781351679800 (epub) | ISBN 9781351679794 (mobi/kindle)
Subjects: | MESH: Bacterial Infections--etiology | Bacterial Infections--pathology | Bacteria--pathogenicity | Bacteria--cytology
Classification: LCC RC115 | NLM WC 200 | DDC 616.9/2--dc23
LC record available at https://lccn.loc.gov/2018043418

Visit the Taylor & Francis Web site at
http://www.taylorandfrancis.com

and the CRC Press Web site at
http://www.crcpress.com

Contents

Series Preface

This volume, *Pocket Guide to Bacterial Infections*, is a collection of concise descriptions of our current knowledge on bacterial infections, including the pathogenesis, clinical features, diagnosis, epidemiology, prophylaxis and therapeutics of the most important bacterial pathogens that infect the gastrointestinal tract, respiratory tract, urinary tract, oral cavity, eyes, skin, blood, and the neonatal brain. In addition, several emerging areas of study, such as the role of the microbiome in bacterial pathogenesis, the role of bacterial infection in atherosclerosis, and the development of lactic acid bacteria as probiotics are also included. The book chapters are organized based on the different host organ systems that bacterial pathogens affect, and each chapter is self-contained. With contributions from 29 experts in the field, this book serves as a remarkable and succinct state-of-the-art teaching and research tool, as well as a practical resource for medical practitioners, microbiologists, pathologists, and graduate students in the biomedical fields of study.

Lijuan Yuan
Blacksburg, Virginia, U.S.A

Editors

K. Balamurugan is endowed professor at the Department of Biotechnology, Alagappa University, Karaikudi, India. His research aims at understanding host-pathogen interaction using *Caenorhabditis elegans* as an *in vivo* model system, a popular and efficient alternative to animal models. The extensive areas of his research using *C. elegans* as a model system include genomic and proteomic level analysis of various pathogens mediated immune (both cellular and neuronal) response, characterization of anti-aging genes, identification of post-translational modifications during host-pathogen interactions, and investigation of wound healing. He has published widely, with 83 peer-reviewed journal publications, 11 patents, numerous book chapters, and hundreds of conference proceedings in his field of research including in the area of host-pathogen interaction using *C. elegans* as the model.

U. Prithika currently focuses her research on exploring and enhancing the stability of different probiotic bacteria present in fermented foods. She is also investigating the effects of probiotics on irritable bowel syndrome. Her previous research experience, which includes working under the mentorship of Prof. K. Balamurugan on model system *Caenorhabditis elegans*, has built a strong foundation in molecular and immunobiology. She has been actively involved in understanding the role of specific genes/proteins during heat-shock and immune-compromised host conditions.

Contributors

B. Agieshkumar
Senior Scientist
Central Inter Disciplinary Research Facility (CIDRF)
Sri Balaji Vidyapeeth Deemed University
Puducherry, India

Charles Solomon Akino Mercy
Medical Microbiology Laboratory (MML)
Department of Microbiology
Centre of Excellence in Life Sciences
Bharathidasan University
Tiruchirappalli, India

Kumarasamy Anbarasu
Department of Marine Biotechnology
Bharathidasan University
Tiruchirappalli, India

Kannan Balaji
Department of Biotechnology
Alagappa University
Karaikudi, India

K. Balamurugan
Department of Biotechnology
Alagappa University
Karaikudi, India

Boopathi Balasubramaniam
Department of Biotechnology
Alagappa University
Karaikudi, India

José Manuel Cabeda
Health Sciences Faculty
Fernando Pessoa University
and
Fernando Pessoa Energy
Environment and Health Research Unit (FP-ENAS)
Porto, Portugal

Gnanasekaran JebaMercy
Department of Biotechnology
Alagappa University
Karaikudi, India

Christopher Karen
Behavioural Neuroscience Laboratory
Department of Animal Science
School of Life Sciences
Bharathidasan University
Tiruchirappalli, India

Shunmugiah Karutha Pandian
Department of Biotechnology
Alagappa University
Karaikudi, India

Emil Kozarov
Center for Molecular Microbiology
Department of Oral Biology
University of Florida Health Science Center
School of Dentistry
Gainesville, Florida

M. Krishna Raja
Department of Biochemistry
JIPMER
Puducherry, India

P. S. Manoharan
Department of Prosthodontics and Crown & Bridge
Indira Gandhi Institute of Dental Sciences
Puducherry, India

Kalimuthusamy Natarajaseenivasan
Medical Microbiology Laboratory (MML)
Department of Microbiology
Centre of Excellence in Life Sciences
Bharathidasan University
Tiruchirappalli, India

and

Neuroscience
Lewis Katz School of Medicine
Temple University
Philadelphia, Pennsylvania

Galina Novik
Belarusian Collection of Microorganisms
Institute of Microbiology
National Academy of Sciences of Belarus
Minsk, Republic of Belarus

Chidambaram Prahalathan
Department of Biochemistry
Bharathidasan University
Tiruchirappalli, India

U. Prithika
Department of Biotechnology
Alagappa University
Karaikudi, India

Arumugam Priya
Department of Biotechnology
Alagappa University
Karaikudi, India

Ann Progulske-Fox
Center for Molecular Microbiology
Department of Oral Biology
University of Florida Health Science Center
School of Dentistry
Gainesville, Florida

Koilmani Emmanuvel Rajan
Behavioural Neuroscience Laboratory
Department of Animal Science
School of Life Sciences
Bharathidasan University
Tiruchirappalli, India

Praveen Rajesh
Department of Conservative Dentistry and Endodontics
Indira Gandhi Institute of Dental Sciences
Puducherry, India

Victoria Savich
Belarusian Collection of Microorganisms
Institute of Microbiology
National Academy of Sciences of Belarus
Minsk, Republic of Belarus

Duraisamy Senbagam
Department of Marine Biotechnology
Bharathidasan University
Tiruchirappalli, India

Balakrishnan Senthilkumar
Department of Medical Microbiology
College of Health and Medical Sciences
Haramaya University
Harar, Ethiopia

Thirukannamangai Krishnan Swetha
Department of Biotechnology
Alagappa University
Karaikudi, India

Ana Moura Teles
Health Sciences Faculty
Fernando Pessoa University
and
Fernando Pessoa Energy
Environment and Health Research Unit (FP-ENAS)
Porto, Portugal

B. Vinoth
Associate professor
Department of Medical Gastroenterology
Aarupadai Veedu Medical College
Puducherry, India

1

Bacteria Causing Gastrointestinal Infections

Overview of Types of Bacteria Causing Gastrointestinal Infections, Mode of Infection, Challenges in Diagnostic Methods, and Treatment

B. Vinoth, M. Krishna Raja, and B. Agieshkumar

Contents

Gastrointestinal (GI) infections are bacterial, parasitic, or viral infections that cause gastroenteritis, an inflammation of the GI tract. A wide range of GI diseases are caused by bacteria, when bacteria or its associated toxins are ingested through contaminated food or water. Though most bacterial GI illness is short lived and self-limiting, it can be fatal, if not treated properly.

This remains a common problem in both primary care and emergency centers in the developing world. Some of the major bacteria that causes GI illness include *Escherichia coli*, *Salmonella* spp., *Shigella*, *Campylobacter* spp., *Clostridium* spp., *Yersinia*, and *Bacillus cereus*. This chapter discusses different types of causative bacteria, mode of infection, and mechanism that contributes to pathophysiology of the disease. This review also adds a note on various methods and challenges in diagnosis and treatment of bacteria causing GI infections.

1.1 Bacterial infections of the gut

1.1.1 Introduction

GI infection caused by bacteria is common throughout the global population, causing significant morbidity and mortality (Ternhag et al. 2008). The disease burden is to such an extent that it stands second in mortality of children younger than 5 years of age. Epidemic outbreak of some bacteria often resulted in high mortality and liability to the community. For instance, around 0.7 million children younger than age of 5 lost their lives among the 1.7 billion diarrheal episodes in the year 2010–2011 (Walker et al. 2013). A changing burden of GI infections around the world is being reported with several newly recognized organisms causing GI disease in the last three decades (Eslick 2010).

The gastrointestinal tract (GIT) is the home for several commensal bacteria living in harmony with the host. This is however being disrupted by some organisms like *Escherichia coli*, *Clostridium perfringens*, and *Clostridium difficile*, which can cause infection of the GIT. Apart from these, many pathogenic bacteria can enter the GIT through contaminated food (i.e., food poisoning) or through some other route (i.e., tuberculosis) and cause infection. Foodborne diarrheal illness is one of the leading causes of acute diarrheal illness worldwide and commonly seen even in developed countries. Bacteria causing GIT infections hence can be grouped into two types (i) primary enteropathogens, which includes bacteria that are gram-negative with specific affinity to enterocytes and has specific pathogenic factors against intestinal epithelial cells. (ii) The second group are those which cause gastroenteritis as a part of foodborne illness with many of them having affinity and pathogenic targets toward other organ systems in addition to GIT. The GI infections manifest either primarily with symptoms like nausea, vomiting, abdominal pain, diarrhea, and fever or along with systemic symptoms like arthritis, hepatitis, renal failure, and so on, depending upon the severity (Shaheen et al. 2006; Goh 2007).

Based on the pathogenesis, these organisms produce either watery or bloody diarrhea. Most of the infections are mild and are associated with only watery diarrhea; however, few organisms are known to produce bloody diarrhea like, *Shigella*, enterohemorrhagic *E. coli* (EHEC), *Salmonella*, *Campylobacter*, and *Yersinia*. Gut infections are mostly foodborne, and the leading cause is *Salmonella*, followed by *Campylobacter*, *Shigella*, and enterotoxigenic *E. coli* (ETEC) according to an US data in 2014. The type of food associated with these infections is also different for different organisms. The frequencies of this infection differ between countries and depend on the health status of the community, weather conditions, food habits, and so on. Despite these differences, it is practically difficult to diagnose a specific organism as a cause primarily based on clinical grounds. The usual way of diagnosing a specific infection is only by isolating the organism in most of these cases. But isolation of certain organisms in this group is difficult and involves either a transport medium or a highly selective medium or application of special techniques. In addition certain organisms like *E. coli* requires molecular methods for knowing the subtypes which are not routinely available and are restricted to reference laboratories. Hence, evaluating patients with diarrheal illness is cumbersome.

Most of these illness are self-limiting lasting only for few days and do not require antibiotics. Only severe and complicated cases require antibiotics, and in fact in some cases, antibiotics might even be harmful as with EHEC. Bacteria developing resistance toward antibiotics makes it mandatory to know the sensitivity pattern before selecting an antibiotic. This makes the isolation of organism even more important. As mentioned before, there has also been a changing trend of bacteria with falling incidence of well-known older infections (i.e., EHEC O157, *Vibrio cholerae*) and increasing incidence of other organisms (i.e., *Campylobacter, Yersinia*) with few maintaining a constant trend (i.e., *Salmonella*), and finally with lots of emerging organisms (i.e., *Arcobacteria, Edwardsiella tarda, Plesiomonas shigelloides, Aeromonas hydrophila, Listeria monocytogenes*, and *Laribacter hongkongensis*). This chapter discusses the general biology and transmission means of more common and emerging bacterial pathogens causing GI infection and describes various diagnostic and treatment methods. The order of the pathogens represents their contribution toward disease severity.

1.2 *Escherichia coli*

E. coli is the foremost gram-negative facultative anaerobic bacilli residing in the intestines of humans and other mammals. Most of the strains remain harmless, but few cause mild to severe diarrheal illness. Six major types of *E. coli* based on their pathogenesis are ETEC, enteropathogenic *E. coli* (EPEC), enteroinvasive *E. coli* (EIEC), EHEC, or Shiga toxin producing *E. coli* (STEC), enteroaggregative *E. coli* (EAEC), and diffusely enteroadherent *E. coli* (DAEC). A comprehensive overview of these groups of *E. coli* is presented in Table 1.1.

Diagnosing the causative agent in individuals is challenging (other than conditions like epidemic outbreaks) because the disease is mostly self-limiting within days and the procedures for detecting the toxins, pathogenic genes, and cytotoxicity through immunoassays, DNA hybridization, cell culture, and adherence assays are done mostly for severe cases.

1.2.1 Enteropathogenic *E. coli* (EPEC)

EPEC is one of the most important causes of acute diarrheal illness in infants and children worldwide. EPEC was found to be a major cause of pediatric diarrhea both in community and hospitalized patients (Lanata et al. 2002). The bacteria spread through fecal-oral route, either directly or indirectly, by consuming contaminated food or water. The infective dose is low and is less than 10,000 colony-forming units (cfu).

Table 1.1 Comprehensive Overview of the Clinical Profiles of Each type of *Escherichia coli*

Types	Site Involved	Pathogenesis	Population Affected	Type of Diarrhea	Antibiotic
EPEC	Small intestine	Attaching effacing lesions I: Typical – Bundle forming pilus, attachment effacement lesion II: Atypical – atypical adherence pattern	Children	Osmotic (watery, persistent diarrhea in few)	Indicated in traveler's diarrhea
ETEC	Small intestine	LT, ST, and cell adherence	Children, travelers	Secretory (watery)	Not indicated
EIEC	Large intestine	Cell invasion	Children and adults	Inflammatory (watery, bloody)	Not indicated
EHEC/STEC a. O157:H7 b. Non O157 c. O154:H4	Large intestine	ST1 and ST2	Children and adults	Inflammatory (watery, bloody)	Not indicated Might precipitate HUS
EAEC	Large intestine	Aggregative adherence to Hep 2 cells Stacked brick appearance	Children	Inflammatory (watery, persistent diarrhea in few)	Indicated
DAEC	Large intestine	Diffuse adherence to Hep 2 cells	Children	Inflammatory (watery, persistent diarrhea in few)	Not indicated

DAEC, diffusely enteroadherent *E. coli*; EAEC, enteroaggregative *E. coli*; EHEC, enterohemorrhagic *E. coli*; EIEC, enteroinvasive *E. coli*; EPEC, enteropathogenic *Escherichia coli*; ETEC, enterotoxigenic *E. coli*; HUS, hemolytic uremic syndrome; LT, labile toxin; ST, stable toxin; STEC, Shiga toxin producing *E. coli*.

1.2.1.1 Pathogenesis – EPEC possesses a localized adherence pattern with the help of their fimbriae. This property is governed by a unique pathogenicity associated island (PAI) termed as locus for enterocytes effacement (LEE). Activation of LEE during contact with enterocytes forms type III secretary system (T3SS) through which bacterial effector molecules is injected into the enterocytes via a pilus-like structure. These effector molecules then initiate a cascade of events like disruption of gap junction, release of interleukin-8 (IL-8), adenosine and disruption of sodium chloride (NaCl)-mediated transport mechanisms leading to diarrhea. Based on the pathogenesis, they are classified into two groups namely typical EPEC (tEPEC) and atypical EPEC (aEPEC). tEPEC has classical pilus formation with attaching effacing lesions, whereas the aEPEC form has atypical adherence pattern. Atypical form appears to be more important and common pathogen than the typical form (Nguyen et al. 2006).

1.2.1.2 Clinical features – EPEC causes watery diarrhea with vomiting and dehydration in infants and children. The incubation period is usually 1–2 days. The diarrhea is often self-limiting, but EPEC are strongly associated with severe and persistent diarrhea in some cases.

1.2.1.3 Diagnosis – Typically targeting the virulence genes by polymerase chain reaction (PCR) could detect EPEC. DNA probing or adherence assays are required to diagnose these *E. coli* types, and it can be done only in labs that harbors molecular biological facilities.

1.2.1.4 Treatment – As most cases recover spontaneously without major complication, antibiotics are not routinely recommended, and the mainstay of management is supportive care in the form of rehydration therapy because it is extremely important in pediatric population. The role of antibiotics in persistent diarrhea is not clearly known (Ochoa and Contreras 2011). Effective breastfeeding have declined the incidence of EPEC and could be preventive against it.

1.2.2 Enterotoxigenic *E. coli* (ETEC)

These organisms were first described from Calcutta in India by De et al. in the year 1956 and were later reported from other areas of the world. In developing countries, they are one of the leading cause of diarrhea in children younger than 5 years of age, with an estimated annual incidence of 280 million cases with 380,000 deaths (Wennerås and Erling 2004; Ochoa and Contreras 2011). They were the most common cause of community-acquired diarrhea (14.1%) and was associated with 9.5% of hospital-acquired diarrhea according to data between 1990 and 2000 (Lanata et al. 2002). It is also the leading cause of traveler's diarrhea.

1.2.2.1 Pathogenesis – The infection is acquired by consuming contaminated food and the infective dose is 10^6 cfu. They reach the small intestine and colon and adhere to the enterocytes through adhesive fimbriae or pili antigens called colonization factor antigens (CFA) and are now renamed as coli surface antigens. There is no penetration of the enterocytes, and the diarrhea is mediated through the production of enterotoxins. ETEC produces two types of toxins namely a heat labile toxin (LT) and a heat stable toxin (ST). There is marked variation in expression of CFA, LT, and ST among ETEC isolated from various parts of the world. CFA/I (17%) expression is more commonly seen among the CFA types and LT expression (60%) is more common than ST.

The LT is structurally and functionally similar to the cholera toxin and is destroyed by heat and acid. It is a combination of A subunit and pentameric ring of 5B subunits. The B subunit binds to GM1 gangliosides of enterocytes and the A subunit causes G protein coupled activation of adenylate cyclase inside the enterocytes, leading to increased production of cyclic adenosine monophosphate (cAMP). These result in increased chloride secretion via cystic fibrosis transmembrane conductance regulator (CFTR) from the intestinal crypt cells of the small intestine resulting in voluminous diarrhea. The ST is not destroyed by heating even at 100°C, and their action is mediated through activation of guanylate kinase resulting in an increase of cyclic guanosine monophosphate (cGMP), leading to increased intestinal secretion from both small and large intestines.

1.2.2.2 Clinical features – ETEC causes large volume watery diarrhea, leading to severe dehydration. The incubation period is 1–2 days, and illness usually lasts for 1–5 days. Fever is rare; however, vomiting is seen in children. In patients with malnutrition, the illness can last up to 3 weeks. Dehydration is more common in adults than children. In children in developing countries, recurrent episodes of diarrhea can cause malnutrition and growth retardation.

1.2.2.3 Diagnosis – Diagnosing ETEC requires DNA hybridization study or an enzyme immune assay (EIA) or cell culture assay for detecting its virulence factor like CFA, LT, and ST.

1.2.2.4 Treatment – The mainstay of treatment is fluid resuscitation because patients can develop severe dehydration following large volume diarrhea. Antibiotics have shown to be effective in reducing the duration of illness in traveler's diarrhea. The antibiotics are chosen based on susceptibility pattern and the preferred drugs are fluoroquinolones, azithromycin, and rifaximin. Because of the disease burden, vaccination is being tried to prevent traveler's diarrhea. An oral inactivated whole-cell vaccine combined with B subunit of cholera toxin (Duckoral) is being used, with modest response of 67% reported from Bangladesh and 52% from Morocco (Clemens et al. 1988; Peltola et al. 1991).

1.2.3 Entero hemorrhagic *E. coli* (EHEC) (STEC/VTEC)

They were first reported from diarrheal outbreaks that occurred in Michigan and Ohio during 1982 (Riley et al. 1983). Among the dysentery cases in emergency departments in the United States, EHEC accounted for 2.6% of the cases (Talan et al. 2001). In the year 2011, a large outbreak of 3842 cases were reported in Germany with a new strain O154:H4. EHEC are associated with hemolytic uremic syndrome (HUS) and hemorrhagic colitis. Cattle, sheep, goat, dog, cat, and pig are the common animal reservoir for EHEC. Consumption of undercooked meat, unpasteurized milk, or open water contamination are some of the ways it is transmitted from animals. Person-to-person transmission is quite common in nursing homes, daycare centers, and so on, and precaution is required because even 100 cfu is enough to be an infective dose. EHEC is also known as Shiga toxin-producing *E. coli* (STEC) or verotoxin producing *E. coli* (VTEC) due to the production of Shiga toxin (Stx) or verotoxin. EHEC is classified as O157 and Non O157 group-based on the serotype. O154:H4 a newer serotype originated from enteroaggregative *E. coli* (EAEC) and has acquired the Shiga toxin gene (Stx2a). Centers for Disease Control and Prevention (CDC) reports in the year 2010 showed that the O157 contributed to 25 outbreaks, and 9 outbreaks were by Non O157 serotypes. ("FOOD Tool | CDC" 2017).

1.2.3.1 Pathogenesis – EHEC pathogenesis is mainly by their toxin production and its effects. Stx (AB5 toxin) has one A subunit and five B subunit The B subunit binds to the receptor in the enterocytes (globotriaosylceramide) and is then internalized along with the A subunit by endocytosis. Then the A subunit which has catalytic activity leads to cascade of events and causes cytotoxicity by inhibiting protein synthesis. The destruction of enterocytes impairs intestinal absorption contributing to diarrhea. Stx are particularly toxic to the endothelial cells, and in the submucosa, they cause microvasculature damage, leading to platelet aggregation and microvascular fibrin thrombi formation mimicking ischemic colitis. After entering the circulation, Stx can damage renal endothelial cells in a similar manner and can cause HUS similar to *Shigella* infection (Bennish et al. 2006; Popoff 2011). EHEC share similar pathogenic profile with EPEC by possessing LEE for adherence and also production of attachment effacement lesions.

1.2.3.2 Clinical features – The incubation period varies between 1 and 14 days, with watery diarrhea after about 1–3 days, later becoming bloody. In patients who are infected with O157, bloody diarrhea develops in 80% of the cases. The other symptoms include nausea, vomiting, and abdominal pain. The abdominal pain may be severe and can precede diarrhea. EHEC causes segmental colitis, and the most common part affected is ascending colon. The involved intestine shows characteristic thumbprint appearance on radiograph due to severe inflammation of mucosa with mucosal edema. Colonoscopy will reveal friable inflamed mucosa with patchy erythema, edema, and superficial ulcerations, mimicking ischemic colitis.

The most important clinical manifestation of EHEC is HUS developing in 6%–9% of cases according to different outbreaks (Bell et al. 1994; Boyce et al. 1995; Bender et al. 1997; Tarr et al. 2005). Children younger than 5 years of age are commonly affected (15%). It is the most common cause of HUS in pediatric population accounting for 90% of the cases (Ardissino et al. 2016). The highest incidence of HUS among EHEC infection was seen in the German epidemic of 2011 where 22% of cases developed HUS; interestingly, adults and females were more affected than children, and the strain was found to be O154:H4 (Frank et al. 2011). The German outbreak resulted from consumption of contaminated sprouts. The classical triad of HUS is microangiopathic hemolytic anemia, thrombocytopenia, and acute renal failure. The patients start developing features of HUS between 5 and 13 days after the onset of diarrhea, and the initial clue is thrombocytopenia that occurs by day 8 or 9 of the illness. The extraintestinal complications commonly occur in patients with HUS and include appendicitis, intussusceptions, intestinal perforation, pancreatitis, stroke, coma, seizures, intestinal strictures, and sudden death. Usually renal function improves following the hematological improvement, and patients recover in 1–2 weeks. The mortality in patients with HUS is about 5%. Some of the patients develop irritable bowel syndrome (IBS) following an episode of HUS.

1.2.3.3 Diagnosis – Bloody stool and elevated leucocytes in the stool should raise the suspicion of EHEC infection. However, the diarrhea is most often watery in the initial stages. The CDC recommends combination of a culture method for O157 and a nonculture toxin assay (i.e., EIA) for identifying Shiga toxin to diagnose EHEC infections. O157 strains do not ferment sorbitol, and hence, sorbitol MacConkey agar is used for screening this organism. Serotyping is done using O157:H7 antisera and simultaneously the colonies should be sent to research laboratories for confirmation. The yield in the culture is better when the stools are tested between 2 and 7 days; however, the fecal shedding might continue for 17 to 29 days. A positive EIA for the toxins in the absence of O157 culture should be confirmed by further isolation of the organism due to possibility of false-positive results. PCR-based Shiga toxin tests are also available for detecting these toxins.

1.2.3.4 Treatment – Duration of diarrhea is 3–8 days, and most recover with no major complications, except few who develops HUS. Many studies have shown that the antibiotics do not have any role on the duration of illness and does not reduce the incidence of HUS. Moreover, few studies have shown that antibiotic usage is associated with increased risk of developing HUS (Zhang et al. 2000; Mølbak et al. 2002; Safdar et al. 2002). Hence at present, antibiotics are not routinely recommended for patients with EHEC infections. The mainstay of management is supportive care with fluid resuscitation. EHEC infections are prevented by consuming properly cooked beef, avoiding unpasteurized milk, and maintaining good hygiene if there is possibility of contact with farm animals.

1.2.4 Enteroinvasive *E. coli* (EIEC)

EIEC are very similar to *Shigella* and possess the same toxin and virulence factors as that of *Shigella*. Both EIEC and *Shigella* appear to have evolved from nonpathogenic *E. coli* by acquisition of invasion plasmid (pINV) at different times. The pathogenesis and clinical features are similar to *Shigella* infection (Refer *Shigella*); however, the incidence is less when compared to *Shigella*. Children younger than 5 years of age are more commonly affected. They are transmitted through contaminated food and water and person-to-person spread also occurs. They are diagnosed by the presence of nonlactose fermenting colonies on MacConkey agar and later by nucleic acid hybridization technique and phenotypic assay for identifying the specific virulence genes and their cytotoxicity pattern. Antibiotics are not routinely recommended for EIEC infections.

1.2.5 Enteroaggregative *E. coli* (EAEC)

EAEC were first identified during the evaluation of diarrhea in Chilean children in 1987 when *E. coli* with a specific adherence pattern of stacked-brick appearance was found when compared to controls. Because of the aggregative clumping of the organism to the enterocytes, they are named enteroaggregative *E. coli*. EAEC causes acute and persistent diarrhea in children and adults in developing and developed countries. They are commonly associated with diarrhea in patients with HIV and also causes traveler's diarrhea. The major pathogenic factors are its characteristic adherence, production of enterotoxins and cytotoxins, and mucosal inflammation. EAEC adheres to the intestinal epithelial cells by aggregative adherence fimbriae (AAF) in a stacked-brick pattern. The toxins produced by EAEC are Shiga enterotoxin 1 (ShET1), enteroaggregative heat-stable toxin 1 (EAST1), and plasmid encoded toxin termed Pet. ShET1 and EAST1 causes increased conductance across enterocytes leading to ion secretion and subsequent diarrhea. Pet toxin is involved in dissolution of the cytoskeleton of the epithelial cell. EAEC causes mucosal inflammation by inducing the release of IL-8 from the epithelial cells. Diagnosis can be made only by epithelial cell adherence assay with Hep 2 cells to study the stacked-brick appearance or by DNA hybridization or PCR assays to detect specific target genes. The role of antibiotic in EAEC is not well evaluated because of the rarity of its diagnosis. However, antibiotics have been shown to be effective for traveler's diarrhea and in patients with HIV. The preferred antibiotics are ciprofloxacin, azithromycin, and rifaximin. Rifaximin are particularly useful in traveler's diarrhea. In developing countries, EAEC produces long-term sequelae like stunted growth and development of IBS.

1.2.6 Diffusely Adherent *E. coli* (DAEC)

DAEC are named because of their characteristic adherence pattern of diffusely adhering to the intestinal epithelial cells. The exact epidemiological pattern, pathogenesis, and their complete clinical profile are not clearly known and are still emerging. The major virulence factor of DAEC is Dr Family of fimbrial adhesins that are responsible for binding of these organisms to the epithelial cells. After adhesion, a part of the organism is internalized by microtubule-dependent pathway, and they lead to cascade of events resulting in a proinflammatory response. DAEC are associated with diarrhea in children (especially older than 2 years of age) in a few parts of both developed and developing countries. They usually cause nonbloody diarrhea. The diagnosis requires cell adhesion assay and DNA hybridization for detecting adherence pattern and virulence factor genes. The exact roles of antibiotics are not clearly known and are not routinely recommended.

1.3 *Shigella*

Shigella is a gram-negative aerobic bacilli belonging to Enterobacteriaceae family. Humans are the only known reservoir, despite which it is difficult to control because of the low infective dose (10–100 cfu), their ability to resist gastric acid, and because of increasing antibiotic resistance (DuPont et al. 1989). They are the leading cause of bloody diarrhea in the world. *Shigella* has four major species namely *S. dysenteriae*, *S. flexneri*, *S. boydii*, and *S. sonnei*. Each of the species, except *S. sonnei*, is further divided into various serotypes according to the on the O antigen. Earlier *S. dysenteriae* type I was the leading cause of bloody diarrhea in developing countries with substantial morbidity and mortality (5%–15%) (Khan et al. 1985), but later

S. flexneri appears to be more common. In an Indian study, the predominant species is *S. flexneri* (60%) followed by *S. sonnei* (24%), *S. dysenteriae* (10%), and *S. sonnei* (6%) (Pazhani et al. 2005). *S. sonnei* outbreaks are occasional, and it causes sporadic diarrhea. In contrast, the predominant species in the United States is *S. Sonnei* (80%) followed by *S. flexneri* (15%–20%), *S. boydii* (1%–2%), and *S. dysenteriae* (<1%) (Scallan et al. 2011).

The mode of transmission is by fecal-oral route. *Shigella* is highly contagious and can spread from person to person through contaminated hands either directly or by the way of contaminated food and water. Sexual transmission and transmission through fomites and houseflies also occurs.

1.3.1 Pathogenesis

Shigella exerts their effect through various mechanisms, which include enterotoxin production, direct invasion, and cytotoxicity. The principal toxins are the Shiga toxin (Stx) primarily produced by *S. dysenteriae* type 1 and Shiga enterotoxin (ShET). ShET 1 is produced only by *S. flexneri* type 2a and ShET2 is produced by most of the serotypes, and both are responsible for the early watery diarrhea of their respective species. The pathogenesis of Stx-induced diarrhea and HUS is similar to that of EHEC as described previously. However, the risk is less when compared to EHEC (Bennish et al. 2006).

The direct invasion and cytotoxicity appear to be more important factor of *Shigella* infection than their toxin production. This organism first invades the M cells which are specialized epithelial cells situated over the lymphoid follicles. Then they reach the gut-associated lymphoid tissue (GALT) in the terminal ileum and colon. There, they are engulfed by the macrophages and induce apoptosis of the macrophages releasing IL-1B that recruits neutrophils, resulting in an inflammatory response and causing destruction of the gap junctions. Once the gap junctions are breached, the organisms present in the lumen enter the subepithelial space and gain access to the basolateral membrane of the enterocytes and eventually invade the enterocytes. They then multiply inside the enterocytes, and this results in destruction of the enterocytes by activating proinflammatory transcription factor, NF-κB and spreads to adjacent enterocytes. The bacterial gene products (approximately 30) involved in the invasion and cytotoxicity are carried on large virulence plasmid present in all the *Shigella* species (Parsot 2005).

1.3.2 Clinical features

The cardinal symptoms of *Shigella* infection are fever, abdominal pain, diarrhea (watery, mucoid or bloody), and vomiting. The incubation period is 1–7 days (average 3 days) (Hui 1994). The first to develop is usually fever

(48 hours), followed by abdominal pain, diarrhea (72 hours), and dysentery (120–144 hours) (DuPont et al. 1969). Though the classical symptom is bloody diarrhea, it is seen only in 40%–60% of cases, and most often the diarrhea is mucoid (50%–99%) and sometimes watery (30%–60%) (Khan et al. 2013). Dysentery or bloody diarrhea is more common with *S. dysentriae* type 1 (80%) and is less common with *S. sonnei* (20%) and other species (Khan et al. 2013).

The clinical course depends on the serotype and host factors like age and immune status. In children the nonbloody diarrhea is usually mild and lasts 1–3 days. In adults, the milder forms of illness usually lasts 7 days, and adults recover without major complications. The more severe cases are usually seen in children with malnutrition and patients with *S. dysenteriae* infection. The severe illness may last for 3–4 weeks and is associated with complications. In prolonged cases, the disease might be confused with ulcerative colitis because even the endoscopic findings and the histology looks similar to ulcerative colitis. HUS is one of the frequent and important complications of EHEC; children infected with type 1 *S. dysenteriae* younger than 5 years of age are also susceptible. The classical triad of HUS is microangiopathic hemolytic anemia, thrombocytopenia, and renal failure. The Stx through its toxicity on the vascular endothelium plays a role in the pathogenesis of HUS. Early antibiotics appears to prevent HUS in *S. dysenteriae* type 1 infection, but it aggravates the chances of HUS in EHEC infection (Wong et al. 2000; Zhang et al. 2000; Bennish et al. 2006; Butler 2012).

The intestinal complications of *Shigella* infection include, proctitis, rectal prolapse, toxic megacolon (3%), intestinal obstruction (2.5%), and perforation (1.7%) (Azad et al. 1986; Bennish 1991; Bennish et al. 1991; Khan et al. 2013). Proctitis and rectal prolapse are usually seen in younger children. Toxic megacolon and obstruction are commonly seen with *S. dysenteriae* type 1 infection. Perforation is seen in *S. dysenteriae* or *S. flexneri* infection especially in infants and patients who are malnourished. Bacteremia is an uncommon complication (<7%) and is associated with increased risk of death. It is commonly seen in children younger than 5 years and people older than 65 years (Martin et al. 1983; Struelens et al. 1985; Davies and Karstaedt 2008). They present with leukocytosis, hypothermia, temperature above 39.5°C, severe dehydration, and lethargy.

Central nervous systems (CNS) complications are commonly encountered in *Shigella* infection. Seizure is the most common CNS manifestation occurring in 10% of patients. Children younger than 15 years are most commonly affected. The other CNS manifestations include encephalopathy with lethargy, confusion, headache, obtundation, and coma (Avraham et al. 1982).

Ekiri syndrome is a rare CNS manifestation of *S. sonnei* infection reported from Japan in pre-World War II era and was characterized by rapid onset of seizure and coma with very high fever and fewer diarrhea symptoms but with very high mortality (15,000 deaths per year) (Bennish 1991). Leukemoid reaction with a white blood cell (WBC) count of more than 50,000 has been reported in about 4% of *Shigella* infection from Bangladesh and are more commonly seen in children between 2 and 10 years with a high mortality (Butler et al. 1984). Sterile inflammatory arthritis (reactive arthritis) is seen in few Shigella infections especially following *S. flexneri* infection. They are uncommon (0.5%) and occurs 1–2 weeks following the onset of dysentery (Porter et al. 2013). Patients positive with HLA B27 are commonly affected with reactive arthritis (70%). This form of arthritis when associated with conjunctivitis and urethritis is termed Reiter's syndrome, which occurs in 1%–2% of *Shigella* infections and commonly seen in young men between 20 and 40 years of age with HLA B27 (Simon et al. 1981; Finch et al. 1986; Keat 2010). The joint involvement is asymmetrical, and lower extremities are commonly affected. The treatment is usually symptomatic with nonsteroidal anti-inflammatory drugs (NSAIDs). Hyponatremia (29%), hypoglycemia, and protein-losing enteropathy are some of the other complications seen with *Shigella* infection.

1.3.3 Diagnosis

Isolation of the organism by culture is required for making a confirmatory diagnosis. Because the organisms are fastidious, the stool should be inoculated in to a specific medium at the bed side, or it has to be transported in a special medium (i.e., Cary-Blair medium or Buffered glycerol saline). The World Health Organization (WHO) has recommended a diagnostic algorithm for diagnosing *Shigella* infection according to which the stool samples are inoculated on MacConkey agar and a more selective xylose lysine deoxycholate agar (XLD agar). *Shigella* is nonlactose fermenter and hence appears pale pink on MacConkey agar and appears pink on XLD medium. The suspected colonies are then inoculated in to Kligler iron agar (KIA) and motility indole urea (MIU) medium. Based on the specific characteristic features on these medium, they can be differentiated from *E. coli* (i.e., *Shigella* are nonmotile, produce alkaline slant, indole positive, urease and oxidase negative with no gas or hydrogen sulfide production, and ferment glucose). Molecular studies like PCR-based assays are currently available to detect the invasion-associated locus of specific species and can detect as few as 10–100 organisms (Avraham et al. 1982; Nataro et al. 1995). The advantage of culture over the molecular studies is that antibiotic susceptibility can also be tested, which is important in deciding appropriate antibiotics, especially in the current scenario of growing antibiotic resistance.

1.3.4 Treatment

The principle management of *Shigella*-induced diarrheal illness includes rehydration, correction of electrolyte imbalance, appropriate antibiotic, and management of complications. Antibiotic therapy reduces the duration of symptoms, duration of excretion of organisms in stool, and also reduces complications; hence, appropriate antibiotics are indicated. If *Shigella* infection is suspected, early empirical antibiotic may be considered if the patient is severely ill, immunocompromised, a healthcare provider, food handler, or a child or adult who is associated with a day care. Drug resistance appears to be a major problem in deciding appropriate antibiotics. Currently ciprofloxacin is the first-line regimen, and pivmecillinam, ceftriaxone, and azithromycin are the second-line regimens. However, the choice of antibiotics should always be decided based on the local drug resistance and sensitivity pattern.

1.3.5 Prevention

Because the organism is highly contagious, frequent hand washing, proper sanitation, and adequate stool disposal are important to prevent the spread of this infection. Unfortunately, there are no vaccines commercially available for this infection.

1.4 *Salmonella*

1.4.1 Nontyphoidal salmonellosis

The most common bacteria that causes GI infections globally is the nontyphoid *Salmonella*. A study conducted in 2010 reported that the global burden of nontyphoid *Salmonella* gastroenteritis went up to 93.8 million cases with mortality of around 155,000 (Majowicz et al. 2010; Eng et al. 2015). The average incidence was estimated to be around 1.14 episodes per 100 person each year. Out of these, nearly 88% of the cases were reported from Asian countries. Ingestion of *Salmonella* through contaminated water or food causes acute gastroenteritis with an incubation period from 4 to 72 hours. Though the majority of *Salmonella* cases occurs through ingestion of contaminated water or food, *Salmonella* can also be acquired by contact with animals and direct person-to-person contact via the fecal-oral route. Salmonellosis may be self-limited or severe and persistent with fever, bloody diarrhea, and weight loss. Diarrhea is usually self-limiting, lasting for 3 to 7 days, and may be grossly bloody. *Salmonella* is reported to be excreted in feces up to 5 weeks after infection in young children and up to 8 weeks in older children. Bacteremia occurs in 5%–10% of infected cases, and some may develop focal infections such as meningitis and bone and joint infections. The mechanism that regulates the disease pathogenicity is not completely understood; however, the disease severity of *Salmonella* depends on factors like the serotype of the infecting bacteria, inoculating dose, and predisposing host factors. Before gaining access to the small intestine, the bacteria can survive in a gastric acid barrier, thus children younger than 1 year of age, advanced age, and patients with reduced gastric acid production were reported to have higher risk of infection. Prolonged or recurrent infection was also reported in impaired cellular immunity patients. Nontyphoid *Salmonella* are able to induce the production of IL-8 by enterocytes, which is a potent neutrophil chemotactic cytokine. The neutrophils attracted in this way lead to further tissue damage by degranulation and release of oxidative toxic substances that render the bacteria to resist the unfavorable environment and survive.

Salmonella can easily be recovered from fecal samples by direct plating, and most laboratories use one medium with low selectivity such as MacConkey agar. Different strategies followed for the laboratory diagnosis of *Salmonella* on routine stool culture was reviewed by Humphries and Linscott (2015). Hektoen enteric agar media is the recommended appropriate selective media for *Salmonella*. XLD, *Salmonella-Shigella*, Hektoen enteric, brilliant green, or bismuth sulfite medium or a chromogenic medium designed for the recovery and detection of specific enteropathogens can also be used. Chromogenic media has been reported to have improved sensitivity and specificity over traditional selective media. Apart from conventional methods, techniques such as molecular assays and MALDI-TOF are also reported to detect *Salmonella*; however, the latter technique is not optimized yet to detect the serotypes.

Antibiotics are not routinely recommended for uncomplicated salmonellosis and are indicated only in patients who are immunocompromised and in severe cases with complications (Sirinavin and Garner 2000). Quinolones are currently the drug of choice for these infections, but there is high level of antibiotic resistance developing to quinolones; hence, judicious usage of antibiotics is required.

1.4.2 *Salmonella typhi*

Salmonella typhi cause illness called *typhoid fever* or *enteric fever*. They are more of a systemic illness rather than a primary GI infection. Humans are the only known reservoir of *S. typhi*. The organism penetrates the intestinal epithelial cells and reaches the lymphatics with very little inflammation; hence, GI symptoms are less in the initial space. They then proliferate in the reticuloendothelial organs, leading to hepatosplenomegaly and are released systemically in large numbers in the next phase with resultant

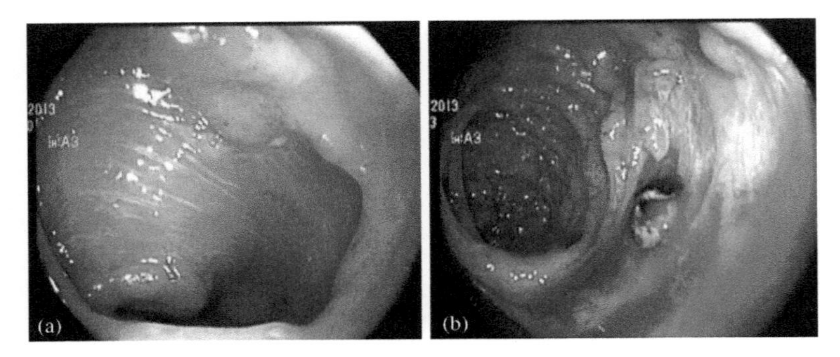

Figure 1.1 Colonoscopy pictures from a patient with typhoid fever. (a) Large deep ulcer in the terminal ileum. (b) Multiple ulcers in the caecum and ascending colon with bleeding from one of the ulcer.

involvement of various systemic organs. During this phase, the GI system might be exposed with heavy bacteremia especially in the payer's patches of terminal ileum. The organisms are also seen abundantly in gall bladder. The lymphoid follicles in the payer's patches might then ulcerate and lead to complications like bleeding and ulcerations, which are seen commonly by the third week of the illness (Figure 1.1). Patients usually recover by 4 weeks, but a few continue to harbor the organism in the gallbladder or other organs and become carriers. They continue shedding the bacteria in feces and cause recurrent outbreaks of infection in the community.

1.4.2.1 Clinical features – The organisms are acquired by ingestion of contaminated food products or contaminated water. The infection characteristically lasts for 4 weeks and then either recovers or ends with complications. The symptoms usually develop after the 7- to 14-day incubation period. During the first week. the patient presents with high-grade fever, myalgia, headache, and abdominal pain. Altered bowel habits are less common during this phase. The patient becomes sick, and the fever continues in the second week. In the third week, patient develops symptoms of toxemia, and GI manifestations might occur in the form of pea soup diarrhea. Only during this phase GI complications secondary to development of large ulcerations like perforation and bleeding occur (Figure 1.1). Altered mental status, often referred to as typhoidal state, happens during this phase. During the fourth week, the patient gradually improves and recovers from illness if the prior storming course is survived.

1.4.2.2 Diagnosis – Typhoid illness is often diagnosed by isolating the organism through culture from the blood, stool, or urine during the first week, second, or third week, respectively. Serological tests in the form of agglutination test (Widal test) to detect the O and H antigen is usually positive by the second week. Serial rising titers of O and H antibodies are more important than an isolated elevation of the titers at a given time. PCR-based assays are now available for the detection of *S. typhi* infection.

1.4.2.3 Treatment – Quinolones and third-generation cephalosporins are currently used for the treatment of *S. typhi* infections, and they are given for at least 7–14 days for a better response. However, because of emerging drug resistance, antibiotics should be decided based on the local sensitivity pattern.

1.5 *Campylobacter*

Campylobacter are gram-negative, motile, comma-shaped bacilli with polar flagellum. They were once grouped under *Vibrio*, and in 1973, they were named a separate genus. Since then, they have emerged as one of the leading cause of acute diarrhea and systemic illness worldwide (Vernon and Chatlain 1973; Skirrow 2006). Along with genus *Arcobacter* and *Helicobacter*, it is phylogenetically grouped in a separate family called "rRNA superfamily VI." All these organisms share common features of colonizing the mucosa of alimentary tract and reproductive tract. Currently there are 18 species recognized, and the most important species causing human infections are *C. jejuni*, *C. coli*, and *C. fetus* (in patients who are immunocompromised). The other species, which rarely cause human infection, includes *C. hyointestinalis*, *C. upsaliensis*, and *C. laridis*. Though *C. jejuni* is the most important in this genus; *Campylobacter* enteritis as a whole is discussed here and wherever appropriate *C. jejuni* is mentioned separately.

Campylobacter species are one of the leading causes of foodborne disease worldwide (Scallan et al. 2011). Children younger than 5 years of age are commonly affected. *Campylobacter* has an extensive animal reservoir including cattle, sheep, pig, birds, and dogs. Human infection occurs from these animals by taking improperly cooked meats or by consuming contaminated foods and water. In the United States, drinking unpasteurized milk and undercooked poultry are the leading cause of *Campylobacter* infection. Person-to-person transmission is relatively low. The infective dose is relatively higher (9000 bacteria); however, a dose as low as 500 can cause clinical illness (Robinson 1981; Black et al. 1988).

1.5.1 Pathogenesis

Adhesion and invasion of the intestinal cells are the important features of *Campylobacter* infection. Because of their spiral shape and the polar flagella, they easily penetrate the mucus layer and reach the surface of epithelial cells. Large-molecular weight plasmids encoding certain virulence factors (pVir in *C. jejuni*) play a role in pathogenesis (Tracz et al. 2005). CadF adhesion factor is responsible for microtubule-mediated internalization of these organisms along the basolateral membrane of M cells resulting in mucosal damage. Though an enterotoxin called cytolethal distending toxin (CDT) is produced by *C. jejuni*, which causes cell cycle arrest and DNA damage, its exact role in the pathogenesis is not clearly known (Pickett et al. 1996; Whitehouse et al. 1998).

1.5.2 Clinical features

The incubation period is 1–3 days and can extend up to 10 days. The illness typically present with acute onset abdominal pain and diarrhea. The pain is colic and periumbilical. Diarrhea is the most common symptom of *Campylobacter* and is seen in almost 90% of cases. It is bloody in 50% of cases, and the frequency may be up to 10 or more times a day. Prodromal illness is seen in one-third of cases, usually lasting 1–3 days before the onset of abdominal pain and diarrhea and includes fever with rigors, headache, dizziness, myalgia, lassitude, anorexia, and vomiting. Patients with prodrome have a more severe illness. The illness is usually self-limited, and patients recover within 7 days with a mortality as low as less than 0.1%. Relapses are seen in 25% of the cases.

Campylobacter infection can mimic acute appendicitis and inflammatory bowel disease (IBD). The severe abdominal pain secondary to ileocecitis causes tenderness in the right iliac fossa (RIF) mimicking appendicitis. In a series of 533 cases with clinical diagnosis of acute pancreatitis, ultrasound showed mesenteric adenitis with thickening of ileum and caecum without visualization of appendix in 61 cases (Puylaert et al. 1989). Out of these 61 patients, 21 were found to have *Yersinia enterocolitica* and 15 had *C. jejuni* infection. Rarely, *Campylobacter*-induced appendicitis has also been reported (Van Spreeuwel et al. 1987). Some of the *Campylobacter* infection present with severe acute colitis with bloody diarrhea; these cases might be confused with IBD, and one has to depend on histology wherein there is acute inflammation with no chronic changes like crypt distortion (Van Spreeuwel et al. 1985). The other complications seen with *Campylobacter* infection are erythema nodosum, urticaria, acute cholecystitis, pancreatitis, septic arthritis, HUS, toxic megacolon, GI hemorrhage, bacteremia, nephritis, meningitis, pericarditis, and myocarditis. Histology of a few patients with immune proliferative small intestinal disease has shown evidence of *C. jejuni* infection; hence, it has been found to be associated with development of intestinal lymphomas (Lecuit et al. 2004).

The late manifestations of *Campylobacter* infections include reactive arthritis and Guillain-Barré syndrome (GBS). The reactive arthritis is similar to *Salmonella*, *Shigella*, and other bacterial diarrheal illness and presents 1–2 weeks after the onset of diarrhea. The ankle, knee, wrist, and small joints of the hand are usually affected and occur commonly in cases positive with HLA B27. The prognosis is good and usually responds to NSAID therapy within 6 months. GBS is one of the common late complications and in the United States: 1 in 1000 cases of *Campylobacter* enteritis (especially *C. jejuni*) develops GBS (Nachamkin et al. 1998). GBS occurs 1–2 weeks following the onset of diarrhea. It has been shown that 30%–40% of GBS were due to *Campylobacter* infection and the prognosis of *C. jejuni*–induced GBS appears to be worse when compared to others (Rees et al. 1995; Nachamkin et al. 1998). The pathogenesis is secondary to the development of antibodies against the epitopes present in *Campylobacter* cross reacting with the GM1 gangliosides of the neurons.

1.5.3 Diagnosis

A rapid presumptive diagnosis can be made by examining fecal smear by Gram stain on dark field microscopy, which shows these organisms as faint gram-negative motile rods with a typical seagull wing appearance. But the gold standard for the diagnosis of *Campylobacter* infection is isolation of the organism by culture. These organisms are fastidious and transport medium, like Cary-Blair medium, should be used for transportation. Generally a selective antibiotic-containing medium is used and incubated at 42°C under carbon dioxide (CO_2) and reduced oxygen conditions. Rapid diagnostic tests, like a latex agglutination test for detecting specific *Campylobacter* antigens and PCR-based test to detect *Campylobacter* DNA in stool samples, are available (Kulkarni et al. 2002; Granato et al. 2010).

1.5.4 Treatment

Antibiotics are not routinely needed because most of infections are mild and recover spontaneously. However, antibiotics are indicated in severe cases with dysenteric symptoms, systemic infection, and immune deficiency; additionally, they should be used during outbreaks in a daycare setting to prevent spreading. *Campylobacter* is resistant to many drugs, and the current choices are ciprofloxacin or azithromycin. In view of increasing fluoroquinolone resistance, azithromycin is the preferred drug (DuPont 2007; Tribble et al. 2007), especially for traveler's diarrhea. In a heavy systemic infection, a more effective drug in the form of carbapenem or aminoglycosides are usually preferred, and an oral macrolide is also used for eradicating intestinal infection (Lachance et al. 1991; Sjögren et al. 1992).

1.5.5 Prevention

Effective pasteurization of milk and consumption of purified drinking water and adequately cooked meat are the important means of preventing *Campylobacter* infection.

1.6 *Yersinia*

The genus *Yersinia* consists of 11 species out of which 3 are important causes of human infections, namely *Y. enterocolitica*, *Y. pseudotuberculosis*, and *Y. pestis*. Yersiniosis, a foodborne GI infection is primarily caused by *Y. enterocolitica* and less commonly by *Y. pseudotuberculosis*. *Y. pestis* is the causative organism for pulmonic and bubonic plague. *Y. enterocolitica* are gram-negative coccobacilli with peritrichous flagella and are facultative anaerobes belonging to Enterobacteriaceae family (Bottone 1997). They are psychrophilic and often require a cold enrichment step for isolation. Infections are common during winter and are frequent in temperate countries. They have an extensive animal reservoir, and the most frequent mode of infection is by eating undercooked pork and drinking raw milk and contaminated water. Yersiniosis is the third-most common zoonosis in the European Union after *Campylobacter* and nontyphoidal salmonellosis (Hoffmann et al. 2012). Lithuania and France have the highest rate at 12.9 and 9.8 cases per 100,000 population, respectively. Children younger than 5 years of age are commonly affected. Diabetes, iron overload, blood transfusion, malnutrition, alcoholism (Rabson et al. 1975; Bouza et al. 1980) are some of the predisposing factors for Yersiniosis. Iron is required for the virulence of these organisms.

1.6.1 Pathogenesis

The incubation period is 1–11 days, and they are excreted in the stools for 14–97 days. The main pathogenesis is cell invasion mediated by various virulence factors. The organism adheres by *Yersinia* adhesion A protein (YadA) binding to the epithelial cells (M cells). Chromosomal genes *inv* and *ail* encode the ability to invade the M cells. Then through plasmid (pYV)-encoded T3SS, bacterial virulence products are secreted into the enterocytes. Only strains carrying the pYV plasmids are virulent and are the primary virulence factor of *Y. enterocolitica*. The variants of these plasmids are also seen in *Y. pseudotuberculosis* and *Y. pestis*. One of the virulent products, Yersinia outer membrane proteins (Yops) is important in suppressing inflammation and phagocytosis, leading to increased survival inside the cells for long period (Boland and Cornelis 1998; Cornelis 2002; Goebel 2012).

Yersinia also has a separate iron uptake system called Yersiniabactin (Ybt) and is involved in efficient uptake of iron even from iron-deprived sites. The expression of certain pathogenic genes is thermoregulated and is the reason for its increased prevalence in cold climates. After invasion, they reach the submucosa and then phagocytosed by the macrophages and taken to payer's patches of terminal ileum and mesenteric lymph nodes where they form microcolonies and subsequently cause systemic disease. Following the primary infection by an autoimmune mechanism, *Yersinia* can cause erythema nodosum, reactive arthropathy, glomerulonephritis, and thyroiditis in few of the cases.

1.6.2 Clinical features

Abdominal pain is the most predominant symptom; fever and diarrhea are less frequent. Diarrhea is usually watery and bloody and seen in one-fourth cases. The other symptoms include nausea, vomiting, headache, pharyngitis, arthralgia, oral aphthous ulcers, and erythema nodosum. The involvement of mesenteric lymph nodes (i.e., mesenteric lymphadenitis) causes pain in the RIF and can mimic appendicitis (pseudo-appendicitis). This form of mesenteric adenitis with pseudo-appendicitis is commonly seen in teenagers and young adults. The other form of enteritis and colitis with diarrhea is commonly seen with children. Some of the complications with severe disease include bacteremia, hepatic and splenic abscess, peritonitis, polymyositis, and osteomyelitis. Septicemia with septic shock is commonly seen in infants, patients who are immunocompromised, and iron overload states. Rapid onset of septic shock has been reported in patients following blood transfusion with a mortality of 50% (Guinet et al. 2011). The common late manifestations presenting after 2 weeks due to autoimmune mechanisms are reactive arthritis and are seen commonly in patients positive with HLA-B27. The others are glomerulonephritis, thyroiditis, and erythema nodosum. *Y. pseudotuberculosis* infections are rare and are associated with granulomatous colitis.

1.6.3 Diagnosis

Yersinia should be suspected in any patient with bloody diarrhea or if the stools have leucocytes (inflammatory diarrhea). The only mode of diagnosing *Yersinia* is by isolating the organism from stool, mesenteric lymph nodes, pharyngeal exudates, peritoneal fluid, or blood. Bacterial identification is difficult, and hence, laboratories should be intimated if *Yersinia* infection is suspected. *Y. enterocolitica* can grow easily in ordinary medium and are nonlactose fermenters on MacConkey agar. However, isolation of the organism needs inoculation in MacConkey agar at 25°C–30°C or using a selective medium. The selective medium is cefsulodin-irgasan-novobiocin (CIN) agar. Enzyme-linked immunosorbent assay (ELISA) and immunoblotting can be used to detect IgG, IgA, and IgM class antibodies. Detection of IgM antibody indicates acute infection and a fourfold increase in the titer form, acute to convalescence stage increases the probability of diagnosis. Agglutination tests and complement fixation tests are also available for diagnosing *Yersinia* infection. However, the results of these serological tests should be interpreted with caution because cross-reactions with antigens of other organisms are possible leading to false-positive results.

1.6.4 Treatment

Yersiniosis is usually a self-limiting illness and mortality is extremely rare (1.2%) (Long et al. 2010). However, antibiotics are indicated in severe infections and patients who are immunocompromised. The preferred antibiotics are fluoroquinolones for adults and septran for children for a period of 5 days. In areas where *Yersinia* is resistant to fluoroquinolones, the alternative choice is septran or doxycycline. In septicemia and more severe infections, a combination of third-generation cephalosporin along with an aminoglycoside is preferred, and the duration of treatment is for 3 weeks. Because *Yersinia* is almost foodborne (90%), proper cooking of the meat, adequate pasteurization of milk, and drinking safe water can prevent such infections.

1.7 *Clostridium difficile*

C. difficile is most commonly implicated in *Clostridium difficile* infection (CDI), which is a leading cause of hospital-associated GI illness (Johnson et al. 1990). It is noninvasive and present in two forms, an antibiotic-resistant latent spore and nonresistant vegetative form. The latter, when in the human body can prioritize growth over toxin production and colonize the intestine causing the disease.

They produce two toxins, namely toxin A (TcdA, 205 kDa) and toxin B (TcdB, 308 kDa), which are involved in the destruction of the intestinal cell cytoskeleton leading to cell death (Carey-Ann and Carroll 2013). *C. difficile*, though a normal commensal, do not cause illness because of its lesser number and competition from other bacterial commensals to assess direct contact with the epithelial cells. However with the use of antibiotics and reduction in the number of other commensals, the condition becomes favorable for *C. difficile* to cause illness (Wilson 1993). *C. difficile* infections may have varying presentations with asymptomatic carriage, mild diarrhea, or a severe condition called *pseudomembranous colitis*.

1.7.1 Incidence

The incidence of *C. difficile* is on the rise due to increasing usage of antibiotics. In the United States, the incidence is 346,805 among hospitalized patients during the year 2010 (Lessa et al. 2015). A meta-analysis including 51 studies from Asia has shown an incidence of 14.8% among all patients with diarrhea, with incidence higher in hospitalized patients when compared to daycare patients at a rate of 5.3% per 10,000 patient days (Brazier and Berriello 2000; Borren et al. 2017). Ramakrishnan and Sriram (2015) who worked on CDI and antibiotic abuse in India concluded an incidence of 1.67% in India, which is similar to the United States.

1.7.2 Clinical and microbiological properties

Morphologically *C. difficile* is seen as irregular rods with elongated spores, which are slightly bulged at the terminal end. *C. difficile* is cultured on cycloserine cefoxitin fructose agar (CCFA) medium and identification on other culture medium is mentioned in Table 1.2. The colonies have a ground glass appearance under microscope and exhibit a yellow-green fluorescence under ultraviolet (UV) illumination. A typical odor of horse manure is also used to identify the microbe. A gas liquid chromatography profile of large amounts of butyric and iso-caproic acid is considered the best method for identification (Fedorko and Williams 1997).

1.7.3 Clinical features

The symptoms of *C. difficile* infection include watery diarrhea, fever, loss of appetite, nausea, and abdominal pain or tenderness. Based on the severity of symptoms, the disease is classified as mild to moderate, severe and complicated disease, and recurrent CDI. Symptomatic illness is characterized by

Table 1.2 Identification of *Clostridium difficile* on Culture Medium

Media	Observation
Gelatin	No liquefaction
Agar	Minute, flat, opaque discs
Egg yolk agar	Surface colonies are dry, irregular, flattened, dry, roughened, somewhat granular, with little or no color No precipitate in the agar and no luster on the colony
Coagulated albumin	No liquefaction
Blood agar	Irregular, flat Hemolysis absent
Blood serum	No liquefaction

Source: Sneath, P.H.A. et al., *Bergey's Manual of Systematic Bacteriology*, Lippincott Williams & Wilkins, Baltimore, MD, 1986.

watery diarrhea occurring often within 2 days of antibiotic usage; 96% of illness occurs within 14 days of antibiotic usage and almost all of them within prior 3 months (Olson et al. 1994). The disease is mild when there are no signs or symptoms of colitis. In moderate disease, patients present with diarrhea and colitis, characterized by fever and abdominal cramps in the lower quadrant.

Fulminant *C. difficile* causes toxic megacolon in which colon is distended more than 6 cm and is prone for perforation. It occurs in less than 5% of patients and is characterized by severe abdominal pain, guarding, rigidity, and high fever with either profuse or absent diarrhea (Bartlett and Gerding 2008; McCollum and Rodriguez 2012).

1.7.4 Pathogenesis

The organism is found freely in soil, in variety of animals, and in raw and processed food products. They are highly prevalent in healthcare facilities, resulting in horizontal transmission through both environmental surface contamination and direct contact by hospital workers. The mode of infection is primarily through person-to-person and through the fecal-oral route. There have been studies showing the increase in transmission of *C. difficile* during hospitalizations, especially in elderly patients in acute care facilities (McFarland et al. 1989; Samore et al. 1994; Chang and Nelson 2000; Vaishnavi 2010).

1.7.5 Diagnosis

In 2010, guideline for the management of patients with CDI was published by Infectious Diseases Society of America (IDSA) and Society for Healthcare Epidemiology of America (SHEA). Based on these guidelines, CDI cases are defined by stool test positivity for *C. difficile* toxins, toxigenic *C. difficile*, or colonoscopic or histopathologic findings, revealing pseudomembranous colitis. Standard reference tests such as cell culture neutralization assay (CCNA) and toxigenic culture methods have being used for the past 30 years for detection of CDI (Planche and Wilcox 2011). EIA-based toxin detection methods are being carried out and they are less expensive, easy, and quick to perform when compared to CCNA. Currently PCR-based identification is widely preferred because nucleic acid amplification methods have a greater sensitivity and reduced turnaround time. Nucleic acid amplification tests (NAATs) are the most recent methods for detection of *C. difficile* and have been used since 1990 (Wilcox et al. 2000; Snell et al. 2004; Zheng et al. 2004; Berg et al. 2007).

1.7.6 Non-laboratory-based tests

Endoscopy and computed tomography (CT) scan may be done in certain cases for diagnosis of pseudomembranous colitis (Kazanowski et al. 2014).

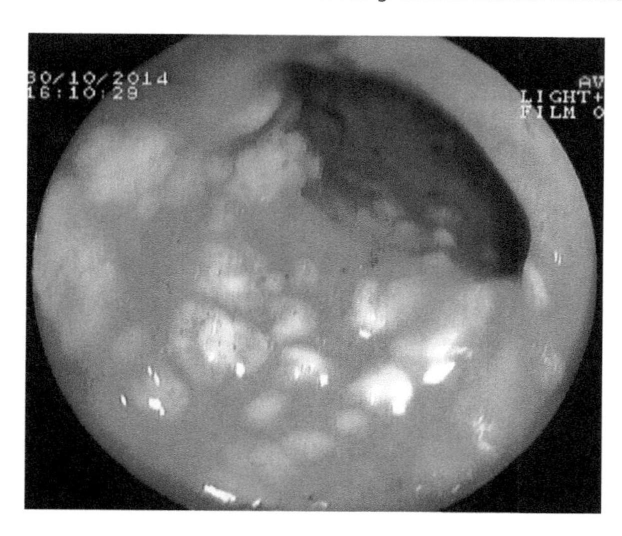

Figure 1.2 Colonoscopy picture showing multiple yellowish plaques of pseudomembranes distributed along the sigmoid colon in a patient with pseudomembranous colitis.

Colonoscopy will reveal pseudo-membranes that are yellow exudative plaques of about 2–5 mm (Figure 1.2).

1.7.7 Treatment

The antibiotic, which caused the CDI, has to be stopped, and few of the patients (15%–25%) respond spontaneously without the need for further specific antibiotics. Antibiotics are only needed in patients who are critically ill and those with severe symptoms. The drugs of choice are metronidazole and vancomycin with a success of about 87% and 97%, respectively (Farrell and LaMont 2000; Zar et al. 2007). Metronidazole may be given orally or intravenously, but vancomycin can only be given orally to be effective. Relapse is common after successful treatment and is seen in 20% of the cases.

1.8 *Clostridium perfringens*

1.8.1 General microbiology

Clostridium perfringens (*C. welchii*) is gram-positive, spore-forming, nonmotile, rod-shaped, obligate anaerobic bacterium with an optimum growth at 43°C–47°C and generation time of less than 10 minutes.

1.8.2 Incidence

According to the CDC, *C. perfringens* is the second-most common bacteria that causes foodborne illnesses in the United States, infecting nearly million people annually. Outbreaks occur regularly and cause substantial morbidity.

1.8.3 Symptoms

After an incubation period of 8 to 16 hours, the symptoms start and last for 24 hours. Symptoms include profuse watery diarrhea, abdominal pain, nausea, and vomiting.

1.8.4 Pathogenesis

C. perfringens are usually involved in food poisoning and open wound infections. The GI infection varies from mild enteric disease to necrotizing enteritis. It is an intestinal commensal and remains dormant until their number increases because of contaminated foods. *C. perfringens* is classified into five types (A–E) based on the production of four exotoxins (i.e., alpha, beta, epsilon, and iota; Table 1.3). In addition to this, several other toxins including enterotoxin (Cpe) and neuraminidase are also produced. Chromosomally coded Cpe are highly resistant to food-preservation procedures compared to the plasmid-encoded isolates (Li and McClane 2006). Because it could not synthesize vital amino acids, it depends on the breakdown of the host tissue for survival. This damage to the intestinal cells results in diarrhea, flushing out the spores and toxins to prevent

Table 1.3 ***Clostridium perfringens*** **Toxins and Its Illness**

Types of *C. perfringens* Toxins	Diseases
A	Gas gangrene (myonecrosis), foodborne illness, and infectious diarrhea in humans; enterotoxemia of lambs, cattle, goats, horses, dogs, alpacas, and others; necrotic enteritis in fowl; equine intestinal clostridiosis; acute gastric dilation in nonhuman primates, various animal species, and humans
B	Lamb dysentery; sheep and goat enterotoxemia (Europe, Middle East); guinea pig enterotoxemia
C	Darmbrand (Germany) and pig-bel (New Guinea) in humans; "struck" in sheep; enterotoxemia in lambs and pigs; necrotic enteritis in fowl
D	Enterotoxemia of sheep; pulpy kidney disease of lambs
E	Enterotoxemia in calves; lamb dysentery; guinea pig enterotoxemia; rabbit "iota" enterotoxemia

severity of infection (Doyle et al. 2007). Heat-resistant spores ingested from the contaminated food survive, proliferate, and produce toxins, leading to the disease.

Alpha toxin is produced by type A strain and present frequently in humans. Zinc activates the toxin after which it interacts with the host cell receptors, and through a series of pathways, the permeability in blood vessels is increased and blood supply is reduced to tissues. Beta toxin is lethal and produced by type B and type C strains. Necrotizing enteritis (enteritis necroticans or pigbel), which is rare, is caused by toxins of type C strains and is often obtained by ingestion of undercooked pork. Epsilon toxin is most commonly isolated from animals. which are produced by type B and type D strains. Potassium ions and fluid leakage occurs due to perforation in tissues. Iota toxin, known as AB toxin, is produced by type E strain. B component interacts with the host cell surface receptor to facilitate the uptake of the toxin, while A component inhibits actin polymerization, thereby breaking down the cytoskeleton. Bacteremia and clostridial sepsis is uncommon, but both are fatal occurring with an infection of the uterus, colon, or biliary tract.

1.8.5 Diagnosis

Culture identification is used for diagnosis with notable colony characteristics like double-zone hemolysis or circular opaque zone on McClung-Toabe egg yolk agar. PCR-based identification of toxin-producing genes, colorimetric assay detecting lecithinase, (Dave 2017), and reverse passive latex agglutination (RPLA) test are available for enterotoxin identification.

1.8.6 Treatment

No specific treatment is needed for milder causes, and even in necrotizing enteritis (Pigbel), only symptomatic and supportive treatment is required.

1.9 *Vibrio*

Vibrio cholerae is a comma-shaped gram-negative bacterium that causes an acute GIT infection known as cholera. Outbreaks of cholera are mainly reported from the underdeveloped and developing world where there are no proper sources for clean drinking water and sewage disposals. However, people can get infected with the bacteria through ingestion of contaminated shellfish or seafood products containing higher concentration of *V. cholerae*. WHO reports that more than 1.4 billion people are at risk of developing cholera every year with over 130,000 deaths worldwide.

Currently the highest incidence of cholera is reported in Africa and south Asian countries. The bacterium *V. cholerae* that causes infections only in humans is a free-living inhabitant of freshwater. More than 200 serotypes of *V. cholerae* have been reported, with each serogroup showing distinct antigenicity based on its O-side chains. *V. cholerae* O1 subgroup and O139 are the strains commonly associated with disease epidemics. Serogroups other than O1 and O139 often produce self-limiting gastroenteritis because they do not have the cholera toxin gene. The cholera toxin gene encodes for the cholera toxin which acts by stimulating the adenylate cyclase system of the GI tract and causes life-threatening secretory diarrhea. Numerous factors contribute to bacterial colonization and disease progression. However, the important part is the ADP-ribosylating cholera toxin, which is not directly required for bacterial colonization, but accounts for the secretory diarrhea as mentioned previously (Sánchez and Holmgren 2008; Millet et al. 2014).

Laboratory diagnosis for *V. cholerae* involves biochemical or serological tests for the identification of the presence of O1 serogroup antigens. The subtyping of O1 serogroup as Inaba, Ogawa, and Hikojima could be done by agglutination test. Slide agglutination test is done by treating the cultures grown on heart infusion agar, Kligler's iron agar, and triple sugar iron agar with the antiserum to detect the specific O antigen. Other than this, some of the biochemical tests like oxidase test, ring test, triple sugar iron test, carbohydrate test, decarboxylase test, and Voges-proskauer test are done infrequently based on the necessity. The hemolytic activity of *V. cholerae* is used to distinguish the biotypes such as classical and E1T; classical types show negative hemolytic activity, whereas E1T or of Australian or the US gulf coast strain shows strong positive hemolytic activity.

Clinical symptoms begin with the sudden onset of painless watery diarrhea after a 24- to 48-hour incubation period. In most of the cases, the diarrhea may quickly become voluminous and is often followed by vomiting with or without abdominal cramps. Most *V. cholerae* infections are asymptomatic with mild to moderate diarrhea, which may not be clinically distinguishable from other causes of gastroenteritis. An estimated 5% of infected patients will develop severe watery diarrhea, vomiting, and dehydration. Stool volume during cholera is more than that of any other infectious diarrhea. The characteristic cholera stool is an opaque white liquid, and it resembles the remnant water that has been used to cooking of rice (i.e., rice water diarrhea). If untreated, the diarrhea and vomiting due to bacterial infection lead to isotonic dehydration, which can lead to acute tubular necrosis and renal failure. Patients with severe disease may develop vascular collapse shock, leading to death.

Intravenous and oral hydration remains the main treatment for cholera. However, antibiotics have also been used as an adjunct to hydration therapy. Studies have shown that antibiotic treatment could greatly reduce the volume of stool output, duration of diarrhea, and duration of positivity for bacterial culture (Roy et al. 1998; Kaushik et al. 2010). Tetracyclines, quinolones, and septran are usually given for the treatment of severe cholera cases.

1.10 *Aeromonas*

Aeromonas are gram-negative ubiquitous bacteria distributed in fresh and brackish water. Based on their temperature preference, they are classified into two groups (Murray and Baron 2007). The psychrophilic species grows at 22°C–25°C are non-motile and cause infection only in fishes. The mesophilic species grows at 35°C–37°C, are motile, and causes disease in humans. Currently, there are eight species in mesophilic group that are known to cause human infection and the common species include *A. hydrophilia, A. caviae,* and *A. veronii.* They cause acute diarrheal illness (i.e., most common manifestation) after ingestion of contaminated drinking water or foods and wound infections following swimming in fresh or brackish waters. Children younger than 2 years of age and adults older than 80 years of age are commonly affected. Though they have been isolated from few of the outbreaks, some studies have failed to show its exact role in gastroenteritis.

The *Aeromonas* adhere to the intestinal epithelial cells and secrete various toxins like heat labile enterotoxin, hemolysin, and cytotoxins. The other virulence factors include VacB, enolase, and the ability to produce Type VI secretory system. Cell invasion also occurs and leads to colitis and dysentery. The affected persons usually present with watery diarrhea (75%–89%), but it may be bloody in few cases. The diarrhea usually lasts for 3 to 10 days. Fever and abdominal pain are less common. Rare complications include segmental colitis, ischemic colitis, and HUS. The treatment is often conservative in the form of rehydration therapy because the illness is self-limited and does not require routine antibiotics. Antibiotics are only indicated in chronic diarrhea and in patients with systemic complications. The preferred antibiotics include fluoroquinolones or a third-generation cephalosporin or a carbapenem.

1.11 *Plesiomonas*

Plesiomonas are gram-negative facultative anaerobic rods belonging to Enterobacteriaceae family and present in freshwater (MacDonell and Colwell 1985). They are acquired by drinking contaminated water, by eating raw seafoods (Oysters and shellfish), or by contact with reptiles. Acute diarrheal illness is the usual manifestation of *Plesiomonas* infections, but they also cause cellulitis and skin abscess during trauma in freshwater as with *Aeromonas*. They also cause traveler's diarrhea in tropical and subtropical countries. Few diarrheal outbreaks have been reported in Japan, China, Cameroon, and Bangladesh (Hori et al. 1966; Tsukamoto et al. 1978; Rutala et al. 1982; Bai et al. 2004; Wouafo et al. 2006). They produce several toxins, including a cholera-like toxin, and they also cause invasion of the epithelial cells, leading to colitis and bloody diarrhea. After an incubation period of 24–48 hours, patients develop watery diarrhea with severe crampy abdominal pain. Sometimes patients develop severe colitis and can present with bloody diarrhea. Pain is severe and is the prominent symptom of *Plesiomonas* colitis; the other symptoms are fever and vomiting. The diarrheal illness is usually self-limiting and recovers without major complications in healthy individuals; however, in some patients, the illness might last from 2 weeks to 3 months. Complications are seen in patients who are immunocompromised and in children and includes septicemia, meningoencephalitis, pneumonia, and so on (Terpeluk et al. 1992; Schneider et al. 2009). Antibiotics are not routinely given, and they are indicated only in chronic diarrhea and for extraintestinal complications. The antibiotics currently used are fluoroquinolones, third-generation cephalosporin, or carbapenem.

1.12 *Bacteriodes fragilis*

B. fragilis are anaerobic commensal bacteria found in human intestine. Only during 1980s, it was found to be associated with diarrheal illness in humans. They produce an enterotoxin called *B. fragilis* enterotoxin (BFT), also called as *fragilysin*, which causes secretion of IL-8 and leads to inflammation. They also cause increased intestinal permeability. The usual clinical feature is development of nonbloody diarrhea in children and adults. It is diagnosed by selective culture under anaerobic condition. Genetic and biological assays are required to diagnose the presence of the enterotoxin and are not routinely available. Because the diagnosis is rarely done, the exact role of antibiotics is not clearly known.

1.13 *Helicobacter pylori*

They are gram-negative, microaerophilic, helical-shaped, motile (Lophotrichous flagella) organisms. Nearly two-third of the global population is infected with *H. pylori*; however, they are common in developing nations. They are acquired by ingestion of contaminated water because of poor sanitation and hygiene. Most often, the infection remains asymptomatic, but few people develop symptomatic acute gastritis or chronic gastritis leading to peptic ulcer disease; they are also associated with development of mucosa-associated lymphoid tissue (MALT) lymphomas and adenocarcinomas primarily of the stomach and rarely of the duodenum.

1.13.1 Pathogenesis

These organisms possess certain special features to survive in the gastric acidic pH. They have urease enzymes that can make the pH alkaline by producing ammonium ions and form an ammonia cloud that protects the organisms from acidic pH. Secondly with the help of their flagella, they move toward the less acidic mucous layer and come in close proximity to the epithelial layers. And lastly, they have various virulence genes that encodes for various virulence factors responsible for causing inflammation. Adhesin called BabA helps in binding to the Lewis b antigen on gastric epithelial cells and SabA to the sialyl-Lewis x antigen on gastric mucosa. Immune responses are triggered by HcpA one of the *Helicobacter* cysteine-rich proteins, cag Pathogenicity Island (PAI). and peptidoglycan causing inflammation (Viala et al. 2004; Dumrese et al. 2009). Expression of vacuolating toxin A (VacA) and the cytotoxin-associated gene A (CagA) proteins causes severe damage to the host cell, which may also pave way for ulceration and gastric cancer (Miehlke et al. 2000; Dossumbekova et al. 2006). CagA is a cag PAI-encoded protein (120–140 kDa), which is also a potential carcinogen that disrupts adherence to adjacent cells, cytoskeleton, cell polarity, intracellular signaling, and other cellular activities (Backert and Selbach 2008). Type IV secretion system helps in injecting the CagA into the gastric epithelial cells. Tyrosine residues of the CagA at four distinct glutamate-proline-isoleucine-tyrosine-alanine (EPIYA) motifs are phosphorylated by the host cell membrane associated Abl and Src kinases, leading to the activation of proto-oncogene Shp2 (Hatakeyama 2004). Signal transduction and gene expression of the host are altered by the activation of epidermal growth factor receptor (EGFR). In perigenetic mechanism, tumor necrosis factor-alpha (TNF-α) alters gastric epithelial cell adhesion, leading to dispersion and migration of mutated epithelial cells. Peptidoglycan triggers NF-κB-dependent pro-inflammatory pathway and IL-8 cytokine secretion by binding with Nodl (Viala et al. 2004). Peptidoglycan also initiates PI3K-Akt pathway leading to migration, cell proliferation, and prevention of apoptosis (Nagy et al. 2009).

1.13.2 Symptoms

Symptoms include abdominal pain (epigastric region), diarrhea (bloody or black), burping, bloating, halitosis, heartburn, and weight loss; less common symptoms are vomiting, nausea, and anorexia. The pain is often related to food and is usually relieved by food in case of duodenal ulcers and aggravated in case of gastric ulcers.

1.13.3 Diagnosis

Noninvasive modalities of diagnosing *H. pylori* include IgG antibody detection, antigen detection in stool sample, and urea breath test. Serology is not useful because it does not identify an active infection. The invasive method of diagnosing *H. pylori* is by performing an endoscopy and taking biopsies from the stomach, and by subjecting the biopsy specimen for rapid urease test (RUT), histopathological examination (HPE), culture and sensitivity, or PCR study. Classical endoscopic finding in patient with *H. pylori* infection is the presence of benign-appearing ulcers either in the stomach, usually along the lesser curve or in the first part of the duodenum (Figure 1.3). Duodenal ulcers are more common than gastric ulcers. Often special stains are used to demonstrate *H. pylori* on HPE (Figure 1.3). The common method

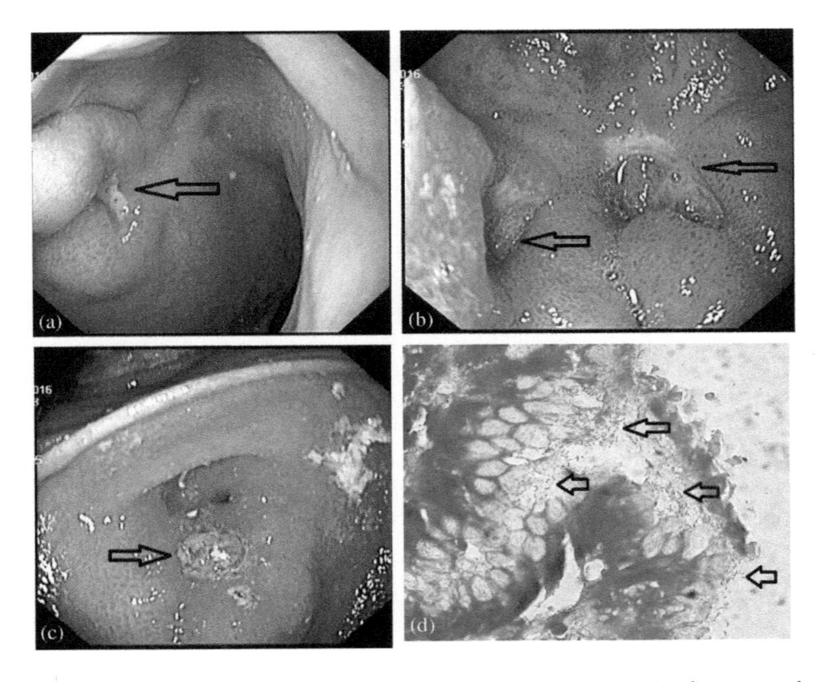

Figure 1.3 Single active ulcer (a) and two active ulcers (b) in the first part of duodenum, (c) an active ulcer in the prepyloric region of stomach causing obstruction, (d) Alcian blue stain of the biopsy from stomach showing shaped *Helicobacter pylori* in the mucous layer of stomach.

of diagnosing *H. pylori* in clinical practice is by endoscopy and biopsy followed by rapid urease test and HPE.

1.13.4 Treatment

All patients with symptomatic ulcer disease need anti–*H. pylori* therapy. Eradication of *H. pylori* requires combination therapy with at least two antibiotics and one proton pump indicator (PPI; i.e., triple therapy) given for 10–14 days. The success rate is about 70% to 85%. The most commonly used combination is one PPI with clarithromycin and amoxicillin. Because of the development of resistance, the antibiotic regimens keep changing, and it is better to select antibiotics based on the local sensitivity pattern. Quadruple therapy with a PPI, bismuth, tetracycline, and metronidazole is also used for difficult cases, and it has a success for about 75% to 90%. Sequential therapy in the form of PPI plus amoxicillin for 5–7 days followed by PPI plus clarithromycin and tinidazole for 5–7 days has a success rate of about 75% to 95%.

1.14 Foodborne illness and food poisoning

1.14.1 Introduction

Foodborne illness is common in developed and developing countries and contributes to significant burden to the society even in developed countries. The most common organisms responsible for foodborne disease vary among different parts of the world. In South Korea, *E. coli*, including EHEC, is the most common organism followed by nontyphoidal *Salmonella* and *S. aureus* (Park et al. 2015). In Japan, the most common organism is *Campylobacter* followed by *Salmonella* species and EHEC (Kumagai et al. 2015), and in United States, the most common is nontyphoidal *Salmonella* followed by *Campylobacter* and *Shigella* (Crim et al. 2015). In most of the developing countries, the exact epidemiology and the etiologic organisms of food poisoning are not available because of improper reporting and difficulties faced in diagnosing the organisms. In the United States, 48 million out of 350 million diarrheal illness is as a result of food poisoning and accounts for 125,000 hospitalization and 3000 deaths annually (Mead et al. 1999; Scharff 2012). Common organisms associated with food poisoning are shown in Tables 1.4 and 1.5, and most of the important organisms causing food poisoning have already been described. The other organisms are briefly described here.

1.14.2 *Staphylococcus aureus*

S. aureus are acquired by eating foods, like dairy products, meats, egg, and salads, prepared by food handlers (Balaban and Rasooly 2000; Centers for Disease Control and Prevention (CDC) 2013). The organisms multiply

Table 1.4 Common Foodborne Illness Causing Bacteria and Its Preliminary Identification Details

Bacteria	Type	Disease
Bacillus cereus	Gram-positive, rod shaped, motile, facultative anaerobic, spore-forming and beta hemolytic bacterium	Foodborne illness
Clostridium difficile	Gram-positive, long irregular with bulge at the ends, motile, anaerobic, and spore forming	Colitis
Clostridium perfringens, or C. welchii, or Bacillus welchii	Gram-positive, rod-shaped, motile, anaerobic, spore-forming	Foodborne illness
Helicobacter pylori or Campylobacter pylori	Gram-negative, helix-shaped, motile, microaerophilic	Stomach ulcer, stomach cancer
Salmonella spp.	Gram-negative, rod-shaped, motile, facultative anaerobe, non-spore-forming	Foodborne illness/ Salmonellosis
Staphylococcus spp. (Staphylococcus aureus)	Gram-positive, round, nonmotile, facultative anaerobes, non-spore forming	Foodborne illness
Yersinia spp. (Yersinia enterocolitica)	Gram-negative, rod-shaped, motile (22°C–29°C)	Foodborne illness/yersiniosis

in the food at room temperature and produce enterotoxins. The toxin is heat stable, and symptoms occur within 1 to 6 hours after food ingestion. Patients present with nausea, vomiting, and abdominal cramps. Fever and diarrhea is less common and occurs only in few cases. Management is usually conservative and supportive.

1.14.3 *Bacillus cereus*

Species of *Bacillus* and related genera remains a large threat because of their resistant endospores. The food-poisoning episode usually occurs because spores survive normal cooking or pasteurization and then germinate and multiply when the food is inadequately refrigerated. *B. cereus* is a large, gram-positive, spore forming, rod-shaped bacteria distributed widely as spore former or vegetative cell in the environment such as freshwater and marine water, decaying organic matters, vegetables, and as vegetative cell in intestinal tract of invertebrates (Scheider et al. 2004; Tewari and Abdullah 2015). On ingestion, the contaminated food or soil products transfer the viable spore or vegetative cells bacteria, leading to the colonization inside the human intestine.

Table 1.5 Diagnosis, Treatment, and Complications of Common Foodborne Pathogens

Bacteria	Diagnosis	Treatment and Complications
Bacillus cereus	Isolation of more than 105 colonies per g of food	Self-limiting rarely septicaemia occurs
Clostridium difficile	Stool culture, stool cytotoxin test. PCR, ELISA, EIA for presence of causative bacterial genetic material or toxins or glutamate dehydrogenase enzyme and its production. Flexible sigmoidoscopy and colonoscopy help in visualizing the pseudo-membranes in colitis	ORT, antibiotics: metronidazole, vancomycin, fidaxomicin. Fecal microbiota transplantation Ion exchange resin: Cholestyramine. Complications: Sepsis, toxic megacolon, pseudomembranous colitis, and perforation of the colon
Clostridium perfringens, or C. welchii, or Bacillus welchii	Stool culture test, >106 spores of this bacteria per gram of stool, Nagler's reaction	Prevent dehydration. Complications: bacteremia, tissue necrosis, cholecystitis, emphysematous, and gas gangrene
Helicobacter pylori or Campylobacter pylori	Histological examination, urine ELISA, stool antigen test or urea breath test, blood antibody test	Triple therapy – proton pump inhibitors (lansoprazole, esomeprazole, pantoprazole, or rabeprazole) and the antibiotics clarithromycin, metronidazole, and amoxicillin
Salmonella spp.	Blood and stool culture, PCR	Electrolytes replacement. For typhoid and paratyphoid fever antibiotics may be prescribed if needed Complications: reactive arthritis, bacteremia
Staphylococcus spp. (Staphylococcus aureus)	Culture identification, PCR	Antibiotic resistance, biofilm formation, osteomyelitis, bacteremia, and septic arthritis

(*Continued*)

Table 1.5 (*Continued*) Diagnosis, Treatment, and Complications of Common Foodborne Pathogens

Bacteria	Diagnosis	Treatment and Complications
Yersinia spp. (*Yersinia enterocolitica*)	Stool culture, tube agglutination, ELISA, radioimmunoassays, imaging studies, colonoscopy	Self-limiting but for severe cases antibiotics may be prescribed. Third-generation cephalosporins, trimethoprim-sulfamethoxazole, tetracyclines, fluoroquinolones (adults >18 years old), aminoglycosides. Complications: sepsis, focal infection, bowel necrosis

EIA, enzyme immune assay; ELISA, enzyme-linked immunosorbent assay; ORT, oral rehydration therapy; PCR, polymerase chain reaction.

Milk and rice are the two most commonly contaminated food products because of paddy soil bacteria. In addition, food poisoning by *B. cereus* can also be contributed to by a variety of contaminated foods like meat dishes, frozen food and food products, ready-to-eat chicken products, and dessert mixers. The organism is also present contaminated with several spices and food additives. The spores of *B. cereus* are heat resistant, and they can be destroyed by steaming under pressure, frying, and grilling. Inactivation of bacterial enteric toxins can be done by heating for 5 minutes at 133°F, and the emetic toxins can be removed by heating to 259°F for more than 90 minutes. Because the spores can strongly adhere to hydrophobic surfaces, it is unlikely to remove the bacteria completely, but tracing spores from farmer to package can reduce the occurrence.

1.14.3.1 Clinical symptoms – The two clinical syndromes caused by *B. cereus* food poisoning are diarrheal form and emetic form. Diarrheal form is the result of a heat labile enterotoxin produced inside the GIT. The patient develops diarrhea and abdominal pain about 8 to 16 hours after consumption of the contaminated food. Foods associated with the diarrheal forms are meat, vegetables, and sauces. The emetic form is due to the emetic toxin (cereulide) ingested along with the contaminated food that induces nausea, vomiting, and abdominal cramps within 1 to 5 hours after ingestion. Fried rice is the most common food associated with emetic form of illness in the United States. The Emetic syndrome is more severe and acute compared to diarrheal syndrome; however, with both forms of illness, patients usually recover within 24 hours with no major complications.

1.14.3.2 Laboratory test – Diagnosis of *B. cereus* infection causing food poisoning is usually clinical and most often an attempt to confirm by a laboratory method is not usually undertaken because the illness is short lived and self-limiting; further, it is not cost effective and is not available in routine laboratories. However, when a large outbreak of food poisoning occurs, the diagnosis is made using certain special reference laboratories. *B. cereus* can be cultured from stool using special culture medium. Commercial assays are being used to detect the diarrheal toxin like reverse passive latex agglutination test (Granum et al. 1993), immunochromatographic tests, and PCR, which targets the gene itself (Fricker et al. 2007; Das et al. 2009). PCR assays have also been used for diagnosing emetic food-borne *B. cereus* infection.

1.14.4.3 Treatment – The treatment of *B. cereus* food poisoning is usually supportive care and antibiotics are not routinely recommended.

1.14.4 *Listeria monocytogenes*

Listeria monocytogenes are gram-negative bacilli, usually causing invasive illness in neonates, patients who are immunocompromised, older, and pregnant. Febrile gastroenteritis is another manifestation of *L. monocytogenes*, which is almost always (99%) foodborne. They are responsible for less than 1% of reported cases of foodborne illness (MacDonell and Colwell 1985). The diarrheal illness is often sporadic, but a few outbreaks have been reported. The incubation period for gastroenteritis is less (usually 24 hours) ranging from 1 to 10 days, when compared to invasive illness which is usually 11 days (i.e., 90% occurs within 28 days). *Listeria* is seen contaminated in a wide range of foods including rice, vegetables, chocolate milks, cheese, smoked trout, corned beef, ham, and delicatessen meat. The infective dose is about 10,000 organisms per gram of food.

Fever is the prominent characteristic clinical feature, and diarrhea is usually watery. Only a few people develop bloody diarrhea. Other symptoms include headache, myalgia, nausea, and vomiting. Most are self-limited and resolve spontaneously, but few can develop dissemination and lead to more invasive illness. Diagnosing *Listeria* gastroenteritis is challenging because they require a highly selective culture medium. Hence, the laboratories should be intimated if *Listeria* is suspected. Serological method for detection of listeriolysin O antigen is available for retrospective diagnosis, and recently DNA-based methods are also available but are not routinely used. Antibiotics are not routinely recommended for uncomplicated illness; however, they are given to patients at high risk (e.g., those who are pregnant, immunocompromised, and older) during outbreaks and in cases with dissemination. The preferred antibiotics in uncomplicated cases are septran, ampicillin, or amoxicillin and in complicated cases, are aminoglycoside in combination with ampicillin or amoxicillin. These infections are best prevented by following strict food-safety measures.

1.14.5 *Vibrio vulnificus*

Vibrio vulnificus is a gram-negative bacilli belonging to the family Vibrionaceae. It is highly lethal opportunistic organism affecting patients who are immunocompromised and those with chronic liver disease. In the United States, *V. vulnificus* is a leading cause of seafood-associated fatality. According to the CDC, there were about 96 cases of *V. vulnificus* infection with a 91% hospitalization rate and mortality rate of 34.8% (Scallan et al. 2011). They are acquired by ingestion of raw seafoods causing gastroenteritis and skin infections requiring amputations. Septicemia and necrotizing fasciitis are the common complications leading to high fatality. Special culture medium is needed for isolation of these organisms. According to the CDC, doxycycline and ceftazidime are the drugs of choice in adults, and septran and an aminoglycoside in children. However, the final decision on selection of the antibiotic should be based on the local susceptibility pattern.

1.14.6 *Cronobacter sakazaki*

C. sakazaki is a gram-negative bacilli belonging to enterobacteriaceae family and is an emerging cause of necrotizing enterocolitis, sepsis, and meningitis in low-birthweight infants and in children younger than 4 years of age (Bowen and Braden 2006; Hunter et al. 2008). It spreads through ingestion of dry foods like infant feeding formula. The incidence is rare but has a high mortality of 40% to 80%. Diagnosis is made from blood culture and cerebrospinal fluid (CSF) culture. Antibiotics are generally administered because of its higher mortality and includes a combination of ampicillin and gentamicin, but in view of resistance, a combination of carbapenem or a newer cephalosporin (cefepime) with an aminoglycoside is also preferable.

1.14.7 *Mycobacterium tuberculosis*

M. tuberculosis infection of the GIT, though not common, is often seen in developing countries where tuberculosis (TB) is endemic, like India, Bangladesh, and Pakistan. Abdominal TB comprises about 5% of all cases of TB (Sharma and Mohan 2004). Intestinal TB is one of the forms of abdominal TB, and the other three are peritoneal, lymph nodal, and solid organ TB. The intestinal form often presents with chronic illness with varying manifestations, and acute diarrheal illness is rare. The disease is acquired by reactivation of latent infection or by ingestion of *Mycobacterium* through intake of unpasteurized milk and undercooked meat. In active pulmonary TB or disseminated TB, it can spread to the GIT by hematogenous route, directly from adjacent structures, or through lymphatics. The organism reaches the submucosal lymphoid tissue and causes inflammatory reaction, leading to lymphangitis, endarteritis, granuloma formation, caseation necrosis, mucosal ulceration, and scarring.

Intestinal TB has two forms, ulcerating or stricturizing form and hypertrophic form, and in a few patients, both the forms can coexist (Figure 1.4), The ulcerative or stricturizing form usually affects jejunum and ileum, and the ulcers are transverse as when compared to longitudinal ulcers of typhoid and tend to be circumferential. Ulcers are often surrounded by inflammatory mucosa and can lead to perforation, bleeding, fistula formation, and strictures upon healing. They present with abdominal pain, distension, chronic diarrhea, nausea, vomiting, constipation, and bleeding. The hypertrophic form commonly affects ileocecal region and presents with abdominal pain, mass in the abdomen, or with signs of obstruction. In addition,

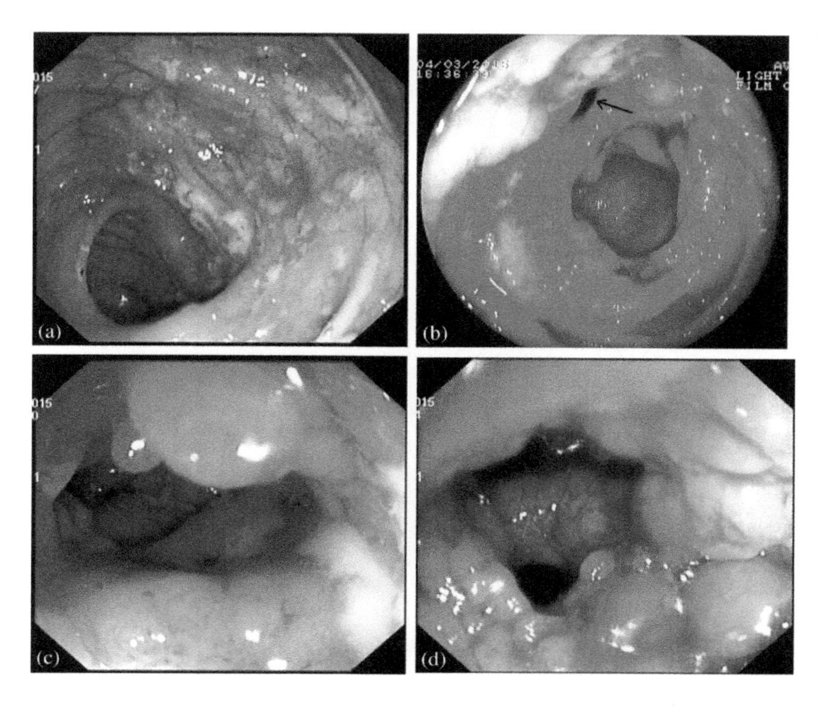

Figure 1.4 Shows various endoscopic findings of intestinal tuberculosis. (a) Ulcer involving hepatic flexure, (b) multiple ulcers involving caecum and ileo-cecal (IC) value with deformed IC value (*arrow*). (c) Proliferative polypoidal lesions in the ascending colon. (d) Proliferative lesion causing obstruction in the ascending colon.

both the forms will have constitutional symptoms like loss of appetite, loss of weight, evening rise of temperature, and night sweats. Abdominal pain is the most common presentation of intestinal TB. The frequent complications are fistula, stricture, and bowel obstruction. Both small and large bowels are involved, and multiple areas of the bowel may be involved. The ileocecal region is the most common site involved and is seen in 75% of the patients, followed in frequency by ascending colon, jejunum, appendix, duodenum, stomach, esophagus, sigmoid, colon, and rectum (Rathi and Gambhire 2016). RIF mass is seen in 25% to 50% of patients (Marshall 1993; Horvath and Whelan 1998).

Diagnosing intestinal TB is often challenging, and it is difficult to differentiate it from Crohn disease. This is extremely important in TB-endemic countries because immunosuppressants given for Crohn disease will aggravate TB if wrongly treated. The definitive diagnosis of TB is only possible by demonstrating the acid-fast bacilli in the biopsy specimen (Figure 1.5), from the resected bowel specimen, or by culturing the organisms from these specimens. Techniques of NAAT, like nested PCR for identifying a particular genetic sequence of TB bacilli, is being added as an acceptable modality of diagnosing TB; however, it will not differentiate active disease from an older infection. The initial evaluation in any patient with suspected intestinal TB would be an abdominal imaging (i.e., preferably contract-enhanced computed tomography [CECT] of the abdomen), which gives information as to the site involved and the presence of complications if any. The samples are usually obtained by performing colonoscopy when the colon or ileocecal region is involved. Obtaining samples in cases with small bowel involvement is challenging and might require enteroscopy. In TB, both the sides of the ileocecal valve is usually involved and is often destroyed, giving an appearance of fish mouth opening. The usual colonoscopy findings are ulcers, strictures, nodules, pseudo-polyps, fibrous bands, and fistulas (Alvares et al. 2005).

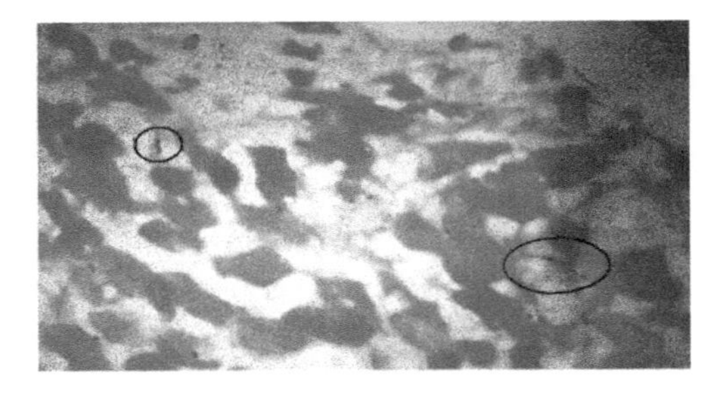

Figure 1.5 Ziehl Neelson stain showing the acid fast *Mycobacterium tuberculosis* bacilli (*round arrows*).

The organisms are commonly seen in the submucosa, and hence, deeper biopsies are required from both the ulcer and the margins for a better yield. The characteristic histopathological features of intestinal TB include presence of caseation granulomas and presence of acid-fast bacilli. The features of TB granuloma that helps in differentiating them from other etiologies include confluent granulomas, granulomas >400 micrometers in diameter, more than five granulomas in biopsies from one segment, and granulomas located in the submucosa or granulation tissue (Pulimood et al. 2005; Rathi and Gambhire 2016). Since the yield of smear and culture is less, the diagnosis sometimes depends on the combinations of clinical, radiological, endoscopic, histopathological findings, and the NAAT results. The other supporting evidence of pulmonary involvement (i.e., chest X-ray [CXR]), if present, may be helpful. The Mantoux test and the QuantiFERON gold assay has only a limited value because they cannot differentiate an active disease from previous infections.

The treatment of the intestinal TB is the anti-tubercular therapy (ATT) regimen similar to pulmonary TB (Makharia et al. 2015; Jullien et al. 2016). Patients show clinical improvement usually after 2 weeks, and complete resolution of ulcers and erosions on follow-up colonoscopy is seen after 2–3 months of anti-tubercular therapy. Surgery is indicated in patients with complications like stricture, obstruction, perforation, abscess, fistula, and bleeding (Aston 1997; Horvath and Whelan 1998; Kapoor 1998).

1.15 Future perspective

Larger number of bacterial organisms are believed to be responsible for GI infections. Some are true pathogens, whereas others are merely commensal in nature without causing any pathological conditions. The molecular strategies used by these bacteria to interact with the host can be unique to specific pathogens or may possess a conserved pattern across several different species. Effective treatment and key to fight the bacterial disease rely on the precise and rapid identification and characterization of these bacteria. Conventional culture-based methods that are widely used in many laboratories can detect only viable bacteria, but not the ones that are non-culturables that are metabolically active but not reproducing (Figure 1.6). Current advancements in molecular biology plays a vital role in clinical diagnostic settings by identifying a large number of bacteria directly from stool samples (e.g., metagenomics allows rapid retrieval of genetic material and screening directly from environmental sources). Thus, these high-throughput methods that are described elsewhere may add knowledge to metabolic function of bacteria and their role in causing GI infections.

Figure 1.6 Different diagnostic methods used in identification of bacteria and its analytes.

References

Alvares, J. F., H. Devarbhavi, P. Makhija, S. Rao, and R. Kottoor. 2005. Clinical, colonoscopic, and histological profile of colonic tuberculosis in a tertiary hospital. *Endoscopy* 37 (4):351–356.

Ardissino, G., S. Salardi, E. Colombo, S. Testa, N. Borsa-Ghiringhelli, F. Paglialonga, V. Paracchini et al. 2016. Epidemiology of haemolytic uremic syndrome in children. Data from the north Italian HUS network. *European Journal of Pediatrics* 175 (4):465–473. doi:10.1007/s00431-015-2642-1.

Aston, N. O. 1997. Abdominal tuberculosis. *World Journal of Surgery* 21 (5):492–499. doi:10.1007/PL00012275.

Avraham A., C. Maayan, and K. J. Goitein. 1982. Incidence of convulsions and encephalopathy in childhood *Shigella* infections: Survey of 117 hospitalized patients. *Clinical Pediatrics* 21 (11):645–648. doi:10.1177/000992288202101101.

Azad, M. A. K., M. Islam, and T. Butler. 1986. Colonic perforation in *Shigella dysenteriae* infection. *The Pediatric Infectious Disease Journal* 5 (1):103.

Backert, S. and M. Selbach. 2008. Role of type IV secretion in *Helicobacter pylori* pathogenesis. *Cellular Microbiology* 10 (8):1573–1581. doi:10.1111/j.1462-5822.2008.01156.x.

Bai, Y., Y.-C. Dai, J.-D. Li, J. Nie, Q. Chen, H. Wang, Y.-Y. Rui, Y.-L. Zhang, and S.-Y. Yu. 2004. Acute diarrhea during army field exercise in southern China. *World Journal of Gastroenterology* 10 (1):127–131. doi:10.3748/wjg.v10.i1.127.

Balaban, N. and A. Rasooly. 2000. Staphylococcal enterotoxins. *International Journal of Food Microbiology* 61 (1):1–10.

Bartlett, J. G. and D. N. Gerding. 2008. Clinical recognition and diagnosis of *Clostridium difficile* infection. *Clinical Infectious Diseases* 46 (Supplement_1):S12–S18. doi:10.1086/521863.

Bell, B. P., M. Goldoft, P. M. Griffin, M. A. Davis, D. C. Gordon, P. I. Tarr, C. A. Bartleson, J. H. Lewis, T. J. Barrett, and J. G. Wells. 1994. A multistate outbreak of *Escherichia coli* O157:H7-Associated bloody diarrhea and hemolytic uremic syndrome from hamburgers. The washington experience. *JAMA* 272 (17):1349–1353.

Bender, J. B., C. W. Hedberg, J. M. Besser, D. J. Boxrud, K. L. MacDonald, and M. T. Osterholm. 1997. Surveillance for *Escherichia coli* O157:H7 infections in minnesota by molecular subtyping. *New England Journal of Medicine* 337 (6):388–394. doi:10.1056/NEJM199708073370604.

Bennish, M. L. 1991. Potentially lethal complications of shigellosis. *Reviews of Infectious Diseases* 13 (Supplement_4):S319–S324. doi:10.1093/clinids/13.Supplement_4.S319.

Bennish, M. L., A. K. Azad, and D. Yousefzadeh. 1991. Intestinal obstruction during shigellosis: Incidence, clinical features, risk factors, and outcome. *Gastroenterology* 101 (3):626–634. doi:10.1016/0016-5085(91)90518-P.

Bennish, M. L., W. A. Khan, M. Begum, E. A. Bridges, S. Ahmed, D. Saha, M. A. Salam, D. Acheson, and E. T. Ryan. 2006. Low risk of hemolytic uremic syndrome after early effective antimicrobial therapy for *Shigella dysenteriae* type 1 infection in Bangladesh. *Clinical Infectious Diseases* 42 (3):356–362. doi:10.1086/499236.

Berg, R. J. 2007. Molecular diagnosis and genotyping of *Clostridium difficile*. Doctoral Thesis. November 21, 2007. https://openaccess.leidenuniv.nl/handle/1887/12445.

Black, R. E., M. M. Levine, M. L. Clements, T. P. Hughes, and M. J. Blaser. 1988. Experimental *Campylobacter jejuni* infection in humans. *The Journal of Infectious Diseases* 157 (3):472–479. doi:10.1093/infdis/157.3.472.

Boland, A. and G. R. Cornelis. 1998. Role of YopP in suppression of tumor necrosis factor alpha release by macrophages during *Yersinia* infection. *Infection and Immunity* 66 (5):1878–1884.

Borren, N. Z., S. Ghadermarzi, S. Hutfless, and A. N. Ananthakrishnan. 2017. "The emergence of *Clostridium difficile* infection in Asia: A systematic review and meta-analysis of incidence and impact. *PLoS One* 12 (5):e0176797. doi:10.1371/journal.pone.0176797.

Bottone, E. J. 1997. *Yersinia enterocolitica*: The charisma continues. *Clinical Microbiology Reviews* 10 (2):257–276.

Bouza, E., A. Dominguez, M. Meseguer, L. Buzon, D. Boixeda, M. J. Revillo, L. De Rafael, and J. Martinez-Beltran. 1980. *Yersinia enterocolitica* Septicemia. *American Journal of Clinical Pathology* 74 (4):404–409. doi:10.1093/ajcp/74.4.404.

Bowen, A. B., and C. R. Braden. 2006. Invasive enterobacter sakazakii disease in infants. *Emerging Infectious Diseases* 12 (8):1185–1189. doi:10.3201/eid1208.051509.

Boyce, T. G., D. L. Swerdlow, and P. M. Griffin. 1995. *Escherichia coli* O157:H7 and the hemolytic–uremic syndrome. *New England Journal of Medicine* 333 (6):364–368. doi:10.1056/NEJM199508103330608.

Brazier, J. S. and S. P. Borriello. 2000. Microbiology, epidemiology and diagnosis of *Clostridium difficile* infection. In *Clostridium Difficile*, pp. 1–33. Current Topics in Microbiology and Immunology. Springer, Berlin, Germany. doi:10.1007/978-3-662-06272-2_1.

Carey-Ann B. D. and K. C. Carroll. 2013. Diagnosis of *Clostridium difficile* infection: An ongoing conundrum for clinicians and for clinical laboratories. *Clinical Microbiology Reviews* 26 (3):604–630. doi:10.1128/CMR.00016-13.

Butler, T. 2012. Haemolytic uraemic syndrome during shigellosis. *Transactions of the Royal Society of Tropical Medicine and Hygiene* 106 (7):395–399. doi:10.1016/j.trstmh.2012.04.001.

Butler, T., M. R. Islam, and P. K. Bardhan. 1984. The leukemoid reaction in shigellosis. *American Journal of Diseases of Children* 138 (2):162–165. doi:10.1001/archpedi.1984.02140400044010.

Centers for Disease Control and Prevention (CDC). 2013. Outbreak of staphylococcal food poisoning from a military unit lunch party–United States, July 2012. *MMWR. Morbidity and Mortality Weekly Report* 62 (50):1026–1028.

Chang, V. T. and K. Nelson. 2000. The role of physical proximity in nosocomial diarrhea. *Clinical Infectious Diseases* 31 (3):717–722. doi:10.1086/314030.

Clemens, J. D., D. A. Sack, J. R. Harris, J. Chakraborty, P. K. Neogy, B. Stanton, N. Huda et al. 1988. Cross-protection by B subunit-whole cell cholera vaccine against diarrhea associated with heat-labile toxin-producing enterotoxigenic *Escherichia coli*: Results of a large-scale field trial. *The Journal of Infectious Diseases* 158 (2):372–377. doi:10.1093/infdis/158.2.372.

Cornelis, G. R. 2002. *Yersinia* type III secretion: Send in the effectors. *Journal of Cell Biology* 158 (3):401–408. doi:10.1083/jcb.200205077.

Crim, S. M., P. M. Griffin, R. Tauxe, E. P. Marder, D. Gilliss, A. B. Cronquist, M. Cartter, M. Tobin-D'Angelo, D. Blythe, and K. Smith. 2015. Preliminary incidence and trends of infection with pathogens transmitted commonly through food—Foodborne diseases active surveillance network, 10 US sites, 2006–2014. *MMWR Morbidity Mortality Weekly Report* 64 (18):495–499.

Das, S., P. K. Surendran, and N. Thampuran. 2009. PCR-based detection of enterotoxigenic isolates of *Bacillus cereus* from tropical seafood. http://lcmr. Nic.in/ljmr/2009/March/0316.Pdf, March. http://imsear.hellis.org/handle/123456789/135791.

Dave, G. A. 2017. A rapid qualitative assay for detection of *Clostridium perfringens* in canned food products. *Acta Biochimica Polonica* 64 (2). http://ojs.ptbioch.edu.pl/index.php/abp/article/view/1641.

Davies, N. E. C. G. and A. S. Karstaedt. 2008. *Shigella* bacteraemia over a decade in Soweto, South Africa. *Transactions of the Royal Society of Tropical Medicine and Hygiene* 102 (12):1269–1273. doi:10.1016/j.trstmh.2008.04.037.

Dossumbekova, A., C. Prinz, M. Gerhard, L. Brenner, S. Backert, J. G. Kusters, R. M. Schmid, and R. Rad. 2006. *Helicobacter pylori* outer membrane proteins and gastric inflammation. *Gut* 55 (9):1360–1361.

Doyle, M. P. (Ed.), Beuchat, L. R., and Inc ebrary. 2007. *Food Microbiology: Fundamentals and Frontiers*. Third edition. Washington, DC: ASM Press. https://trove.nla.gov.au/version/52019080.

Dumrese, C., L. Slomianka, U. Ziegler, S. S. Choi, A. Kalia, A. Fulurija, W. Lu et al. 2009. The secreted *Helicobacter* cysteine-rich protein a causes adherence of human monocytes and differentiation into a macrophage-like phenotype. *FEBS Letters* 583 (10):1637–1643. doi:10.1016/j.febslet.2009.04.027.

DuPont, H. L. 2007. Azithromycin for the self-treatment of traveler's diarrhea. *Clinical Infectious Diseases: An Official Publication of the Infectious Diseases Society of America* 44 (3):347–349. doi:10.1086/510594.

DuPont, H. L., R. B. Hornick, A. T. Dawkins, M. J. Snyder, and S. B. Formal. 1969. The response of man to virulent *Shigella flexneri* 2a. *The Journal of Infectious Diseases* 119 (3):296–299. doi:10.1093/infdis/119.3.296.

DuPont, H. L., M. M. Levine, R. B. Hornick, and S. B. Formal. 1989. Inoculum size in shigellosis and implications for expected mode of transmission. *The Journal of Infectious Diseases* 159 (6):1126–1128.

Eng, S. K., P. Pusparajah, N. S. Ab Mutalib, H. L. Ser, K. G. Chan, and L. H. Lee. 2015. Salmonella: A review on pathogenesis, epidemiology and antibiotic resistance. *Frontiers in Life Science* 8 (3): 284–293. doi: 10.1080/21553769.2015.1051243.

Eslick, G. D. 2010. Future perspectives on infections associated with gastrointestinal tract diseases. *Infectious Disease Clinics* 24 (4):1041–1058. doi:10.1016/j.idc.2010.08.002.

Farrell, R. J. and J. T. LaMont. 2000. Pathogenesis and clinical manifestations of *Clostridium difficile* diarrhea and colitis. In *Clostridium difficile*, 109–125. Current Topics in Microbiology and Immunology. Springer, Berlin, Heidelberg. doi:10.1007/978-3-662-06272-2_6.

Federko, D. P., and E. C. Williams. 1997. Use of cycloserine-Cefoxitin-fructose agar and L-proline-aminopeptidase (PRO Discs) in the rapid identification of Clostridium difficile. *Journal of Clinical Microbiologyl* 35 (5):1258–1259.

Finch, M., G. Rodey, D. Lawrence, and P. Blake. 1986. Epidemic reiter's syndrome following an outbreak of shigellosis. *European Journal of Epidemiology* 2 (1):26–30. doi:10.1007/BF00152713.

FOOD Tool|CDC. 2017. Government. Centers for Disease Control and Prevention. October 10, 2017. https://wwwn.cdc.gov/foodborne outbreaks/.

Frank, C., D. Werber, J. P. Cramer, M. Askar, M. Faber, M. an der Heiden, H. Bernard et al. 2011. Epidemic profile of shiga-toxin–producing *Escherichia coli* O104:H4 outbreak in Germany. *New England Journal of Medicine* 365 (19):1771–1780. doi:10.1056/NEJMoa1106483.

Fricker, M., U. Messelhäußer, U. Busch, S. Scherer, and M. Ehling-Schulz. 2007. Diagnostic real-time PCR assays for the detection of emetic *Bacillus cereus* strains in foods and recent food-borne outbreaks. *Applied and Environmental Microbiology* 73 (6):1892–1898. doi:10.1128/AEM.02219-06.

Goebel, W. 2012. *Genetic Approaches to Microbial Pathogenicity*. Berlin, Germany: Springer Science & Business Media.

Goh, K. L. 2007. Changing trends in gastrointestinal disease in the Asia–pacific region. *Journal of Digestive Diseases* 8 (4):179–185. doi:10.1111/j.1751-2980.2007.00304.x.

Granato, P. A., L. Chen, I. Holiday, R. A. Rawling, S. M. Novak-Weekley, T. Quinlan, and K. A. Musser. 2010. Comparison of premier CAMPY enzyme immunoassay (EIA), ProSpecT campylobacter EIA, and immunoCard STAT! CAMPY tests with culture for laboratory diagnosis of *Campylobacter* enteric infections. *Journal of Clinical Microbiology* 48 (11):4022–4027. doi:10.1128/JCM.00486-10.

Granum, P. E., S. Brynestad, and J. M. Kramer. 1993. Analysis of enterotoxin production by *Bacillus cereus* from dairy products, food poisoning incidents and non-gastrointestinal infections. *International Journal of Food Microbiology* 17 (4):269–279.

Guinet, F., E. Carniel, and A. Leclercq. 2011. Transfusion-transmitted *Yersinia enterocolitica* sepsis. *Clinical Infectious Diseases* 53 (6):583–591. doi:10.1093/cid/cir452.

Hatakeyama, M. 2004. Oncogenic mechanisms of the *Helicobacter pylori* CagA Protein. *Nature Reviews Cancer* 4 (9):688. doi:10.1038/nrc1433.

Hoffmann, S., M. B. Batz, and J. G. Morris. 2012. Annual cost of illness and quality-adjusted life year losses in the United States due to 14 foodborne pathogens. *Journal of Food Protection* 75 (7):1292–1302. doi:10.4315/0362-028X.JFP-11-417.

Hori, M., K. Hayashi, K. Maeshima, M. Kigawa, T. Miyasato, Y. Yoneda, and Y. Hagihara. 1966. Food poisoning caused by *Aeromonas shigelloides* with an antigen common to *Shigella dysenteriae* 7. *Journal of Japanese Association of Infectious Disease* 39:433–441.

Horvath, K. D. and R. L. Whelan. 1998. Intestinal tuberculosis: Return of an old disease. *The American Journal of Gastroenterology* 93 (5):692. doi:10.1111/j.1572-0241.1998.207_a.x.

Hui, Y. H. 1994. *Foodborne Disease Handbook: Diseases Caused by Viruses, Parasites, and Fungi; 3. Diseases Caused by Hazardous Substances.* Marcel Dekker: New York.

Humphries, R. M. and A. J. Linscott. 2015. Laboratory diagnosis of bacterial gastroenteritis. *Clinical Microbiology Reviews* 28 (1):3–31. doi:10.1128/CMR.00073-14.

Hunter, C. J., M. Petrosyan, H. R. Ford, and N. V. Prasadarao. 2008. *Enterobacter sakazakii*: An emerging pathogen in infants and neonates. *Surgical Infections* 9 (5):533–539. doi:10.1089/sur.2008.006.

Johnson, S., C. R. Clabots, F. V. Linn, M. M. Olson, L. R. Peterson, D. N. Gerding, S. Johnson, F. V. Linn, L. R. Peterson, and D. N. Gerding. 1990. Nosocomial *Clostridium difficile* colonisation and disease. *The Lancet* 336 (8707):97–100. doi:10.1016/0140-6736(90)91605-A.

Jullien, S., S. Jain, H. Ryan, and V. Ahuja. 2016. Six-month therapy for abdominal tuberculosis. *The Cochrane Database of Systematic Reviews*. doi:10.1002/14651858.CD012163.pub2.

Kapoor, V. K. 1998. Abdominal tuberculosis. *Postgraduate Medical Journal* 74 (874):459–467. doi:10.1136/pgmj.74.874.459.

Kaushik, J. S., P. Gupta, M. M. A. Faridi, and S. Das. 2010. Single dose azithromycin versus ciprofloxacin for cholera in children: A randomized controlled trial. *Indian Pediatrics* 47 (4):309–315. doi:10.1007/s13312-010-0059-5.

Kazanowski, M., S. Smolarek, F. Kinnarney, and Z. Grzebieniak. 2014. *Clostridium difficile*: Epidemiology, diagnostic and therapeutic possibilities-a systematic review. *Techniques in Coloproctology* 18 (3):223–232. doi:10.1007/s10151-013-1081-0.

Keat, A. 2010. Reiter's syndrome and reactive arthritis in perspective. *New England Journal of Medicine*. Review-article. doi:10.1056/NEJM198312293092604.

Khan, M. U., N. C. Roy, R. Islam, I. Huq, and B. Stoll. 1985. Fourteen years of shigellosis in dhaka: An epidemiological analysis. *International Journal of Epidemiology* 14 (4):607–613.

Khan, W. A., J. K. Griffiths, and M. L. Bennish. 2013. Gastrointestinal and extra-intestinal manifestations of childhood shigellosis in a region where all four species of shigella are endemic. *PLoS One* 8 (5):e64097. doi:10.1371/journal.pone.0064097.

Kulkarni, S. P., S. Lever, J. M. J. Logan, A. J. Lawson, J. Stanley, and M. S. Shafi. 2002. Detection of *Campylobacter* species: A comparison of culture and polymerase chain reaction based methods. *Journal of Clinical Pathology* 55 (10):749–753. doi:10.1136/jcp.55.10.749.

Kumagai, Y., S. Gilmour, E. Ota, Y. Momose, T. Onishi, F. Bilano, V. Luanni, F. Kasuga, T. Sekizaki, and K. Shibuya. 2015. Estimating the burden of foodborne diseases in Japan. *Bulletin of the World Health Organization* 93:540–549. doi:10.2471/BLT.14.148056.

Lachance, N., C. Gaudreau, F. Lamothe, and L. A. Larivière. 1991. Role of the beta-lactamase of *Campylobacter jejuni* in resistance to deta-lactam agents. *Antimicrobial Agents and Chemotherapy* 35 (5):813–818. doi:10.1128/AAC.35.5.813.

Lanata, C. F., W. Mendoza, and R. E. Black. 2002. Improving diarrhoea estimates. Monitoring and evaluation team child and adolescent health and development. World Health Organization Geneva, Switzerland. http://www.Who.Int/Child-Adolescent-Health/NewPublications/CHILD-HEALTH/EPI/Improving-Diarrhoea-Estimates. Pdf (Revised in May 2007).

Lecuit, M., E. Abachin, A. Martin, C. Poyart, P. Pochart, F. Suarez, D. Bengoufa et al. 2004. Immunoproliferative small intestinal disease associated with *Campylobacter jejuni*. *New England Journal of Medicine* 350 (3):239–248. doi:10.1056/NEJMoa031887.

Lessa, F. C., Y. Mu, W. M. Bamberg, Z. G. Beldavs, G. K. Dumyati, J. R. Dunn, M. M. Farley et al. 2015. Burden of *Clostridium difficile* infection in the United States. *New England Journal of Medicine* 372 (9):825–834. doi:10.1056/NEJMoa1408913.

Li, J. and B. A. McClane. 2006. Further comparison of temperature effects on growth and survival of *Clostridium perfringens* type A isolates carrying a chromosomal or plasmid-borne enterotoxin gene. *Applied and Environmental Microbiology* 72 (7):4561–4568. doi:10.1128/AEM.00177-06.

Long, C., T. F. Jones, D. J. Vugia, J. Scheftel, N. Strockbine, P. Ryan, B. Shiferaw, R. V. Tauxe, and L. H. Gould. 2010. *Yersinia pseudotuberculosis* and *Y. enterocolitica* infections, foodNet, 1996–2007. *Emerging Infectious Diseases* 16 (3):566–567. doi:10.3201/eid1603.091106.

MacDonell, M. T., and R. R. Colwell. 1985. Phylogeny of the vibrionaceae, and recommendation for two new genera, *Listonella* and *Shewanella*. *Systematic and Applied Microbiology* 6 (2):171–182. doi:10.1016/S0723-2020(85)80051-5.

Majowicz, S. E., J. Musto, E. Scallan, F. J. Angulo, M. Kirk, S. J. O'Brien, T. F. Jones et al. 2010. The global burden of nontyphoidal salmonella gastroenteritis. International Collaboration on Enteric Disease "Burden of Illness Studies". *Clinical Infectious Diseases* 50 (6):882-889. doi: 10.1086/650733.

Makharia, G. K., U. C. Ghoshal, B. S. Ramakrishna, A. Agnihotri, V. Ahuja, S. D. Chowdhury, S. D. Gupta et al. 2015. Intermittent directly observed therapy for abdominal tuberculosis: A multicenter randomized controlled trial comparing 6 Months versus 9 Months of therapy. *Clinical Infectious Diseases* 61 (5):750–757. doi:10.1093/cid/civ376.

Marshall, J. B. 1993. Tuberculosis of the gastrointestinal tract and peritoneum. *The American Journal of Gastroenterology* 88 (7):989–999.

Martin, T., B. F. Habbick, and J. Nyssen. 1983. Shigellosis with bacteremia: A report of two cases and a review of the literature. *The Pediatric Infectious Disease Journal* 2 (1):21.

McCollum, D. L. and J. M. Rodriguez. 2012. Detection, treatment, and prevention of *Clostridium difficile* infection. *Clinical Gastroenterology and Hepatology* 10 (6):581–592. doi:10.1016/j.cgh.2012.03.008.

McFarland, L. V., M. E. Mulligan, R. Y. Y. Kwok, and W. E. Stamm. 1989. Nosocomial acquisition of *Clostridium difficile* infection. *New England Journal of Medicine* 320 (4):204–210. doi:10.1056/NEJM198901263200402.

Mead, P. S., L. Slutsker, V. Dietz, L. F. McCaig, J. S. Bresee, C. Shapiro, P. M. Griffin, and R. V. Tauxe. 1999. Food-related illness and death in the United States. *Emerging Infectious Diseases* 5 (5):607–625.

Miehlke, S., C. Kirsch, K. Agha-Amiri, T. Günther, N. Lehn, P. Malfertheiner, M. Stolte, G. Ehninger, and E. Bayerdörffer. 2000. The *Helicobacter pylori* vacA s1, m1 genotype and cagA is associated with gastric carcinoma in Germany. *International Journal of Cancer* 87 (3):322–327. doi:10.1002/1097-0215(20000801)87:3<322::AID-IJC3>3.0.CO;2-M.

Millet, Y. A., D. Alvarez, S. Ringgaard, U. H. von Andrian, B. M. Davis, and M. K. Waldor. 2014. Insights into *Vibrio cholerae* intestinal colonization from monitoring fluorescently labeled bacteria. *PLoS Pathogens* 10 (10):e1004405. doi:10.1371/journal.ppat.1004405.

Mølbak, K., P. S. Mead, and P. M. Griffin. 2002. Antimicrobial therapy in patients with *Escherichia coli* O157:H7 infection. *JAMA* 288 (8):1014–1016. doi:10.1001/jama.288.8.1014.

Murray, P. R and E. J. Baron. 2007. *Manual of Clinical Microbiology*. Washington, DC: ASM Press.

Nachamkin, I., B. M. Allos, and T. Ho. 1998. *Campylobacter* species and guillain-barré syndrome. *Clinical Microbiology Reviews* 11 (3):555–567.

Nagy, T. A., M. R. Frey, F. Yan, D. A. Israel, D. B. Polk, and R. M. Peek. 2009. *Helicobacter pylori* regulates cellular migration and apoptosis by activation of phosphatidylinositol 3-kinase signaling. *The Journal of Infectious Diseases* 199 (5):641–651. doi:10.1086/596660.

Nataro, J. P., J. Seriwatana, A. Fasano, D. R. Maneval, L. D. Guers, F. Noriega, F. Dubovsky, M. M. Levine, and J. G. Morris. 1995. Identification and cloning of a novel plasmid-encoded enterotoxin of enteroinvasive *Escherichia coli* and *Shigella* strains. *Infection and Immunity* 63 (12):4721–4728.

Nguyen, R. N., L. S. Taylor, M. Tauschek, and R. M. Robins-Browne. 2006. Atypical enteropathogenic *Escherichia coli* infection and prolonged diarrhea in children. *Emerging Infectious Diseases* 12 (4):597–603. doi:10.3201/eid1204.051112.

Ochoa, T. J. and C. A. Contreras. 2011. Enteropathogenic *E. coli* (EPEC) infection in children. *Current Opinion in Infectious Diseases* 24 (5):478–483. doi:10.1097/QCO.0b013e32834a8b8b.

Olson, M. M., C. J. Shanholtzer, J. T. Lee, and D. N. Gerding. 1994. Ten years of prospective *Clostridium difficile* associated disease surveillance and treatment at the minneapolis VA medical center, 1982–1991. *Infection Control & Hospital Epidemiology* 15 (6):371–381. doi:10.2307/30145589.

Park, M. S., Y. S. Kim, S. H. Lee, S. H. Kim, K. H. Park, and G. J. Bahk. 2015. Estimating the burden of foodborne disease, South Korea, 2008–2012. *Foodborne Pathogens and Disease* 12 (3):207–213. doi:10.1089/fpd.2014.1858.

Parsot, C. 2005. *Shigella Spp.* and enteroinvasive *Escherichia coli* pathogenicity factors. *FEMS Microbiology Letters* 252 (1):11–18. doi:10.1016/j.femsle.2005.08.046.

Pazhani, G. P., T. Ramamurthy, U. Mitra, S. K. Bhattacharya, and S. K. Niyogi. 2005. Species diversity and antimicrobial resistance of *Shigella Spp.* isolated between 2001 and 2004 from hospitalized children with diarrhoea in Kolkata (Calcutta), India. *Epidemiology & Infection* 133 (6):1089–1095. doi:10.1017/S0950268805004498.

Peltola, H., A. Siitonen, M. J. Kataja, H. Kyrönseppa, I. Simula, L. Mattila, P. Oksanen, and M. Cadoz. 1991. Prevention of travellers' diarrhoea by oral B-subunit/whole-cell cholera vaccine. *The Lancet* 338 (8778):1285–1289. doi:10.1016/0140-6736(91)92590-X.

Pickett, C. L., E. C. Pesci, D. L. Cottle, G. Russell, A. N. Erdem, and H. Zeytin. 1996. Prevalence of cytolethal distending toxin production in *Campylobacter jejuni* and relatedness of *Campylobacter Sp.* CdtB gene. *Infection and Immunity* 64 (6):2070–2078.

Planche, T. and M. Wilcox. 2011. Reference assays for *Clostridium difficile* infection: One or two gold standards? *Journal of Clinical Pathology* 64 (1):1–5. doi:10.1136/jcp.2010.080135.

Popoff, M. R. 2011. Multifaceted interactions of bacterial toxins with the gastrointestinal mucosa. *Future Microbiology* 6 (7):763–797. doi:10.2217/fmb.11.58.

Porter, C. K., D. Choi, and M. S. Riddle. 2013. Pathogen-specific risk of reactive arthritis from bacterial causes of foodborne illness. *The Journal of Rheumatology* 40 (5):712–714. doi:10.3899/jrheum.121254.

Pulimood, A. B., S. Peter, B. S. Ramakrishna, A. Chacko, R. Jeyamani, L. Jeyaseelan, and G. Kurian. 2005. Segmental colonoscopic biopsies in the differentiation of ileocolic tuberculosis from Crohn's disease. *Journal of Gastroenterology and Hepatology* 20 (5):688–696. doi:10.1111/j.1440-1746.2005.03814.x.

Puylaert, J. B. C. M., R. J. Vermeijden, S. D. J. Van Der Werf, L. Doornbos, and R. J. Koumans. 1989. Incidence and sonographic diagnosis of bacterial ileocaecitis masquerading as appendicitis. *The Lancet* 334 (8654):84–86. doi:10.1016/S0140-6736(89)90322-X.

Rabson, A. R., A. F. Hallett, and H. J. Koornhof. 1975. Generalized *Yersinia enterocolitica* infection. *The Journal of Infectious Diseases* 131 (4):447–451. doi:10.1093/infdis/131.4.447.

Ramakrishnan, N. and K. Sriram. 2015. Antibiotic overuse and *Clostridium difficile* infections: The Indian paradox and the possible role of dietary practices. *Nutrition* 31 (7):1052–1053. doi:10.1016/j.nut.2015.02.002.

Rathi, P. and P. Gambhire. 2016. Abdominal tuberculosis. *The Journal of the Association of Physicians of India* 64 (2):38–47.

Rees, J. H., S. E. Soudain, N. A. Gregson, and R. A. C. Hughes. 1995. *Campylobacter jejuni* infection and guillain–barré syndrome. *New England Journal of Medicine* 333 (21):1374–1379.

Riley, L. W., R. S. Remis, S. D. Helgerson, H. B. McGee, J. G. Wells, B. R. Davis, R. J. Hebert et al. 1983. Hemorrhagic colitis associated with a rare *Escherichia coli* serotype. *New England Journal of Medicine* 308 (12):681–685. doi:10.1056/NEJM198303243081203.

Robinson, D. A. 1981. Infective dose of *Campylobacter jejuni* in milk. *British Medical Journal (Clinical Research Ed.)* 282 (6276):1584.

Roy, S. K., A. Islam, R. Ali, K. E. Islam, R. A. Khan, S. H. Ara, N. M. Saifuddin, and G. J. Fuchs. 1998. A randomized clinical trial to compare the efficacy of erythromycin, ampicillin and tetracycline for the treatment of cholera in children. *Transactions of the Royal Society of Tropical Medicine and Hygiene* 92 (4):460–462. doi:10.1016/S0035-9203(98)91094-X.

Rutala, W. A., F. A. Sarubbi Jr, C. S. Finch, J. N. MacCormack, and G. E. Steinkraus. 1982. Oyster-associated outbreak of diarrhoeal disease possibly caused by *Plesiomonas shigelloides*. *The Lancet* 319 (8274):739.

Safdar, N., A. Said, R. E. Gangnon, and D. G. Maki. 2002. Risk of hemolytic uremic syndrome after antibiotic treatment of *Escherichia coli* O157:H7 enteritis: A meta-analysis. *JAMA* 288 (8):996–1001.

Samore, M. H., P. C. DeGirolami, A. Tlucko, D. A. Lichtenberg, Z. A. Melvin, and A. W. Karchmer. 1994. *Clostridium difficile* colonization and diarrhea at a tertiary care hospital. *Clinical Infectious Diseases* 18 (2):181–187. doi:10.1093/clinids/18.2.181.

Sánchez, J. and J. Holmgren. 2008. Cholera toxin structure, gene regulation and pathophysiological and immunological aspects. *Cellular and Molecular Life Sciences: CMLS* 65 (9):1347–1360. doi:10.1007/s00018-008-7496-5.

Scallan, E., R. M. Hoekstra, F. J. Angulo, R. V. Tauxe, M.-A. Widdowson, S. L. Roy, J. L. Jones, and P. M. Griffin. 2011. Foodborne illness acquired in the United States—major pathogens. *Emerging Infectious Diseases* 17 (1):7–15. doi:10.3201/eid1701.P11101.

Scharff, R. L. 2010. Economic burden from health losses due to food borne illness in the United States. *Journal of Food Protection* 75 (1):123–131. doi: 10.4315/0362-028XJFP-11-058.

Schneider, F., N. Lang, R. Reibke, H. J. Michaely, W. Hiddemann, and H. Ostermann. 2009. *Plesiomonas shigelloides* pneumonia. *Médecine et Maladies Infectieuses* 39 (6):397–400. doi:10.1016/j.medmal.2008.11.010.

Schneider, K. R., M. E. Parish, R. M. Goodrich, and T. Cookingham. 2004. *Preventing Foodborne Illness: Bacillus Cereus and Bacillus Anthracis*. Florida Cooperative Extension Services, EDIS FSHN036. University of Florida, Gainesville, FL.

Shaheen, N. J., R. A. Hansen, D. R. Morgan, L. M. Gangarosa, Y. Ringel, M. T. Thiny, M. W. Russo, and R. S. Sandler. 2006. The burden of gastrointestinal and liver diseases, 2006. *The American Journal of Gastroenterology* 101 (9):2128. doi:10.1111/j.1572-0241.2006.00723.x.

Sharma, S. K., and A. Mohan. 2004. Extrapulmonary tuberculosis. *The Indian Journal of Medical Research* 120 (4):316–353.

Simon, D. G., R. A. Kaslow, J. Rosenbaum, R. L. Kaye, and A. Calin. 1981. Reiter's syndrome following epidemic shigellosis. *The Journal of Rheumatology* 8 (6):969–973.

Sirinavin, S. and P. Garner. 2000. Antibiotics for treating *Salmonella* gut infections. *The Cochrane Database of Systematic Reviews* 2:CD001167.

Sjögren, E., B. Kaijser, and M. Werner. 1992. Antimicrobial susceptibilities of *Campylobacter jejuni* and *Campylobacter coli* isolated in Sweden: A 10-year follow-up report. *Antimicrobial Agents and Chemotherapy* 36 (12):2847–2849. doi:10.1128/AAC.36.12.2847.

Skirrow, M. B. 2006. John McFadyean and the centenary of the first isolation of *Campylobacter* species. *Clinical Infectious Diseases* 43 (9):1213–1217. doi:10.1086/508201.

Sneath, P. H. A., N. S. Mair, M. E. Sharp, and J. G. Holt. 1986. *Bergey's Manual of Systematic Bacteriology*. Lippincott Williams & Wilkins: Baltimore, MD.

Snell, H., M. Ramos, S. Longo, M. John, and Z. Hussain. 2004. Performance of the TechLab C. DIFF CHEK-60 enzyme immunoassay (EIA) in combination with the *C. difficile* Tox A/B II EIA Kit, the Triage *C. difficile* panel immunoassay, and a cytotoxin assay for diagnosis of *Clostridium difficile*-associated diarrhea. *Journal of Clinical Microbiology* 42 (10):4863–4865. doi:10.1128/JCM.42.10.4863-4865.2004.

Struelens, M. J., D. Patte, I. Kabir, A. Salam, S. K. Nath, and T. Butler. 1985. *Shigella* septicemia: Prevalence, presentation, risk factors, and outcome. *The Journal of Infectious Diseases* 152 (4):784–790. doi:10.1093/infdis/152.4.784.

Talan, D. A., G. J. Moran, M. Newdow, S. Ong, W. R. Mower, J. Y. Nakase, R. W. Pinner, and L. Slutsker. 2001. Etiology of bloody diarrhea among patients presenting to United States emergency departments: Prevalence of *Escherichia coli* O157:H7 and other enteropathogens. *Clinical Infectious Diseases* 32 (4):573–580. doi:10.1086/318718.

Tarr, P. I., C. A. Gordon, and W. L. Chandler. 2005. Shiga-toxin-producing *Escherichia coli* and haemolytic uraemic syndrome. *The Lancet* 365 (9464):1073–1086. doi:10.1016/S0140-6736(05)71144-2.

Ternhag, A., A. Törner, Å. Svensson, K. Ekdahl, and J. Giesecke. 2008. Short- and long-term effects of bacterial gastrointestinal infections. *Emerging Infectious Diseases* 14 (1):143–148. doi:10.3201/eid1401.070524.

Terpeluk, C., A. Goldmann, P. Bartmann, and F. Pohlandt. 1992. *Plesiomonas shigelloides* sepsis and meningoencephalitis in a neonate. *European Journal of Pediatrics* 151 (7):499–501. doi:10.1007/BF01957753.

Tewari, A. and S. Abdullah. 2015. *Bacillus cereus* food poisoning: International and Indian perspective. *Journal of Food Science and Technology* 52 (5):2500–2511. doi:10.1007/s13197-014-1344-4.

Tracz, D. M., M. Keelan, J. Ahmed-Bentley, A. Gibreel, K. Kowalewska-Grochowska, and D. E. Taylor. 2005. PVir and bloody diarrhea in *Campylobacter jejuni* enteritis. *Emerging Infectious Diseases* 11 (6):839–843. doi:10.3201/eid1106.041052.

Tribble, D. R., J. W. Sanders, L. W. Pang, C. Mason, C. Pitarangsi, S. Baqar, A. Armstrong, A. et al. 2007. Traveler's diarrhea in Thailand: randomized, double-blind trial comparing single-dose and 3-day azithromycin-based regimens with a 3-day levofloxacin regimen. *Clinical Infectious Diseases* 44 (3):338–346. doi:10.1086/510589.

Tsukamoto, T., Y. Kinoshita, T. Shimada, and R. Sakazaki. 1978. Two epidemics of diarrhoeal disease possibly caused by *Plesiomonas shigelloides*. *Epidemiology & Infection* 80 (2):275–280. doi:10.1017/S0022172400053638.

Vaishnavi, C. 2010. "Clinical spectrum & pathogenesis of *Clostridium difficile* associated diseases. http://Icmr.Nic.in/Ijmr/2010/April/0403. Pdf, April. http://imsear.hellis.org/handle/123456789/135472.

Van Spreeuwel, J. P., G. C. Duursma, C. J. Meijer, R. Bax, P. C. Rosekrans, and J. Lindeman. 1985. *Campylobacter colitis*: Histological immunohistochemical and ultrastructural findings. *Gut* 26 (9):945–951. doi:10.1136/gut.26.9.945.

Van Spreeuwel, J. P., J. Lindeman, R. Bax, H. J. Elbers, R. Sybrandy, and C. J. Meijer. 1987. *Campylobacter*-associated appendicitis: Prevalence and clinicopathologic features. *Pathology Annual* 22:55–65.

Vernon, M. and R. Chatlain. 1973. Taxonomic study of the genus *Campylobacter* (Sebald and Veron) and designation of the neotype strain for the type species, *Campylobacter fetus* (Smith and Taylor) sebald and veron. *International Journal of Systematic and Evolutionary Microbiology* 23:122–134.

Viala, J., C. Chaput, I. G. Boneca, A. Cardona, S. E. Girardin, A. P. Moran, R. Athman et al. 2004. Nod1 responds to peptidoglycan delivered by the *Helicobacter pylori cag* pathogenicity Island. *Nature Immunology* 5 (11):1166. doi:10.1038/ni1131.

Walker, C. L. F., I. Rudan, L. Liu, H. Nair, E. Theodoratou, Z. A. Bhutta, K. L. O'Brien, H. Campbell, and R. E. Black. 2013. Global burden of childhood pneumonia and diarrhoea. *Lancet (London, England)* 381 (9875):1405–1416. doi:10.1016/S0140-6736(13)60222-6.

Wennerås, C. and V. Erling. 2004. Prevalence of enterotoxigenic *Escherichia coli*-associated diarrhoea and carrier state in the developing world. *Journal of Health, Population and Nutrition* 22 (4):370–382.

Whitehouse, C. A., P. B. Balbo, E. C. Pesci, D. L. Cottle, P. M. Mirabito, and C. L. Pickett. 1998. *Campylobacter jejuni* cytolethal distending toxin causes a G2-phase cell cycle block. *Infection and Immunity* 66 (5):1934–1940.

Wilcox, M. H., W. N. Fawley and P. Parnell. 2000. Value of lysozyme agar incorporation and alkaline thioglycollate exposure for the environmental recovery of *Clostridium difficile*. *Journal of Hospital Infection* 44 (1):65–69. doi:10.1053/jhin.1999.0253.

Wilson, K. H. 1993. The microecology of *Clostridium difficile*. *Clinical Infectious Diseases: An Official Publication of the Infectious Diseases Society of America* 16 (Suppl 4):S214–S218.

Wong, C. S., S. Jelacic, R. L. Habeeb, S. L. Watkins, and P. I. Tarr. 2000. The risk of the hemolytic–uremic syndrome after antibiotic treatment of *Escherichia coli* O157:H7 infections. *New England Journal of Medicine* 342 (26):1930–1936. doi:10.1056/NEJM200006293422601.

Wouafo, M., R. Pouillot, P. F. Kwetche, M.-C. Tejiokem, J. Kamgno, and M.-C. Fonkoua. 2006. An acute foodborne outbreak due to *Plesiomonas shigelloides* in yaounde, cameroon. *Foodborne Pathogens and Disease* 3 (2):209–211. doi:10.1089/fpd.2006.3.209.

Zar, F. A., S. R. Bakkanagari, K. M. L. S. T. Moorthi, and M. B. Davis. 2007. A comparison of vancomycin and metronidazole for the treatment of *Clostridium difficile*–associated diarrhea, stratified by disease severity. *Clinical Infectious Diseases* 45 (3):302–307. doi:10.1086/519265.

Zhang, X., A. D. McDaniel, L. E. Wolf, G. T. Keusch, M. K. Waldor, and D. W. Acheson. 2000. Quinolone antibiotics induce Shiga toxin-encoding bacteriophages, toxin production, and death in mice. *The Journal of Infectious Diseases* 181 (2):664–670. doi:10.1086/315239.

Zheng, L., S. F. Keller, D. M. Lyerly, R. J. Carman, C. W. Genheimer, C. A. Gleaves, S. J. Kohlhepp, S. Young, S. Perez, and K. Ye. 2004. Multicenter evaluation of a new screening test that detects *Clostridium difficile* in fecal specimens. *Journal of Clinical Microbiology* 42 (8):3837–3840. doi:10.1128/JCM.42.8.3837-3840.2004.

2

Gateways of Pathogenic Bacterial Entry into Host Cells—*Salmonella*

Balakrishnan Senthilkumar, Duraisamy Senbagam, Chidambaram Prahalathan, and Kumarasamy Anbarasu

Contents

2.1 Introduction

Enteric fever is one of the major tropical bacterial, foodborne disease caused by *Salmonella* species, causing high mortality and morbidity (Marks et al. 2017). The major symptoms include nausea, vomiting, abdominal cramps, loss of appetite, bloody diarrhea, fever, headache, and so on. The incubation period varies from individual to individual but usually is from 6 to 72 hours. The bacteria persists a long time asymptomatically in healthy carriers from months to years in the host organism bowel and is released every day in their feces, but at the same time, they are highly invasive (Ilakia et al. 2015; Prestinaci et al. 2015). Southeast Asia and western Pacific countries are in the top list of high morbidity and mortality from enteric fever. In 2015, it was assessed to cause about 178,000 deaths and 17 million hospitalizations; more than 85% of all incidences occur in three major countries, Bangladesh, India, and Pakistan (Wang et al. 2016). In United States alone, 1.4 million cases of human salmonellosis, 7000 hospitalizations, and almost 600 deaths were chronicled each year (Hendriksen et al. 2004; Voetsch et al. 2004). Globally, *Salmonella enterica* serovar Typhimurium was the frequently isolated serovar in most of the enteric fever cases (Leekitcharoenphon et al. 2016). Other nontyphoidal *Salmonella* (NTS) serovars that usually cause self-limiting diarrhea can also lead to multiple systemic infections and are generally referred to as invasive NTS (iNTS). Globally iNTS is estimated to cause 3.4 million hospitalizations and 681,000 fatalities annually (Ao et al. 2015). Although, *Salmonella* infection contributes substantially to global morbidity and mortality during the pre-antibiotic era, the percentage of case fatality rate has decreased from 30% to less than 1% with the use of effective antibiotics (Crump et al. 2015). But, recently, there has been increasing number of *Salmonella* Typhi/Parathyphi incidents that were reported because of the emergence of antibiotic-resistance behaviors (Kariuki et al. 2015). Multidrug-resistant (MDR) *Salmonella* serovars are resistant to first-line antibiotics, ampicillin, tetracycline, trimethoprim, sulfamethoxazole, and chloramphenicol; however, they are sensitive to fluoroquinolones. In Southeast Asia and African countries, they are also susceptible to fluoroquinolones.

The transfer of antimicrobial resistance (AMR) genes among bacteria is commonly enabled by plasmid or transposon exchange. *Salmonella* pathogens are highly invasive in nature. H58, a dominant haplotype of *S. typhi* mostly identified in the high risky regions, are mostly associated with an IncHI1 plasmid. Such plasmids anchor a composite transposon that can carry multiple resistance genes, including *blaTEM-1* (ampicillin resistance), *dfrA7*, *sul1*, *sul2* (trimethoprim-sulfamethoxazole resistance), *catA1* (chloramphenicol resistance), and *strAB* (streptomycin resistance) genes (Klemm et al. 2018). Since 2016, a large proportion of *S. typhi* resistant to drugs, including chloramphenicol, ampicillin, trimethoprim-sulfamethoxazole, and third-generation cephalosporins, have been reported in Pakistan and the United Kingdom (Feasey et al. 2015; Hendriksen et al. 2015; Wong et al. 2015). Because of the emergence of these extensively drug-resistant

(XDR) isolates, urgent action is needed for the hour to avoid lineages. Understanding their invasive mechanism and their proliferation within the host cells could be an efficient approach to control the spreading of such perilous lineages worldwide. Keeping that in mind, this chapter narrates the mechanism of *Salmonella* pathogenicity manipulates in the host cell using different strategies for successful invasion to establish an intracellular niche.

2.2 *Salmonella*

Salmonella, a gram-negative, rod-shaped, facultative anaerobic, nonsporing bacteria, belongs to the member of Enterobacteriaceae family, causing gastrointestinal (GI) disorders and multisystemic fatal infections. Initially the genus *Salmonella* consists of two species namely *S. bongori* and *S. enterica* (Porwollik et al. 2004). *S. enterica*, the enteric pathogen, is further classified with six subspecies (*S. enterica* subsp. *enterica*, *S. enterica* subsp. *salamae*, *S. enterica* subsp. *arizonae*, *S. enterica* subsp. *diarizonae*, *S. enterica* subsp. *indica*, and *S. enterica* subsp. *houtenae*) and more than 2600 serovers (Porwollik et al. 2004; Guibourdenche et al. 2010). *Salmonella* pathogens can survive in the GI tract of various animals including human and birds. However, the enteric fever causing *Salmonella* serotypes and the disease-causing pathogens spread from asymptomatic human carriers via feces and contaminated food and water. Based on agglutination properties of their outer membrane protein antigens such as somatic O, flagellar H and capsular Vi, they are commonly classified (Guibourdenche et al. 2010). Most of the *Salmonella* human infections are due to strains of *Salmonella enterica* subsp. *enterica*. Now its nomenclature is established on the basis of serotypes name belonging to subspecies. In case of *Salmonella* enterica subsp. Enterica serotype Typhimurium is edited to *Salmonella* Typhimurium (Brenner et al. 2000).

In the view of clinical perspective, *Salmonella* serotypes are grouped on the basis of host range and outcome of the disease. For example, *Salmonella* Typhi is a human-host specific restricted pathogen causing a septicemic typhoid syndrome (enteric fever), and *Salmonella* Gallinarum is restricted to birds and causes a severe systemic disease called *fowl typhoid* (Shivaprasad 2000). Broad host range serotypes such as *Salmonella* Typhimurium and *Salmonella* Enteritidis cause majority of GI salmonellosis (Velge et al. 2005), including in human, domestic livestock, and fowl. In a single host, different *Salmonella* serotypes can bring different pathologies. Oral inoculation of weaned calves with *S.* Dublin causes severe systemic infection, whereas

Salmonella Gallinarum becomes avirulent and *S.* Typhimurium causes acute enteritis (Paulin 2002). Finally, pathogenesis is not restricted to dose and route of inoculation, which is mainly influenced by genesis and immune status of the host organism (Calenge et al. 2010).

2.3 *Salmonella*—Invasion strategies

Salmonella infection is initiated by ingestion of either contaminated food or water followed by passage of the bacteria from the stomach to the intestine. Stomach acids are good physiological barrier for the prevention of salmonellosis; however, patients who are immunocompromised or taking anti-acidic drugs or antibiotics lost their stomach natural acidity balance, which leads to the infection. Once they successfully cross the stomach barriers, they easily adhere to intestinal epithelial cells to invade the stomach, and their passage is initiated by transcytosis (through enterocytes or M cells at the apical side), migration to the basolateral side, and exocytosis into the interstitial space of the lamina propria (Muller et al. 2012). Within the propria, the cells are taken up accidentally by phagocytes and disseminate quickly via the efferent lymph in mesenteric lymph node and bloodstream in spleen and liver. But different routes of entry have been observed in different host and serotypes. In the case of cattle, *S.* Dublin passage was observed through the epithelial layer and associates with MHC class II–positive cells in propria because of their extracellular nature in efferent lymph (Pullinger et al. 2007).

The successful invasion of *Salmonella* into host cells is known to be critical for bacterial survival and onset of disease. In general, the intracellular pathogenic bacteria have evolved in two strategies such as trigger and zipper mechanism. Previously Takeuchi (1967) who first reported the mechanism of entry via trigger mechanism using Type III secretion system (T3SS-1); however, recent studies added zipper mechanism (Coombes et al. 2005; Desin et al. 2009; Rosselin et al. 2010). Among them, most of the *Salmonella* serovars are using zipper mechanism via the Rck invasin, a protein encoded by the *rck* gene located on the large virulence plasmid (Rychlik et al. 2006; Futagawa-Saito et al. 2010; Rosselin et al. 2010). However, the other pathogenic enterobacteria such as *Shigella flexneri* enters via trigger mechanism by secreting T3SS (Schroeder and Hilbi 2008), and *Yersinia pseudotuberculosis* uses zipper mechanism. None of these bacteria, other than *Salmonella*, has been shown to use both mechanisms to invade their host cells (Cossart and Sansonetti 2004), which demonstrates its strong pathogenicity.

2.4 Host invasion pathways

2.4.1 Type III secretion system (T3SS) dependent invasion mechanism

T3SS is one of the best studied entry mechanism of *Salmonella*. It allows the bacteria to enter the host nonphagocytic cells via trigger mechanism and persuades massive actin rearrangements and intense membrane distressing at that site (Cossart and Sansonetti 2004).

2.4.1.1 T3SS structure – *Salmonella* serovars contain a large number of gene clusters, commonly known as *Salmonella* pathogenicity islands (SPIs) that usually encode virulence factors. Until recently, SPI-1 is thought to have genes required for bacterial entry (Galan 1996), whereas SPI-2 is essential for intracellular survival (Shea et al. 1996). However, both SPI-1 and SPI-2 encode T3SS conserved among many gram-negative pathogens (Tampakaki et al. 2004) and consisted more than 20 proteins. Subsets of these proteins forms a needle-like complex structure that spans both inner and outer membrane and inserts into the host plasma membrane (Kubori and Galan 2003) by energy-dependent (i.e., adenosine triphosphate [ATP]) manner (Galan and Wolf-Watz 2006). They can cross over the inner and outer membrane of a host cell and create a pore in the membrane on contact with the host cells. Their structure bears a resemblance to basal body of flagella, proposing an evolutionary relationship of these two organelles.

The needle complex of T3SS consists of a multiring base constituted with SPI-1–encoded proteins such as PrgK, PrgH, and InvG and a slight needle-like single protein PrgI. The detailed protein system encoded by SPI-1 is tabulated with their specific activity (Table 2.1). The assembly of all these proceed in an orderly manner (Schraidt and Marlovits 2011). The needle projects single protein PrgI arrives from the outer membrane as a filament of 50-nm length, a complex of three proteins, namely SipB, SipC, and SipD, commonly known as translocon and located at the edge of needle, which enables it to make a pore in the target cell and allows the secretion of effector proteins (Mattei et al. 2011). Hydrophilic domains of SipD make direct interaction with PrgI needle protein and hydrophobic domains of SipB and SipC directly involved in pore formation (Miki et al. 2004; Rathinavelan et al. 2011). The interaction between SipB and cellular cholesterol is essential for effector translocation. In conclusion, the translocation of T3SS-1 effector proteins involves an apparatus present at the inner membrane and are made up of highly conserved proteins namely SpaP, SpaQ, SpaR, SpaS, InvA, InvC, and OrgB.

Table 2.1 Functional System of *Salmonella* Pathogenicity Island (SPI-1)

SPI-1 Effector Proteins	Functions
SpaP, SpaQ, SpaR, SpaS InvA, InvC (ATPase) OrgB	Exportation apparatus
PrgH, PrgK InvG, InvH PrgJ InvJ	Needle complex
SipB, SipC, SipD	Translocon
InvF HilA, HilD SirC, SprB	Regulators
SipA, SptP, AvrA	Effectors
SicA InvB SicP	Chaperones
OrgA, OrgC, InvE, InvI IacP, IagB, SpaO	Unclassified

2.4.1.2 Machinery system of T3SS-dependent entry – Among the vast number of SPI-1 effectors, six proteins, SipA, SipC, SopB, SopD, SopE, and SopE2, are playing a key role to invade the cell, whereas others are contributed in postinvasional process, including host cell survival and modulatory inflammatory responses (Patel and Galan 2005). To trigger *Salmonella* invasion into host cells, effectors operate actin cytoskeleton either directly or indirectly and also influence the delivery of vesicles to the entry site to provide additional membrane and allow the extension and ruffling of the plasma membrane to promote invasion. Later, membrane fusion occurs to induce the sealing of the future *Salmonella*-containing vacuole (SCV) and actin filaments are depolymerized, enabling the host cell to recover its normal shape.

2.4.1.3 Actin cytoskeleton rearrangements mediate invasion – *Salmonella* entry into the host cells is directly or indirectly mediated by actin cytoskeleton. Drugs that are interrupting actin dynamics completely prevent bacterial internalization (Finlay et al. 1991). After establishing contact with epithelial cells, *Salmonella* induces actin cytoskeletal rearrangements at the site of bacterial entry that direct bacterial internalization (Finlay et al. 1991; Francis et al. 1993; Ginocchio et al. 1994).

SipC, one of the components of bacterial translocon, consists of two membrane-spanning domain, s120 amino acid N-terminus and 209 amino acid

C-terminus, while the C-terminus extends into the host cytoplasm (Hayward and Koronakis 1999). C-terminal domain of SipC nucleates the gathering of actin filaments, resulting in rapid filament growth from the pointed ends (Hayward and Koronakis 1999). Notably, the SipC C-terminus is also required for translocation of effector protein, suggesting that it modulates both translocon assembly and activities. Chang et al. (2005) has demonstrated that actin nucleation and effector translocation are detachable. A short region proximately next to the second transmembrane domain of SipC (residues 201–220) is responsible for promoting actin nucleation, whereas the C-terminal 88 residues are accountable for effector translocation. Actin filament polymerization has been promoted by a second effector protein, SipA; by sinking the monomer concentration, it is essential for filament assembly by which improving the filament bundling activity of fimbrin, a host protein and influencing nucleation function of SipC (Zhou et al. 1999; McGhie et al. 2001). Besides, SipA binds to assembled filaments and prevents depolymerization (McGhie et al. 2004). Topological analysis indicates that SipA acts as a "molecular staple" using two extended arm domains to truss actin monomers. So, the synergistic activity of SipC and SipA supports the formation of actin filaments in the vicinity to the attached host bacteria, and they become stable over these filaments by host regulatory proteins (Lilic et al. 2003).

Once the bacterial invasion is accomplished, host cell cytoskeleton immediately returns to its basal state, specifically noted in polarized epithelial cells, in which apical microvilli are totally reassembled. This process is mediated by effector protein SptP, and it contains two distinct catalytic segments such as, an N-terminal Rho GAP domain and a C-terminal tyrosine phosphatase domain (Kaniga et al. 1996). GAP domain of SptP clearly mimics eukaryotic RhoGAPs in its overall structure and catalytic mechanism. Stebbins and Galan (2000) observed a reduced actin assembly once the host cells are treated with purified SptP. The role of SptP tyrosine phosphatase is in cytoskeletal recovery after *Salmonella* entry. The known targets of SptP phosphatase activity are tyrosine kinase ACK (activated Cdc42-associated kinase), intermediate filament protein vimentin, and p130Cas (Murli et al. 2001). ACK is a downstream effector protein of Cdc42 and has been involved in stimulation of ERK activation during *Salmonella* infection (Ly and Casanova 2007). The p130Cas becomes transitorily tyrosine phosphorylated in the course of *Salmonella* infection (Shi and Casanova 2006).

2.4.2 T3SS-independent invasion

To onset diseases ranging from typhoid fever to gastroenteritis or to asymptomatic carrier state, *Salmonella* needs to cross several barriers. Until now, SP-1– and SP-2–mediated invasion has been the focus because of their key role in invasion and internalization into various cell types, particularly enterocyte cell lines (Agbor and McCormick 2011). Conversely, the role of outer membrane proteins (OMP) consists of adhesive molecules, and virulence factors have been demonstrated in a number of pathogenic gram-negative bacteria. The association of *Salmonella* OMP with host cells is known to trigger a variety of biological events that include induction of innate and adaptive immunity and stimulate cell invasion (Galdiero et al. 2003).

2.4.2.1 OMP Rck invasion – Rck, a 19-kDa outer membrane protein belongs to Ail/Lom family, consists of five members (Rck, Ail, Lom, OmpX, and PagC). This *rck* gene located on large virulence plasmid contributing the expression of virulence genes such as *spvRABCD* (*Salmonella* plasmid virulence), *pef* (plasmid-encoded fimbriae), *srgA* (SdiA-regulated gene, putative disulphide bond oxidoreductase), or *mig-5* genes (macrophage-inducible gene coding for putative carbonic anhydrase) (Rychlik et al. 2006).

In general, *rck* carries virulence gene (Buisan et al. 1994); it is highly conserved and persist only in serovars of Enteritidis and Typhimurium (Futagawa-Saito et al. 2010), which are frequently associated with human and domestic animals infections causing pathogens; on the other hand, it was not detected in the serovars of Choleraesuis, Gallinarum, and Pullorum (Chu et al. 1999; Rychlik et al. 2006).

Rck is able to promote adhesion and internalization of coated beads (Rosselin et al. 2010) or of noninvasive *Escherichia coli* strains (Heffernan et al. 1994). It was showed that 46 amino acids of Rck were screened as being necessary for entry process. Their binding to the cell surface is generally inhibited by soluble Rck and induces distinct membrane rearrangements due to cell signaling (Rosselin et al. 2010). These demonstrate that Rck induces a Zipper-mode of invasion mechanism, supporting the certainty that *Salmonella* is the first bacterium to be demonstrated as able to induce both zipper and trigger mechanisms for their host cell invasion (Rosselin et al. 2010).

2.4.2.2 OMP PagN invasion – PagN, a 26 KDa outer membrane protein. has been identified as the important protein involved in *Salmonella* invasion mechanism (Lambert and Smith 2008). The gene, *pagN*, is located on the seventh centisome genomic island and is broadly scattered among *Salmonella enterica* serotypes (Folkesson et al. 1999). The *pagN* open reading frame was primarily identified during a Tn*phoA* random-insertion screening in *S.* Typhimurium performed to discover PhoP-activated genes (Belden and Miller 1994). In general, PagN is similar to both the Tia and Hek invasions of *E. coli* and represents 39% and 42% similarity in amino acids with these two invasions, respectively. Tia and Hek are predicted to have eight transmembrane regions, four long exposed extracellular loops, and three short periplasmic turns (Mammarappallil and Elsinghorst 2000; Fagan et al. 2008). Thus, PagN probably adopts a similar conformation as that of Hek and Tia.

The role of PagN in host invasion is supported by the fact that PagN overexpressed in a noninvasive *E. coli* strain-induced cell invasion (Lambert and Smith 2008). However, they also demonstrated that *pagN* deletion in *S.* Typhimurium leads to a three- to fivefold reduction in invasion of enterocytes. Thus, PagN might facilitate interactions between *Salmonella* and mammalian cells in specific conditions that do not allow SPI-1 expression (Lambert and Smith 2008). The PagN protein interacts with cell surface heparin sulfate proteoglycans to enter into the mammalian cell line CHO-K1 (Lambert and Smith 2009). However, because of the inability of proteoglycans to transduce signaling cascade, they might act as coreceptors for invasion and not as the receptor. At the cellular and molecular level, the PagN-mediated invasion mechanism is poorly described; hence, detailed studies are needed to categorize the PagN receptors at the molecular level.

2.4.2.3 Role of the hemolysin HlyE – HlyE, a small 2.3-kb pore-forming hemolysin encoded by SPI-18 island is missing in *S.* Typhimurium but was found in *S.* Typhi and *S.* Paratyphi A. It shares more than 90% similarity with *E. coli* HlyE (ClyA) hemolysin. *S.* Typhi *hlyE* mutants are weak in their ability to invade HEp-2 cells, compared to that of wild-type strain (Fuentes et al. 2008). However, the cellular mechanism of enhancing their invasion is not yet understood.

2.4.3 Caveolae-mediated invasion

Caveolae, a 21–24 kDa integral membrane flask-shaped projection in the plasma membrane of endothelial cells, consist of three caveolin proteins named as caveolin 1, 2, and 3. Caveolin 1 and 2 are expressed together as a hetero oligomer in the plasma membrane, and caveolin 3 was expressed only in muscle tissue (Tang et al. 1996; Smart et al. 1999). Caveolin-1, a scaffolding protein expressed within the caveolar membrane, interacts with signaling proteins, such as epidermal growth factor receptor, G-proteins, *Src*-like kinases, Ha-Ras, insulin receptors, and integrins for regulating their activities (Smart et al. 1999). Caveolar endocytosis is an endocytosis mechanism including nutrients and pathogens in most of the prokaryotes and parasites via endoplasmic reticulum (ER) → golgi → cytoplasm not fuse with lysosome, letting helps the pathogens to escape from lysosomal degradation (Schnitzer et al. 1994).

Salmonella can invade nonphagocytic cells by the effector proteins T3SS, and all are encoded within SPI-1. These proteins activate signaling cascades inside the host cell, leading to a variety of responses, including the formation of actin-rich membrane ruffles, which are ultimately responsible for bacterial internalization. *Salmonella* could be easily invaded by nonphagocytic senescent host cells in which caveolin-1 was also increased. At the time of destruction of caveolae structures by methyl-b-cyclodextrin or siRNA of caveolin-1 in the senescent cells, *Salmonella* invasion was reduced noticeably compared to that in nonsenescent cells (Lim et al. 2010). Besides, caveolin-1 is highly expressed in Peyer's patch and spleen of aged mice that are target sites for *Salmonella* invasion. This is the evidence supporting that increased level of caveolae and caveolin-1 in aged cells might be responsible for their increased susceptibility to microbial infections (Lim et al. 2010).

Lim et al. (2014) described a new cellular and molecular machinery for the caveolin-1–dependent entry of *Salmonella* into host cells via the direct regulation of actin reorganization. The caveolae are not able to form *Salmonella*-containing vacuoles or endosomes within the host cells. Instead, they quickly moved to the apical plasma membrane upon actin condensation during early invasion. The injected bacterial protein SopE counteracted with Rac1 to regulate actin reorganization, and both proteins are directly interacted with caveolin-1 in caveolae during the early stage. After the complete internalization of *Salmonella*, SopE levels decreased both in the caveolae and in the host cytoplasm; Rac1 activity was also reduced. Downregulation of caveolin-1 leads to decreased invasion of *Salmonella*. These results suggested that caveolin-1 might be involved in *Salmonella* entry through their interaction with SopE and Rac1 of T3SS-1, leading to better membrane ruffling for phagocytosis into host cells (Lim et al. 2014).

2.4.4 Unknown factors influencing invasion

Besides all these well-documented entry mechanisms, some other unidentified and unfamiliar factors seems to be involved during cell invasion. Rosselin et al. (2011) demonstrated that *Salmonella* serovars that were not expressing the effectors like Rck, PagN, and T3SS-1 have the ability to invade different host cells by unknown mechanism. This unknown entry mechanism depends on the cell type and cell line. For example, 3T3 fibroblasts and MA104 kidney epithelial cells are most accommodating to these mechanisms to enter independently of T3SS-1, PagN, and Rck. But nonpolarized HT29 enterocytes are not prone to these unknown entry mechanisms. They stated that cytoskeletal and membrane rearrangements similar to zipper or trigger machinery system have been observed during microscopic examination of infected cell types. These factors are found to be involved in inducing the signaling cascade that mediate the *Salmonella* entry into human foreskin fibroblasts (Aiastui et al. 2010). Steffen et al. (2004) and Hanisch et al. (2012) described a new invasion pathway that could be independent of Arp2/3 complex, and it depends on the formation of myosin II-rich stress fiber-like structures at entry sites via the activation of RhoA/Rho kinase signaling pathway.

In fact, a *Salmonella* mutant does not express any of the known invasion factors displayed both local and massive actin accumulations and also in intense membrane rearrangements. These findings showed that invasion factors other than PagN, Rck, and T3SS-1 apparatus were able to induce either a zipper or trigger mechanism (Rosselin et al. 2011). Overall and contrary to the existing theory, *Salmonella* can enter cells through a zipper-like mechanism mediated by Rck and other unknown invasions, in addition to the trigger mechanism mediated by its T3SS-1 apparatus and other unknown determinants. In consequence of these findings, avenues are open for the identification of new and unknown invasion factors.

2.5 Intracellular survival

After host cell invasion, *Salmonella* are internalized within a membrane-bound compartment known as small colony variants (SCVs). (Many bacterial pathogens have developed different strategies to neutralize the host immune response, either by preventing vacuole–lysosome fusion or by escaping into cytosol.) *Salmonella* are classified as a vacuolar pathogen; hence upon invasion, this bacterium resides and multiplies within SCV, a specialized vacuole, as the only intracellular niche of *Salmonella* (Bakowski et al. 2008), and it undergoes different stages of maturation like macrophages. These SCVs consist with markers such as Rab5, EEA1 (early endosome antigen 1), and TfR (transferrin receptor) (Steele-Mortimer et al. 1999; Bakowski et al. 2008). Once the SCV matures and is surrounded by actin, it migrates toward a perinuclear position and initiates formation of *Salmonella*-induced filaments (SIF), which helps transportation of nutrients to the SCV, thereby facilitating bacterial replication (Knodler and Steele-Mortimer 2003; Salcedo and Holden 2003).

Due to the defect in SCV maturation, *Salmonella* are released in nutrient-rich cytosol and replicated in higher rate termed *hyper-replication* (Knodler et al. 2014). These cytosolic *Salmonella* is sensed by the host epithelial cells, leading to an inflammatory response characterized by the activation of caspase 1 and caspases 3/7 and the apical release of IL-18, an important cytokine regulator of intestinal inflammation (Monack et al. 2001).

In the SCV, when *Salmonella* adapts to the intravascular environment, the two component regulatory system PhoP/PhoQ suppresses the expression of SPI-1 genes, which are no longer needed, but through reduction in hilA transcription another component increases SPI-2 T3SS expression (Bijlsma and Groisman 2005; Golubeva et al. 2012). Death of internalized *Salmonella* could be associated either with the complete SCV–lysosome fusion (Viboud and Bliska 2001) or autophagy, a capture mechanism of either cytosol-adapted or vacuolar bacteria that allow them to lysosomal compartment for killing. So, it is concluded that the net intracellular growth of *Salmonella* is the replication in both vacuolar and cytosolic region and of intracellular bacteria destruction.

Salmonella, in their latent period, is fused with SCV-lysosome to proliferate with in the host cell without being destroyed. Increase in the SCV number imbalances in the ratio of number of vacuoles to the number of acidic lysosomes. So, cells deficient in lysosomes are helpful for *Salmonella* by increasing their survival and proliferation with in the host cell (Eswarappa et al. 2010).

2.6 Conclusion

The detailed traits associated with epidemics of *Salmonella* serovars are still not understood. This is specifically evidenced with the recent screening of SPI-1–deficient *Salmonella* associated with human enteric disease and indicating that SPI-1 T3SS is not only the sole effector to cause pathogenesis (Boumart et al. 2014). Some new paradigms discussed in this chapter describe that *Salmonella* serovars could be entered in nonphagocytic cells by various invasion strategies that should adapt our vision and helps to revisit the host-specificity bases. Therefore, further studies are needed to focus on T3SS-independent mechanisms to identify the signaling pathways induced and host receptors involved in it.

Acknowledgments

KA thanks to DBT (Government of India) for financial assistance (BT/PR20721/BBE/117/241/2016) and DS thanks to UGC for Women postdoctoral fellowship program.

References

Agbor, T. A. and B. A. McCormick. 2011. *Salmonella* effectors: Important players modulating host cell function during infection. *Cell Microbiology* 13:1858–1869.

Aiastui, A., M. G. Pucciarelli, and F. G. Del Portillo. 2010. *Salmonella* enterica serovar Typhimurium invades fibroblasts by multiple routes differing from the entry into epithelial cells. *Infection and Immunity* 78:2700–2713.

Ao, T. T., N. A. Feasey, M. A. Gordon, K. H. Keddy, F. J. Angulo, and J. A. Crump. 2015. Global burden of invasive nontyphoidal *Salmonella* disease, 2010. *Emerging Infectious Diseases* 21 (6):941.

Bakowski, M. A., V. Braun, and J. H. Brumell. 2008. *Salmonella*-containing vacuoles: Directing traffic and nesting to grow. *Traffic* 9:2022–2031.

Belden, W. J. and S. I. Miller. 1994. Further characterization of the PhoP regulon: Identification of new PhoP-activated virulence loci. *Infection and Immunity* 62 (11):5095–5101.

Bijlsma, J. J. and E. A. Groisman. 2005. The PhoP/PhoQ system controls the intramacrophage type three secretion system of *Salmonella* enterica. *Molecular Microbiology* 57:85–96.

Boumart, Z., P. Velge, and A. Wiedemann. 2014. Multiple invasion mechanisms and different intracellular Behaviors: A new vision of *Salmonella*–host cell interaction. *FEMS Microbiology Letters* 361:1–7.

Brenner, F. W., R. G. Villar, F. J. Angulo, R. Tauxe, and B. Swaminathan. 2000. *Salmonella* nomenclature. *Journal of Clinical Microbiology* 38:2465–2467.

Buisan, M., J. M. Rodrigues- Pena, and R. Rotger. 1994. Restriction map of the *Salmonella* Enteritidis virulence plasmid and its homology with the plasmid of *Salmonella* Typhimurium. *Microbial Pathogenesis* 16:165–169.

Calenge, F., P. Kaiser, A. Vignal, and C. Beaumont. 2010. Genetic control of resistance to salmonellosis and to *Salmonella* carrier-state in fowl: A Review. *Genetics Selection Evolution* 42:11.

Chang, J., J. Chen, and D. Zhou. 2005. Delineation and characterization of the actin nucleation and effector translocation activities of *Salmonella* SipC. *Molecular Microbiology* 55:1379–1379.

Chu, C. S., S. F. Hong, C. J. Tsai, W. S. Lin, T. P. Liu, and J. T. Ou. 1999. Comparative physical and genetic maps of the virulence plasmids of *Salmonella* enterica serovars Typhimurium, Enteritidis, Choleraesuis, and Dublin. *Infection and Immunity* 67:2611–2614.

Coombes, B. K., M. E. Wickham, M. J. Lowden, N. F. Brown, and B. B. Finlay. 2005. Negative regulation of *Salmonella* pathogenicity island 2 is required for contextual control of virulence during typhoid. *Proceedings of National Academy of Sciences of U S A* 102 (48):17460–17465.

Cossart, P. and P. J. Sansonetti. 2004. Bacterial invasion: The paradigms of enteroinvasive pathogens. *Science* 304:242–248.

Crump, J. A., M. Sjound-Karlsson, M. A. Gordon, and C. M. Parry. 2015. Epidemiology, clinical presentation, laboratory diagnosis, antimicrobial resistance, and antimicrobial management of invasive *Salmonella* infections. *Clinical Microbiology Review* 28:901–937.

Desin, T. S., P. K. Lam, B. Koch, C. Mickael, E. Berberov, A. L. Wisner et al. 2009. Salmonella enterica serovar Enteritidis pathogenicity island 1 is not essential for but facilitates rapid systemic spread in chickens. *Infection and Immunity* 77:2866–2875.

Eswarappa, S. M., V. D. Negi, S. Chakraborty, B. K. Chandrasekhar, and D. Chakravortty. 2010. Division of the *Salmonella*-containing vacuole and depletion of acidic lysosomes in *Salmonella*-infected host cells are novel strategies of *Salmonella* enterica to avoid lysosomes. *Infection and Immunity* 78:68–79.

Fagan, R. P., M. A. Lambert, and S. G. Smith. 2008. The hek outer membrane protein of Escherichia coli strain RS218 binds to proteoglycan and utilizes a single extracellular loop for adherence, invasion, and autoaggregation. *Infection and Immunity* 76 (3):1135–1142.

Feasey, N. A., K. Gaskell, V. Wong, C. Msefula et al. 2015. Rapid emergence of multidrug resistant, H58-lineage *Salmonella* typhi in Blantyre, Malawi. *PLOS Neglected Tropical Diseases* 9:e0003748.

Finlay, B. B., S. Ruschkowski, and S. Dedhar. 1991. Cytoskeletal rearrangements accompanying *Salmonella* entry into epithelial cells. *Journal of Cell Science* 99:283–296.

Folkesson, A., A. Advani, S. Sukupolvi, J. D. Pfeifer, S. Normark, and S. Lofdahl. 1999. Multiple insertions of fimbrial operons correlate with the evolution of *Salmonella* serovars responsible for human disease. *Molecular Microbiology* 33 (3):612–622.

Francis C. L., T. A. Ryan, B. D. Jones, S. J. Smith, S. Falkow. 1993. Ruffles induced by *Salmonella* and other stimuli direct macropinocytosis of bacteria. *Nature* 364:639–642.

Fuentes, J. A., N. Villagra, M. Castillo-Ruiz, and G. C. Mora. 2008. The *Salmonella* Typhi hlyE gene plays a role in invasion of cultured epithelial cells and its functional transfer to *S.* Typhimurium promotes deep organ infection in mice. *Research Microbiol*ogy 159 (4):279–287.

Futagawa-Saito, K., A. T. Okatani, N. Sakurai-Komada, W. Ba-Thein, and T. Fukuyasu. 2010. Epidemiological characteristics of *Salmonella* enterica serovar Typhimurium from healthy pigs in Japan. *Journal of Veterinary Medical Science* 72:61–66.

Galan, J. E. 1996. Molecular and cellular bases of *Salmonella* entry into host cells. *Current Topics in Microbiology and Immunology* 209:43–60.

Galan, J. E. and H. Wolf-Watz. 2006. Protein delivery into eukaryotic cells by type III secretion machines. *Nature* 444 (7119):567–573.

Galdiero, M., M. G. Pisciotta, E. Galdiero, and C. R. Carratelli. 2003. Porins and lipopolysaccharide from *Salmonella* Typhimurium regulate the expression of CD80 and CD86 molecules on B cells and macrophages but not CD28 and CD152 on T cells. *Clinical Infectious Diseases* 9:1104–1111.

Ginocchio, C. C., S. B. Olmsted, C. L. Wells, and J. E. Galan. 1994. Contact with epithelial cells induces the formation of surface appendages on *Salmonella* Typhimurium. *Cell* 76:717–724.

Golubeva, Y. A., A. Y. Sadik, J. R. Ellermeier, and J. M. Slauch. 2012. Integrating global regulatory input into the *Salmonella* pathogenicity island 1 type III secretion system. *Genetics* 190:79–90.

Guibourdenche, M., P. Roggentin, M. Mikoleit, P. I. Fields, J. Bockemuhl, P. A. Grimon, and F. X. Weill. 2010. Supplement 2003–2007 (No. 47) to the White–Kauffmann–Le Minor scheme. *Research Microbiology* 161:26–29.

Hanisch, J., T. E. Stradal, and K. Rottner. 2012. A novel contractility pathway operating in *Salmonella* invasion. *Virulence* 3:81–86.

Hayward, R. D. and V. Koronakis. 1999. Direct nucleation and bundling of actin by the SipC protein of invasive *Salmonella*. *EMBO Journal* 18:4926–4934.

Heffernan, E. J., L. Wu, J. Louie, S. Okamoto, J. Fierer, and D. G. Guiney. 1994. Specificity of the complement resistance and cell association phenotypes encoded by the outer membrane protein genes rck from *Salmonella* Typhimurium and ail from *Yersiniaenterocolitica*. *Infection and Immunity* 62:5183–5186.

Hendriksen, R. S., P. Leekitcharoenphon, O. Lukjancenko, C. Lukwesa-Musyani. et al. 2015. Genomic signature of multidrug-resistant *Salmonella* entericaserovartyphi isolates related to a massive outbreak in Zambia between 2010 and 2012. *Journal of Clinical Microbiology* 53:26272.

Hendriksen, S. W. M., K. Orsel, J. A. Wagenaar, and A. Miko. 2004. *Salmonella* Typhimurium DT104A variant. *Emerging Infectious Diseases* 10:2225–227.

Ilakia, S., D. Senbagam, B. Senthilkumar, and P. Sivakumar. 2015. A prevalence study of typhoid fever and convalescent phase asymptomatic typhoid carriers among the schoolchildren in the northern part of Tamil Nadu. *Journal of Public Health* 23 (6):373–378.

Kaniga, K., J. Uralil, J. B. Bliska, and J. E. Galan. 1996. A secreted protein tyrosine phosphatase with modular effector domains in the bacterial pathogen *Salmonella typhimurium*. *Molecular Microbiology* 21:633–641.

Kariuki, S., M. A. Gordon, N. Feasey, and C. M. Parry. 2015. Antimicrobial resistance and management of invasive *Salmonella* disease. *Vaccine* 33 (3):C21–C29.

Klemm, E. J., A. J. Shakoor, F. N. Page, K. Qamar et al. 2018. Emergence of an extensively drug-resistant *Salmonella* enterica serovar Typhi clone harboring a promiscuous plasmid encoding resistance to fluoroquinolones and third-generation cephalosporins. *mBio* 9 (1):e00105-18.

Knodler, L. A. and S. Mortimer. 2003. Taking possession: Biogenesis of the *Salmonella*-containing vacuole. *Traffic* 4:587–599.

Knodler, L. A., V. Nair, and O. S. Mortimer. 2014. Quantitative assessment of cytosolic *Salmonella* in epithelial cells. *PLoS One* 9:e84681.

Kubori, T. and J. E. Galan. 2003. Temporal regulation of *Salmonella* virulence effector function by proteasome-dependent protein degradation. *Cell* 115 (3):333–342.

Lambert, M. A. and S. G. Smith. 2008. The PagN protein of *Salmonella* enterica serovar Typhimurium is an adhesin and invasin. *BMC Microbiology* 8 (8):142.

Lambert, M. A. and S. G. Smith. 2009. The PagN protein mediates invasion via interaction with proteoglycan. *FEMS Microbiology Letters* 297 (2):209–216.

Leekitcharoenphon, P., R. S. Hendriksen, S. L. Hello, F. X. Weill, D. L. Baggesen, S. R. Jun, D. W. Ussery et al. 2016. Global genomic epidemiology of *Salmonella enterica* Serovar Typhimurium DT104. *Applied and Environmental Microbiology* 82 (8):2516–2526.

Lilic, M., V. E. Galkin, A. Orlova, M. S. VanLoock, E. H. Egelman, and C. E. Stebbins. 2003. *Salmonella* SipA polymerizes actin by stapling filaments with nonglobular protein arms. *Science* 301:1918–1921.

Lim, J. S., H. E. Choy, S. C. Park, J. M. Han, I. K. Jang, and K. A. Cho. 2010. Caveolae-mediated entry of *Salmonella* Typhimurium into senescent nonphagocytotic host cells. *Aging Cell* 9:243–251.

Lim, J. S., M. Shin, H. J. Kim, K. S. Kim, H. E. Choy, and K. A. Cho. 2014. Caveolin-1 mediates *Salmonella* invasion via the regulation of SopE-dependent Rac1 activation and actin reorganization. *Journal of Infectious Diseases* 210 (5):793–702.

Ly, T. K. and J. E. Casanova. 2007. Mechanisms of *Salmonella* entry into host cells. *Cellular Microbiology* 9 (9):2103–2111.

Mammarappallil, J. G. and E. A. Elsinghorst. 2000. Epithelial cell adherence mediated by the enterotoxigenic *Escherichia coli* Tia protein. *Infection and Immunity* 68 (12):6595–6601.

Marks, F., V. V. Kalckreuth, P. Aaby, A. Adu-Sarkodie et al. 2017. Glob health Incidence of invasive *Salmonella* disease in sub-Saharan Africa: A multicentre population-based surveillance study. *Lancet* 5:e310–e323.

Mattei, P. J., E. Faudry, V. Job, T. Izor, I. Attree, and A. Dessen. 2011. Membrane targeting and pore formation by the type III secretion system translocon. *FEBS Journal* 278 (3):414–426.

McGhie, E. J., R. D. Hayward, and V. Koronakis. 2001. Cooperation between actin-binding proteins of invasive *Salmonella*: SipA potentiates SipC nucleation and bundling of actin. *EMBO Journal* 20:2131–2139.

McGhie, E. J., R. D. Hayward, and V. Koronakis. 2004. Control of actin turnover by a Salmonella invasion protein. *Molecular Cell* 13:497–420.

Miki, T., N. Okada, Y. Shimada, and H. Danbara. 2004. Characterization of *Salmonella* pathogenicity island 1 type III secretion-dependent hemolytic activity in *Salmonella* entericaserovarTyphimurium. *Microbial Pathogenesis* 37 (2):65–72.

Monack, D. M., W. W. Navarre, and S. Falkow. 2001. *Salmonella*-induced macrophage death: The role of caspase-1 in death and inflammation. *Microbes and Infection* 3:1201–1212.

Muller, A. J., P. Kaiser, K. E. Dittmar, T. C. Weber, S. Haueter, K. Endt et al. 2012. *Salmonella* gut invasion involves TTSS-2-dependent epithelial traversal, basolateral exit, and uptake by epithelium-sampling lamina propria phagocytes. *Cell Host & Microbe* 11:19–32.

Murli, S., R. O. Watson, and J. E. Gala. 2001. Role of tyrosine kinases and the tyrosine phosphatase SptP in the interaction of *Salmonella* with host cells. *Cell Microbiology* 3:795–810.

Patel, J. C. and J. E. Galan. 2005. Manipulation of the host actin cytoskeleton by *Salmonella*—All in the name of entry. *Current Opinion in Microbiology* 8 (1):10–15.

Paulin, S. M., P. R. Watson, A. R. Benmore, M. P. Stevens, P. W. Jones, B. Villarreal-Ramos. et al. 2002. Analysis of *Salmonella* enterica serotype-host specificity in calves: Avirulence of *S.* enterica serotype Gallinarum correlates with bacterial dissemination from mesenteric lymph nodes and persistence in vivo. *Infection and Immunity* 70:6788–6797.

Porwollik, E. F., C. Boyd, P. Choy, L. Cheng, E. Florea, L. Proctor, and M. McClelland. 2004. Characterization of Salmonella enterica subspecies I genovars by use of microarrays. *Journal of Bacteriology* 186 (17):5883–5898.

Prestinaci F., P. Pezzotti, and A. Pantosti. 2015. Antimicrobial resistance: A global multifaceted phenomenon. *Pathogens and Global Health* 109 (7):309–318.

Pullinger, G. D., S. M. Paulin, B. Charleston, P. R. Watson, A. J. Bowen, F. Dziva et al. 2007. Systemic translocation of *Salmonella* entericaserovar Dublin in cattle occurs predominantly via efferent lymphatics in a cell-free niche and requires type III secretion system 1 (T3SS-1) but not T3SS-2. *Infection and Immunity* 75:5191–5199.

Rathinavelan, T., C. Tang, and R. N. De Guzman. 2011. Characterization of the interaction between the *Salmonella* type III secretion system tip protein SipD and the needle protein PrgI by paramagnetic relaxation enhancement. *Journal of Biological Chemistry* 286 (6):4922–4930.

Rosselin, M., N. Abed, I. V. Payant, E. Bottreau, P. Y. Sizaret, P. Velge et al. 2011. Heterogeneity of type III secretion system (T3SS)-1-independent entry mechanisms used by *Salmonella* Enteritidis to invade different cell types. *Microbiology* 157:839–847.

Rosselin, M., V. I. Payant, C. Roy, E. Bottreau, P. Y. Sizaret, L. Mijouin, P. Germon, E. Caron, P. Velge, and A. Wiedemann. 2010. Rck of *Salmonella* enterica, subspecies entericaserovarenteritidis, mediates zipper-like internalization. *Cell Research* 20 (6):647–664.

Rychlik, I., D. Gregorova, and H. Hradecka. 2006. Distribution and function of plasmids in *Salmonella* enterica. *Veterinary Microbiology* 112:1–10.

Salcedo, S. P. and D. W. Holden. 2003. SseG, a virulence protein that targets *Salmonella* to the Golgi network. *EMBO Journal* 22:5003–5014.

Schnitzer, J. E., P. Oh, E. Pinney, and J. Allard. 1994. Filipin-sensitive caveolae mediated transport in endothelium: Reduced transcytosis, scavenger endocytosis, and capillary permeability of select macromolecules. *Journal of Cell Biology* 127:1217–1232.

Schraidt, O. and T. C. Marlovits. 2011. Three-dimensional model of *Salmonella*'s needle complex at subnanometer resolution. *Science* 331 (6021):1192–1195.

Schroeder, G. N. and H. Hilbi. 2008. Molecular pathogenesis of Shigella spp.: Controlling host cell signaling, invasion, and death by type III secretion. *Clinical Microbiology Review* 21 (1):134–156.

Shea, J. E., M. Hensel, C. Gleeson, and D. W. Holden. 1996. Identification of a virulence locus encoding a second type III secretion system in *Salmonella typhimurium*. *Proceedings of National Academy and Sciences of USA* 93:2593–2597.

Shi, J. and J. E. Casanova. 2006. Invasion of host cells by *Salmonella typhimurium* requires focal adhesion kinase and p130Cas. *Molecular Biology of the Cell* 17:4698–4708.

Shivaprasad, H. L. 2000. Fowl typhoid and pullorum diseases. *Revue scientifique et technique*. 19:406–424.

Smart, E. J., G. A. Graf, M. A. McNiven, W. C. Sessa, J. A. Engelman, P. E. Scherer, T. Okamoto, and M. P. Lisanti. 1999. Caveolins, liquid-ordered domains, and signal transduction. *Molecular and Cell Biology* 19:7289–7304.

Stebbins, C. E. and J. E. Galan. 2000. Modulation of host signaling by a bacterial mimic: structure of the *Salmonella* effector SptP bound to Rac1. *Molecular Cell* 6:1449–1460.

Steele-Mortimer, O., S. Meresse, J. P. Gorvel, B. H. Toh, and B. B. Finlay. 1999. Biogenesis of *Salmonella* Typhimurium-containing vacuoles in epithelial cells involves interactions with the early endocytic pathway. *Cell Microbiology* 1:33–49.

Steffen, A., K. Rottner, J. Ehinger, M. Innocenti, G. Scita, J. Wehland, and T. E. Stradal. 2004. Sra-1 and Nap1 link Rac to actin assembly driving lamellipodia formation. *EMBO Journal* 23:749–759.

Takeuchi, A. 1967. Electron microscope studies of experimental *Salmonella* infection. I. Penetration into the intestinal epithelium by *Salmonella* typhimurium. *American Journal of Pathology* 50 (1):109–136.

Tampakaki, A. P., V. E. Fadouloglou, A. D. Gazi, N. J. Panopoulos, and M. Kokkinidis. 2004. Conserved features of type III secretion. *Cell Microbiology* 6 (9):805–816.

Tang, Z., P. E. Scherer, T. Okamoto, K. Song, C. Chu, D. S. Kohtz, I. Nishimoto, H. F. Lodish, and M. P. Lisanti. 1996. Molecular cloning of caveolin-3, a novel member of the caveolin gene family expressed predominantly in muscle. *Journal of Biological Chemistry* 271:2255–2261.

Velge, P., A. Wiedemann, M. Rosselin, N. Abed, Z. Boumart, A. M. Chausse, O. G. Pinet, F. Namdari, S. M. Roche, A. Rossignol, and I. V. Payant. 2002. Multiplicity of *Salmonella* entry mechanisms, a new paradigm for *Salmonella* pathogenesis. *MicrobiologyOpen* 1 (3):243–258.

Viboud, G. I. and J. B. Bliska. 2001. A bacterial type III secretion system inhibits actin polymerization to prevent pore formation in host cell membranes. *EMBO Journal* 20:5373–5382.

Voetsch, A. C., T. J. Van Gilder, F. J. Angulo, M. M. Farley, S. Shallow, R. Marcus, P. R. Cieslak, V. C. Deneen, and R. V. Tauxe. 2004. Food Net estimate of the burden of illness caused by nontyphoidal *Salmonella* infections in the United States. *Clinical Infectious Diseases* 38 (3):127–134.

Wang, H., M. Naghavi, C. Allen, R. M. Barber, Z. A. Bhutta et al. 2016. Global, regional, and national life expectancy, all-cause mortality, and cause-specific mortality for 249 causes of death, 1980–2015: A systematic analysis for the Global Burden of Disease Study 2015. *Lancet* 388 (10053):1459–1444.

Wong. V. K., S. Baker, D. J. Pickard et al. 2015. Phylogeographical analysis of the dominant multidrug-resistant H58 clade of *Salmonella* Typhi identifies inter- and intra continental transmission events. *Nature Genetics* 47:632–639.

Zhou, D., M. S. Mooseker, and J. E. Galan. 1999. Role of the *S.* typhimurium actin-binding protein SipA in bacterial internalization. *Science* 283 (5410):2092–2095.

3 Prevalence of Bacterial Infections in Respiratory Tract

Boopathi Balasubramaniam, U. Prithika, and K. Balamurugan

Contents

3.1 Introduction

The respiratory tract is mainly responsible for breathing. It is distributed into two major portions, the upper respiratory tract (URT) and the lower respiratory tract (LRT). URT comprises of various organs such as sinuses, nasal cavity, pharynx, and larynx; LRT comprises of trachea, bronchi, lungs, and diaphragm (Figure 3.1). The foremost function of the human respiratory

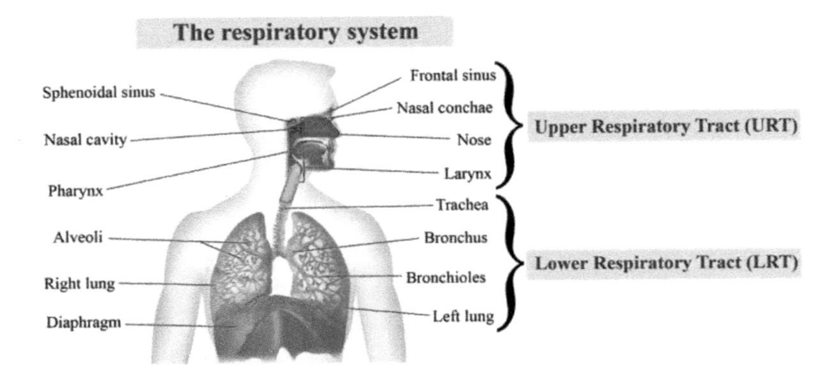

Figure 3.1 The representative figure illustrates the internal organs exist in upper respiratory and lower respiratory tracts of a human being. (Extracted and Modified from Wikiversity contributors, WikiJournal of Medicine/Medical gallery of Blausen Medical 2014, *Wikiversity*, https://en.wikiversity.org/w/index.php?title=WikiJournal_of_Medicine/Medical_gallery_of_Blausen_Medical_2014&oldid=1753045, 2017.)

system is to passage the inhaled air into lungs and to aid the diffusion of oxygen molecules into the bloodstream. Respiratory tract infections (RTIs) represent a common major health problem and also the most frequently reported infections throughout the world because of their ease of transmission, occurrence, and considerable mortality and morbidity by affecting people of all ages. Traditionally, they are divided into two classes: lower respiratory tract infections (LRTIs) and upper respiratory tract infections (URTIs). Generally, bacterial infections may affect the lower or upper respiratory tract and the pathogens causing these symptoms have a tendency to localize to one region. Some of these RTIs are mild (acute), some are chronic (long-lasting), and sometimes some are self-limiting (Richter et al. 2016). The respiratory tract can be infected by various pathogenic bacteria, both gram-positive and -negative (Felmingham and Gruneberg 2000). But by chance, most of these infections can be treated by well-known antibiotic therapies.

3.2 An overview of Respiratory Tract Bacterial Infections

According to World Lung Foundation's Acute Respiratory Infections Atlas, acute respiratory infections are a major cause for 4.25 million deaths every year and also the third-leading cause of deaths (after heart disease and stroke) in the world. Pneumonia, a form of acute RTI frequently caused by both bacteria and viruses, is the sole leading cause of infantile death around the world. According to the World Health Organization (WHO), pneumonia accounts for 16% of all deaths in children younger than the age of 5, killing 920,136 children in 2015 (WHO 2016). Tuberculosis, a chronic RTI caused due to *Mycobacterium tuberculosis*, had at least 6.3 million new cases (up from 6.1 million in 2015), which is equivalent to 61% of the estimated incidence of 10.4 million (WHO 2018), leading to 1.5 million deaths (Table 3.1).

Table 3.1 The Prevalence of Respiratory Tract Infections Caused by the Pathogens

Disease	Frequency/ Prevalence	Bacteria	Reference(s)
Pharyngitis/ tonsillitis	7.5% of people (in any given 3 months)	*Haemophilus influenzae* *Corynebacterium diptheriae* *Streptococcus pneumoniae* *Bordetella pertussis* *Mycoplasma pneumoniae*	Jones et al. (2005)
Laryngitis	Common	*Haemophilus influenzae* *Streptococcus pneumoniae*	Wood et al. (2014)
Acute otitis media	471 million cases (2015)	*Streptococcus pneumoniae* *Staphylococcus aureus* *Haemophilus influenzae* β-Haemolytic *streptococci*	Wood et al. (2014)
Acute rhinosinusitis	10%–30% each year (developed world)	*Streptococcus pneumoniae* *Haemophilus influenzae* *Staphylococcus aureus*	Rosenfeld et al. (2015)
Bronchiolitis	~20% (children younger than age group 2)	*Haemophilus influenzae* *Corynebacterium diptheriae* *Streptococcus pneumoniae* *Bordetella pertussis* *Mycoplasma pneumoniae*	Schroeder and Mansbach (2014); Friedman et al. (2014)

(*Continued*)

Table 3.1 (*Continued*) The Prevalence of Respiratory Tract Infections Caused by the Pathogens

Disease	Frequency/ Prevalence	Bacteria	Reference(s)
Acute bronchitis	~5% of people/ year	*Haemophilus influenzae* *Corynebacterium diptheriae* *Streptococcus pneumoniae* *Bordetella pertussis* *Mycoplasma pneumoniae*	Wenzel and Fowler (2006)
Pneumonia	450 million cases (7%) per year	*Staphylococcus aureus* *Streptococcus pneumoniae* *Haemophilus influenzae* *Klebsiella pneumoniae* *Mycobacterium tuberculosis* *Mycoplasma pneumoniae* *Legionella* spp.	Ruuskanen et al. (2011); Lodha et al. (2013)
Tuberculosis	33% of people	*Mycobacterium tuberculosis*	WHO (2017)

3.3 Upper Respiratory Tract Bacterial Infections (URTBIs) and Lower Respiratory Tract Bacterial Infections (LRTBIs)

Upper RTIs are defined as the infections transmitted and localized in the nasal cavity including, sinuses, pharynx, and larynx. The foremost roles of the URT are to humidify the exposed heat and filter the respired air through several compartments (i.e., nasal cavity/meatus, oropharynx, nasopharynx, and pharynx) before the respired air reaches the lungs (Sahin-Yilmaz and Naclerio 2011). These compartments usually allow for the passage and colonization of bacterial pathogens (Siegel and Weiser 2015). Differences in mucus secretion, temperature, and relative oxygen concentration all over the URT regulate the bacterial colonization (Rigottier-Gois 2003). The leading causes of URTBIs are *Staphylococcus aureus, Streptococcus pneumoniae, Haemophilus influenzae,* β-hemolytic *streptococci, Corynebacterium diptheriae, Neisseria gonorrhoeae, Mycoplasma hominis,* and *Mycoplasma pneumoniae* (Figure 3.2). The LRT is called the tracheobronchial tree or respiratory tree, which starts with the larynx and includes trachea, two

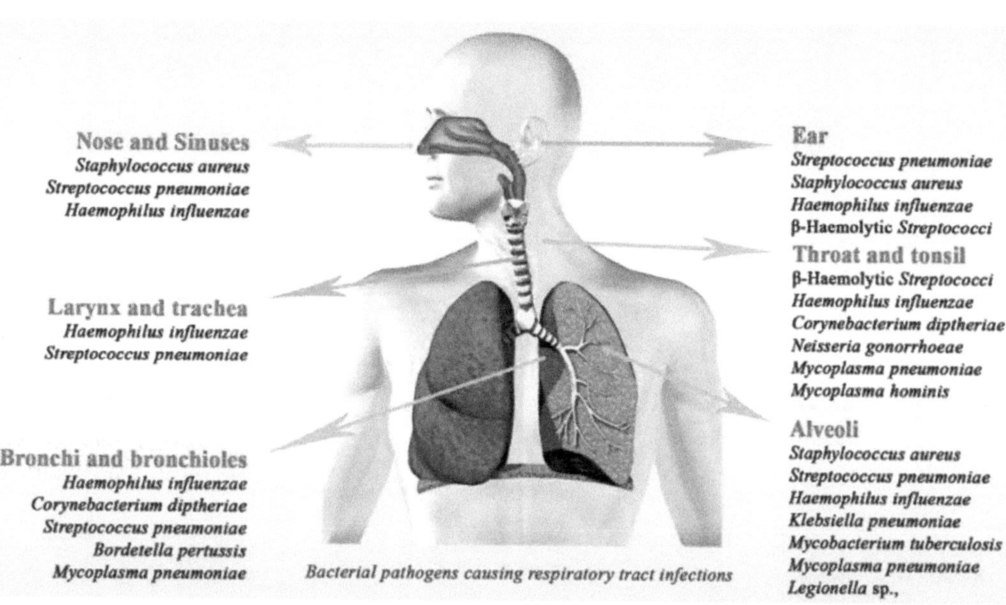

Figure 3.2 The representative figure demonstrates the bacterial infections caused in both the lower respiratory tract and upper respiratory tract. (Extracted and Modified from Dasaraju, P.V. and Liu, C., *Med. Microbial.*, 4, 1996.)

bronchi that divides from the trachea, and the lungs. The lung is the organ where gas exchange generally takes place (Wikibooks contributors 2018). The LTBIs are caused by the colonization of several pathogens such as *H. influenzae, C. diptheriae, S. pneumoniae, Bordetella pertussis, M. pneumoniae, S. aureus, Klebsiella pneumoniae, M. tuberculosis*, and *Legionella* species (Figure 3.2).

3.3.1 *Staphylococcus aureus*

S. aureus is one of the major commensal bacteria present in human beings and preferably colonizes the epithelium of the anterior nares (Foster 2004). *S. aureus* is a gram-positive bacterium that is capable of causing a multitude of diseases ranging from mild to chronic severity. Approximately 20% of the total population are colonized with *S. aureus*; 60% are periodic, and 20% population never carry the bacteria (Von Eiff et al. 2001a). The emergence of methicillin-resistant *S. aureus* (MRSA) is considered to be a major threat in the current scenario. The rates of MRSA carriages differ extensively from 3% to 30% (Kitti et al. 2011; Shaw et al. 2013; Jimenez-Truque et al. 2017). A report by Davis et al. (2004) revealed that 21% of admitted patients were methicillin-susceptible *S. aureus* (MSSA) carriers and 3.4% of patients were MRSA carriers. However, a higher ratio of the carriers of MRSA (19%) developed invasive staphylococcal disease than MSSA carriers (1.5%).

3.3.1.1 Pathogenesis – The pathogenesis of *S. aureus* was referred to classical mode and resembles the pathology of soft-tissue infection and skin (Foster 2005; Tong et al. 2015). Reports revealed that first-line of defense against *S. aureus* infection was the neutrophil response (Figure 3.3). Generally, *S. aureus* invades the cells by means of different methods (e.g., chemotaxis of leukocytes, blocking sequestering host antibodies, resisting destruction after ingestion by phagocytes, and hiding from detection via capsule or biofilm formation). The bacterial cell surface receptors present in the cell walls correspond to microbial surface components recognizing adhesive matrix molecules (MSCRAMMs) or adhesins (Dunyach-Remy et al. 2016). MSCRAMMs assist the bacterial adhesion to epithelial cells. In the intracellular module, *S. aureus* forms small-colony variants (SCVs) (Proctor et al. 2006). So, they would be able to persist in a metabolically inactive state while conserving the integrity of the epithelial cells. SCVs retain several phenotypic and metabolic differences from the usual *S. aureus* clinical isolates (Von Eiff et al. 2001b; Tuchscherr et al. 2010). Certainly, they are comparatively resistant to the usual antibiotics (Baumert et al. 2002; Garcia et al. 2013). Therefore, it is challenging to eradicate *S. aureus* with well-known antibiotic therapies (Proctor and Peters 1998). Furthermore, the synthesis and secretion of glycocalyx matrix

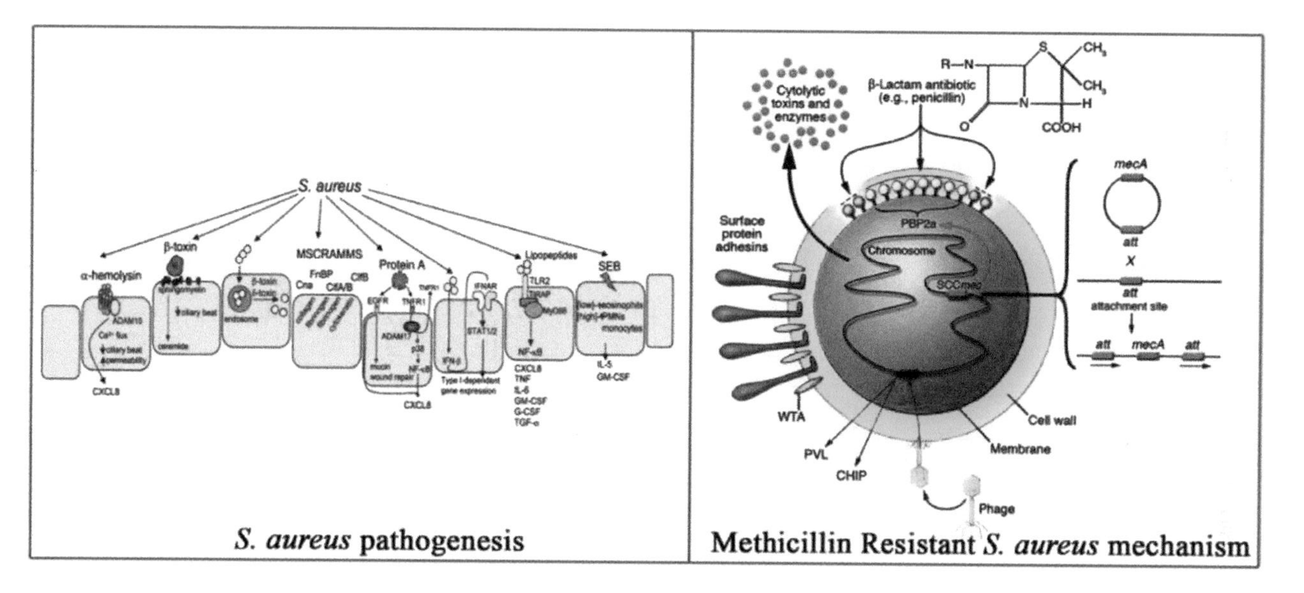

Figure 3.3 Schematic diagram of *Staphylococcus aureus* pathogenesis and methicillin-resistant *S. aureus* (MRSA) emergence. *S. aureus* secretes several virulence factors that evade host immune defenses. On the other hand, the bacterium expresses adhesins and also secretes many toxins (i.e., α-toxins, etc.) and enzymes by the activation of needed chromosomal genes. Methicillin-resistance is attained by the insertion of a DNA element called SCCmec through horizontal gene transfer mechanism. And also, the mecA gene encodes a novel β-lactam–resistant penicillin binding protein, PBP2a, which endures to synthesize a new cell wall peptidoglycan moiety even when the penicillin-binding proteins are repressed. (Extracted and Modified from Parker, D. and Prince, A., *Semin. Immunopathol.*, 34, 281–297, 2012; Foster, T. J., *J. Clin. Invest.*, 114, 1693–1696, 2004.)

with the combination of specific surface adherence mechanisms also plays a crucial role in the *S. aureus* virulence. It was reported that *S. aureus* was able to secrete hazardous toxins, which can lead to tissue necrosis in the host. *S. aureus* toxins also have vital role in the excavating and spreading of the infection in the patients (Dunyach-Remy et al. 2016).

3.3.1.2 Prophylaxis – Bacterial strains lacking cytotoxins are considered to be avirulent in a mice model of *S. aureus* pneumonia; similarly, vaccine-based therapies that provoke the toxins come up with the protection against lethal disease. Studies suggest that disrupting the function of the cytotoxin affords a potent preventive mechanism to treat *S. aureus* pneumonia. Derivatives of β-cyclodextrin are sevenfold symmetrical small molecules that block the assembled α-hemolysin pore, negotiating the functions of toxin (Ragle et al. 2010). In the current research field, a significant problem common with *S. aureus* infections is the rapid progression of antibiotic resistance. In *S. aureus* biofilm-associated infections, this may be provoked by the proliferation of antibiotic minimum inhibitory concentrations (MICs) compared with planktonic bacteria (Howlin et al. 2015). Vancomycin is the most commonly prescribed antibiotic for the *S. aureus* biofilm-associated infections (Liu et al. 2011). Conversely, physicians are extremely cautious about the direction of this antibiotic owing to the tendency of *S. aureus* to develop resistance. Evidence for this tendency is the recent development of vancomycin intermediate *S. aureus* and vancomycin-resistant *S. aureus* (VRSA) strains (Howden et al. 2010). Still, the combined therapy of vancomycin with rifampicin has shown conflicting activities. Also multiple studies show that this combination might be effective against MSSA rather than MRSA biofilm infections (Olson et al. 2010; Salem et al. 2011; Zimmerli 2014). Daptomycin, a cyclic lipopeptide molecule, is a novel drug that has been administrated for vancomycin-unresponsive *S. aureus* conditions. The mode of actions of daptomycin relies with the disruption of cytoplasmic membrane resulting in cessation of DNA, RNA, protein synthesis, and rapid depolarization. Daptomycin was found to be the most effective among the drugs tested (i.e., clindamycin, linezolid, vancomycin, and tigecycline) in clearing-off the prevailing biofilm of *S. aureus* (Bhattacharya et al. 2015).

3.3.2 *Streptococcus pneumoniae*

S. pneumoniae (*Pneumococcus*) is a gram-positive, extracellular bacterial pathogen, and a classic example of highly invasive bacteria. It is reported to be the leading cause of mortality and morbidity globally, causing more deaths than any other communicable diseases. At the highest risk, 1 million children (younger than 5 years of age) are dying annually (Centers for Disease, Control, and Prevention 2008). Pneumococcal diseases vary from mild RTIs such as sinusitis and otitis media to more chronic diseases such as septicemia, pneumonia, and meningitis. Though, *Pneumococcus* can cause mild to lethal diseases, it is more commonly a dormant colonizer of the URT where up to 60% of children could carry *Pneumococci* in the nasal routes asymptomatically (Henriqus Normark et al. 2003; Nunes et al. 2005).

3.3.2.1 Pathogenesis – Be it in the mucosal surface or blood, the bacterial pathogenesis involves the bacterial adherence to cell surfaces of host system followed by cellular invasion (Figure 3.4). The successful invasion of the bacterial cells was due to adherence to the two receptors present on the surface of various cell types (e.g., the platelet activating factor receptor (PAFr) and 37/67 kDa laminin receptor [LR]). Orihuela et al. (2009) reported

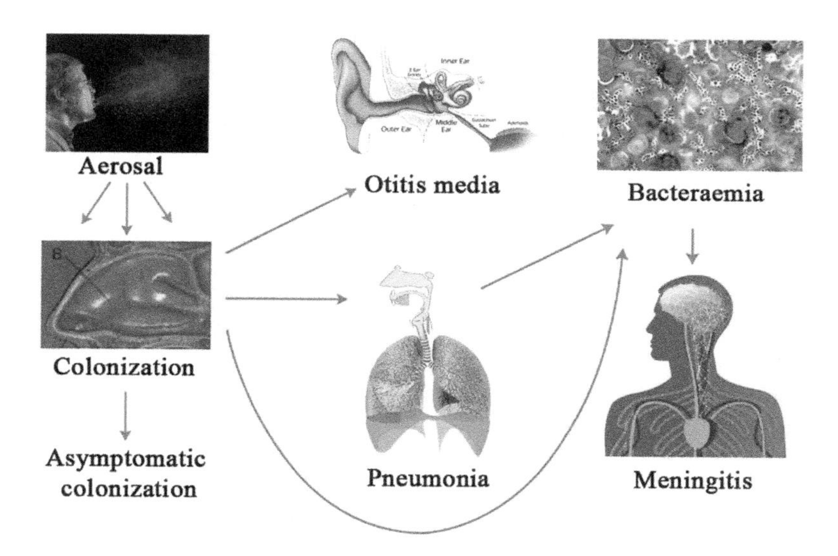

Figure 3.4 Schematic illustration of *Staphylococcus pneumoniae* pathogenesis. *S. pneumoniae* are generally transmitted through air and the attachment followed by colonization is mediated through surface proteins, containing unknown methods of immune evasion. Bacteria dissociate from the matrix formed because of biofilms and rise into the middle ear and to the lung through the pharynx. The interaction most commonly leads to serotype-specific immunity, and the disease progression of otitis media is most common in children. The most severe development in the *S. pneumoniae* pathogenesis is meningitis. This series of aggressive steps is shared by the three major bacterial pathogens (*Pneumonococcus*, *Haemophilus*, and *Meningococcus*) and is driven by interplay between innate immunity and innate invasion. (Extracted and Modified from Henriques-Normark, B. and Tuomanen, E.I., *Cold Spring Harb. Perspect. Med.*, 3, a010215, 2013.)

that the bacterial adhesin protein CbpA primarily bound to endothelial LR. CbpA surface-exposed loop mediated binding to the polymeric immunoglobulin receptor was accountable for the *Pneumococcus* translocation across the nasopharyngeal epithelium (Zhang et al. 2000). It was also notable that CbpA binding mediated adherence of the bacterial cells to LR but not invasion. Significantly, LR also acted as a target for *Neisseria meningitidis* and *H. influenzae* pathogenesis. The antibody of CbpA cross-reacted and blocked the adherence of these meningeal pathogens, representing a shared binding mechanism. These results suggest that a series of pathogens target LR as a major step in the host-pathogen interactions.

3.3.2.2 Prophylaxis – Even though successful conjugate polysaccharide vaccines exist, serotype-independent protein-based pneumococcal vaccines deal a foremost progression for preventing life-threatening pneumococcal infections in terms of cost, particularly in developing countries. IL-17A-secreting CD4+ T cells (TH17) facilitates resistance to the mucosal colonization of multiple pathogens including *S. pneumoniae*. A study by Moffitt et al. revealed that screening of expression library containing >96% of predicted pneumococcal virulence proteins. They have identified antigens recognized by TH17 cells from mice immune system due to pneumococcal colonization. The identified antigens also provoked IL-17A secretion from colonized mice splenocytes and human peripheral blood mononuclear cells (PBMCs) proposed that comparable responses were clued-up during natural contact. The report demonstrated the potential of screening through proteomic approaches to identify the specific antigens for designing the subunit vaccines against mucosal pathogens by means of harnessing TH17-mediated immunity (Moffitt et al. 2011). Extensive use of the 7-valent pneumococcal conjugate vaccine (PCV7) in US infants (23 months of age) was also endorsed for certain children (59 months of age) to prevent the pneumococcal disease distribution. After the PCV7 introduction, vivid decreases in invasive pneumococcal disease (IPD) were observed in children. A significant reduction in adult pneumococcal diseases was also seen. However, other IPD serotypes have not been comprised in the PCV7, which provoked the necessity for the development of a pneumococcal conjugate vaccine with extended coverage (Committee on Infectious, Diseases 2010).

3.3.3 *Haemophilus influenzae*

H. influenzae is one of the significant gram-negative bacteria associated with respiratory infections, ranging from acute otitis media to chronic obstructive pulmonary disease (COPD) and also includes invasive diseases such as sepsis and meningitis (Murphy et al. 2009). The serotype variation among the *H. influenzae* can be differentiated by the encapsulated and nonencapsulated forms. Encapsulated *H. influenzae* is categorized into six diverse serotypes such as a to f, whereas the nonencapsulated *H. influenzae* is labeled as no-typeable *H. influenzae* (NTHI) (Van Wessel et al. 2011). Generally, *H. influenzae* exist in the mucosa, and NTHI is predominantly associated with RTIs, while encapsulated *H. influenzae* serotypes including *H. influenzae* type b (Hib) cause invasive diseases such as meningitis and septicemia. Until the 1990s, Hib was considered to be the most common serotype, but a dramatic decrease was observed in Hib infections was perceived after the development of a conjugate vaccine. However, an accumulative occurrence of invasive diseases caused by nontype b *H. influenzae* has been recently reported throughout the world (Resman et al. 2011; Rubach et al. 2011).

3.3.3.1 Pathogenesis – Once the human airway epithelium is immunocompromised, *H. influenzae* can create a respiratory infection. The first step of pathogenesis is nasopharyngeal colonization (Figure 3.5). The pathogen starts adhering to the respiratory tract epithelial cells and escapes from a number of host defenses, including complement fixation, secretory IgA, mucociliary clearance, lactoferrin, lysozyme, and antimicrobial peptides. Surface adherence of the *H. influenzae* to mucin can assist clearance, and impairment of mucociliary clearance is a crucial stage in the virulence. When mucociliary is damaged, the pathogen is able to interact with nonciliated cells through one or few of its several bacterial adhesins. The pathogen either invades the epithelial cell or enters to the subepithelial layer by means of transcytosis or aggregates to form biofilm. Still, it remains unclear how *H. influenzae* forms carriage to form biofilm. Whereas, *H. influenzae* is not habitually known as an intracellular pathogen, feasible bacteria have been frequently found within host cells (Schaechter 2009). A report by Singh et al. suggest that *H. influenzae* interacts with the host epithelium and *H. influenzae* protein E (PE) intermediates the binding to the epithelial surface, using an unknown receptor (Singh et al. 2013). The bacterial adhesion leads to the induction of a pro-inflammatory response by the epithelial cells. During the pathogenesis, PE binds to laminin (Ln), which contributes *H. influenzae* adhesion to the basement membrane and the host extracellular matrix (ECM). Moreover, *H. influenzae* binds with plasminogen by using PE. Once, the plasminogen is bound to PE of *H. influenzae*, it is transformed into active from called *plasmin* by host urokinase plasminogen activator (uPA) or tissue plasminogen activator (tPA). Active plasmin may possibly help in thebacterial invasion and degradation of the ECM (Singh et al. 2013).

3.3.3.2 Prophylaxis – Vaccines are available that can prevent *H. influenzae* type b (Hib) disease, which is the most common serotype ("strain") of *H. influenzae* bacteria. But Hib vaccine does not prevent the diseases caused

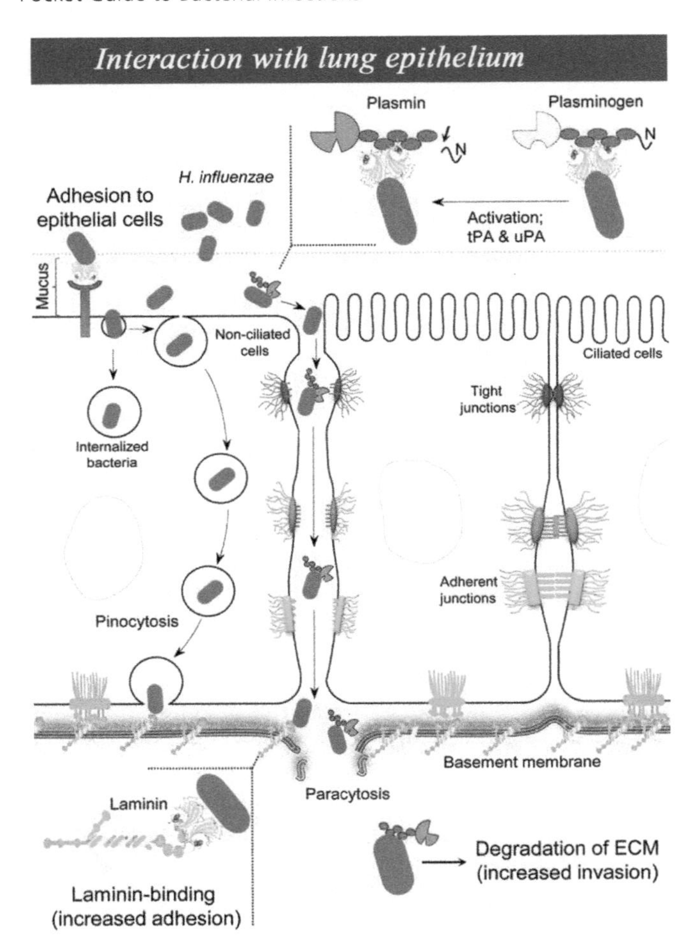

Figure 3.5 Microbial pathogenesis of *Haemophilus influenzae* in the lung epithelial cells. The *H. influenzae* protein E (PE) molecule facilitates binding to the surface of epithelial cell by using an unknown receptor. This interaction induces the bacterial adhesion and elevates a proinflammatory response in the epithelial cells. (Extracted and Modified from Singh, B. et al., *Infect. Immun.*, 81, 801–814, 2013.)

by any other types of *H. influenzae*. Generally, Hib vaccine is endorsed for children younger than 5 years of age in the United States and was also prescribed to 2-month-old infants. In certain circumstances, individuals at increased threat for getting invasive Hib diseases (when *H. influenzae* invades other internal parts of the body and produces septicemia) and who are completely vaccinated may possibly need additional doses of the Hib vaccine. It is also recommended that unimmunized children and adults with certain medical circumstances should also be given the Hib vaccine. A child with *H. influenzae* (including Hib) infection may not develop defending levels of antibodies, providing chances for the individual to be affected with *H. influenzae* disease again. Children whose age is younger than 2 years who have recovered from invasive Hib disease are not be considered safe and must be given Hib vaccine immediately. In certain circumstances, a person in close contact with someone who is infected with Hib should also receive antibiotics to prevent them from receiving the disease (CDC 2016).

3.3.4 *Mycobacterium tuberculosis*

M. tuberculosis arises globally and is reported to kill 2–3 million people per year (Gagneux et al. 2006). Tuberculosis (TB) is considered one of the ancient documented human diseases and still stands as one of the prevalent killers among the other well-known infectious diseases, even with the worldwide usage of a live attenuated vaccine and a number of antibiotics. The discovery of mutilated bones in various neolithic locations in Denmark, Italy, and other countries in the Middle East Europe suggests that TB was found all over the world 4,000 years ago. The origin of *M. tuberculosis* has been the subject of recent investigations, and it is now thought that the bacteria in the genus *Mycobacterium*, like Actinomycetes, were primarily found in the soil and that some species of *Mycobacterium* evolved to live in mammals (Smith 2003).

3.3.4.1 Pathogenesis – Generally, *M. tuberculosis* infection was initiated by the nasopharyngeal inhalation of aerosol droplets that comprises the infectious bacteria. The preliminary stages of TB infection are usually characterized by the innate immune responses that involve in the recruitment of inflammatory immune cells to the alveoli and bronchioles of lung (Figure 3.6). Following bacterial diffusion to the exhausting lymph node, dendritic cell presentation of bacterial cells (including virulence factors) leads to T cell priming and activates the development of antigen-specific T cells, which are gathered to the lung. At this time, the recruitment of B cells, T cells, activated macrophages, and other leukocytes leads to the formation of granulomas which can comprise the *M. tuberculosis* cells (Nunes-Alves et al. 2014). Most of the time, the infected person will persist in a "latent" state, which has no clinical symptoms. A small number of colonies of *M. tuberculosis* is enough to develop an active disease condition, which can lead to the release of the bacteria into the airways. When individuals with active TB cough, they can produce potentially infectious droplets that can transmit the bacterial infection. The "central dogma" of defensive immunity

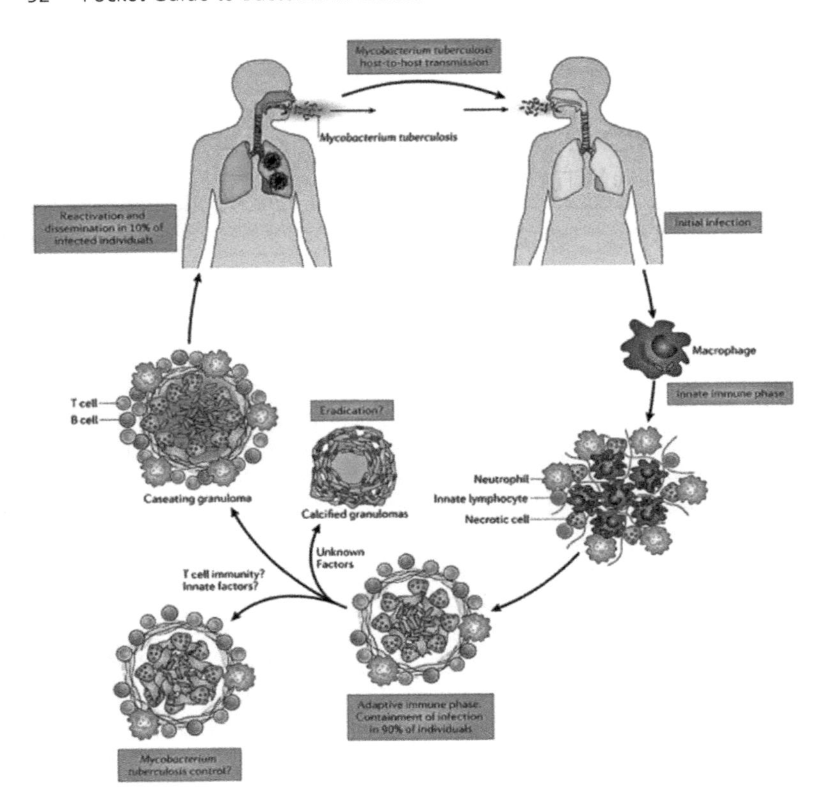

Figure 3.6 Pathogenesis of *Mycobacterium tuberculosis*. Primary infection is commenced by the inhaled aerosol droplets that contain pathogenic bacteria. At the initial stages, infection is characterized by innate immune responses that comprise the recruitment of inflammatory cells such as macrophages, antigen-specific T cells to the lung. The recruitment of B cells, T cells, other leukocytes, and activated macrophages leads to the formation of granulomas, which can have the *Mycobacterium tuberculosis* bacteria. Once individuals acquire an active tuberculosis (TB) cough, they can produce infectious aerosol droplets that transmit the TB infection. (Extracted and Modified from Nunes-Alves, C. et al., *Nat. Rev. Microbiol.*, 12, 289–299, 2014.)

to TB is that CD4+ T cells that produce interferon-γ (IFN-γ), T helper 1 (TH1) cells, which associates with tumor necrosis factor (TNF; produced by the macrophage of T cell), and also these activate antimicrobial activity that is capable of controlling the development of *M. tuberculosis* further. Finally, a number of components of the innate immunity, including vitamins and interleukin-1 (IL-1) can collaborate with cytokines that are produced by T cells to fight against the TB (Nunes-Alves et al. 2014).

3.3.4.2 Prophylaxis – TB is a curable disease and general drug-susceptible TB disease can be treated with a standard 6-month passage of antimicrobial drugs that are prescribed under the supervision of a physician and support to the patient by a trained volunteer. It is estimated that 53 million people were saved through TB treatment and diagnosis between 2000 and 2016. It was found that drug-resistant strains emerge when anti-TB medicines are used improperly, through inappropriate prescriptions of poor-quality medications by physicians, and patients stopping antibiotics prematurely. Multidrug-resistant tuberculosis (MDR-TB) is caused by the *M. tuberculosis* bacteria that do not respond to rifampicin and isoniazid, the two most powerful first-line anti-TB drugs. Second-line drugs can cure the MDR-TB, but second-line treatments are restricted and require an extensive chemotherapy (up to 2 years of treatment). In several cases, extreme drug resistance can develop and is called extensively drug-resistant TB (XDR-TB); it is a more serious form of MDR-TB caused by the *M. tuberculosis* bacteria that do not counter to the second-line anti-TB drugs. WHO estimated that there were 600,000 new cases during 2016 with drug resistance to rifampicin (the most effective first-line drug) of which 490,000 people had MDR-TB. The MDR-TB problem is mainly estimated in three countries (India, China, and the Russian Federation), which collectively account for close to half of the total cases throughout the world. Further, it is estimated that around 6.2% of MDR-TB cases had XDR-TB in 2016. However, WHO approved the use of a standardized short treatment for MDR-TB in patients who were resistant to second-line TB medicines. Individuals with XDR-TB cannot use this regimen; however, one of the new drugs (delamanid and bedquiline) may be used for the regimen to control the prevalence for long time. Reports by WHO suggest that as of June 2017, 54 countries had hosted delamanid and 89 nations had hosted bedaquiline to develop the efficiency of MDR-TB treatments (WHO 2017).

3.3.5 *Klebsiella pneumoniae*

K. pneumoniae was first isolated during the late nineteenth century and was initially recognized as Friedlander's bacterium owing the discoverer's name (Merino et al. 1992). It is a gram-negative, nonmotile, encapsulated bacterium that is found in the environments including water and soil and on medical devices such as catheters (Bagley 1985; Rock et al. 2014). Prominently, reports suggest that *K. pneumoniae* freely colonizes surfaces of human mucosa, including the important organs such as oropharynx and gastrointestinal (GI) tract and where the properties of its establishment seems to be benign (Bagley 1985; Rock et al. 2014; Dao et al. 2014).

From those colonized sites, *K. pneumoniae* may enter into other nearby tissues and cause severe infections. This can be represented by the ability of these bacteria to escape from the immune response and survive at many sites in the hosts, rather than aggressively suppress various mechanisms of the immune system (Paczosa and Mecsas 2016). As a common route of infection, *K. pneumoniae* are typically seen in individuals with an immune-suppressed condition. The individuals with *K. pneumonia* infection is assumed to have weakened respiratory host defenses, including alcoholism, diabetes, liver disease, malignancy, glucocorticoid therapy, COPD, and renal failure. Most of these infections are acquired while a person is in the hospital for some other reasons and is known as a *nosocomial infection*.

3.3.5.1 Pathogenesis – *Klebsiella* infections usually spread through the pathogenic bacteria via respiratory tract, which initially causes pneumonia or infection in the bloodstream (septicemia). *Klebsiella* can spread promptly and easily but not through the air. Healthcare locations are the most vulnerable places for *Klebsiella* infections, owing to the nature of trials that allow tranquil access of the pathogen into the body. Patients who are on ventilators or have catheters or surgery lesions are highly disposed to spreading this deadly nosocomial infection. Infection of *K. pneumoniae* occurs in the lungs, where they cause necrosis, inflammation, and hemorrhage within the lung tissue. This is caused by aspirating oropharyngeal microorganisms into the LRT. About 70% of hospitalized pneumonia cases can be detected as aspiration pneumonia (ASP) based on the description determined by Japanese hospital acquired pneumonia (HAP) and nursing and healthcare-associated pneumonia (NHCAP) guidelines (Teramoto et al. 2008; Kohno et al. 2013). Reports suggest that the ratio of ASP cases to the incidence of hospital-acquired pneumonia increases with age. ASP contains two pathological conditions: dysphagia-associated miss-swallowing and airspace infiltration with bacterial pathogens. Microaspiration of oropharyngeal matters is particularly communal in the elderly, including those who are affected after a stroke and may cause small penetrations to the lungs, which then develop into ASP (Kikuchi 1994; Teramoto 2009; Shimada et al. 2014). Occurrence of pneumonia among outpatients in communication with the healthcare system such as hospitals is characterized as healthcare-associated pneumonia. The incidence of ASP is high in very elderly patients and those with healthcare-associated pneumonia (Teramoto et al. 2009).

3.3.5.2 Prophylaxis – Various reports clearly reveal that *K. pneumoniae* is resistant to a number of well-known antibiotics. Based on the area of infection that has been affected by *K. pneumoniae*, the antibiotic treatment varies. The choice of antibiotic treatment is specifically improved for people affected with confirmed bacteremia. The available antibiotics with high intrinsic activities against *K. pneumoniae* consist of carbapenems, cephalosporin, quinolones, and aminoglycosides. These treatments against *K. pneumoniae* are primarily used as mono-antibiotic therapy or a combination with one or more antibiotic therapy. An initial course of treatment, typically between 48 and 72 hours of combination aminoglycoside therapy, is suggested for the patients who are severely ill with pneumonia and also bacteremia. This would then be followed by other extended-spectrum cephalosporin antibiotic. Carbapenem resistance is an emerging problem and is considered to be remarkable due to the production of *K. pneumoniae* carbapenemase beta-lactamase (KPC) (Arnold et al. 2011). Due to the widespread of KPC-producing bacteria, clinicians rely on tigecycline and polymyxins for the *K. pneumoniae* treatment. It is found that polymyxins have been the only agent among other antibiotics that is active against KPC-producing *K. pneumonia*. However, they were occasionally used due to their association with neurotoxicity and nephrotoxicity. Very few data support the polymyxin as a promising drug for KPC treatment. Swallowing function assessment is significant for the management and diagnosis of pneumonia. When performed in the elderly who require a high level of attention, bedside simple swallowing provocation tests and swallowing function assessments may be desirable (Teramoto et al. 1999, 2004; Osawa et al. 2013).

3.4 Conclusion

The human respiratory system contains a series of organs accountable for the intake of oxygen and dismissing the waste product of cellular function (i.e., carbon dioxide). Whenever something goes wrong with any of the parts of respiratory tract, it makes it harder to intake the oxygen required and to get rid of the excess carbon dioxide. Common respiratory symptoms comprise difficulty in breathing, chest pain, and cough. Numerous pathogens have been found to endure the disease progression from lungs to other organs by a mutual invasion through receptors. There are various antibiotic therapies available to treat RTIs, but the lack of adequate therapeutic options for RTIs caused by emerging bacterial pathogens signifies a main hazard to human health throughout the world. There are only few reports collectively reviewed about the RTIs and bacterial pathogens starting from etiology, treatment, and pathogenesis. This chapter summarizes the outline of RTIs and the pathogenesis followed by prophylaxis of potential RTI bacterial pathogens.

References

Arnold, R. S., K. A. Thom, S. Sharma, M. Phillips, J. Kristie Johnson, and D. J. Morgan. 2011. Emergence of *Klebsiella pneumoniae* carbapenemase-producing bacteria. *Southern Medical Journal* 104 (1):40–45.

Bagley, S. T. 1985. Habitat association of *Klebsiella* species. *Infection Control & Hospital Epidemiology* 6 (2):52–58.

Baumert, N., C. von Eiff, F. Schaaff, G. Peters, R. A. Proctor, and H. G. Sahl. 2002. Physiology and antibiotic susceptibility of *Staphylococcus aureus* small colony variants. *Microbial Drug Resistance* 8 (4):253–260.

Bhattacharya, M., D. J. Wozniak, P. Stoodley, and L. Hall-Stoodley. 2015. Prevention and treatment of *Staphylococcus aureus* biofilms. *Expert Review of Anti-infective Therapy* 13 (12):1499–1516.

CDC. 2016. https://www.cdc.gov/hi-disease/about/prevention.html.

Centers for Disease, Control, and Prevention. 2008. Progress in introduction of pneumococcal conjugate vaccine—worldwide, 2000–2008. *MMWR. Morbidity and Mortality Weekly Report* 57 (42):1148.

Committee on Infectious, Diseases. 2010. Recommendations for the prevention of Streptococcus pneumoniae infections in infants and children: use of 13-valent pneumococcal conjugate vaccine (PCV13) and pneumococcal polysaccharide vaccine (PPSV23). *Pediatrics* 126 (1):186–190.

Dao, T. T., D. Liebenthal, T. K. Tran, B. N. T. Vu, D. N. T. Nguyen, H. K. T. Tran, C. K. T. Nguyen et al. 2014. *Klebsiella pneumoniae* oropharyngeal carriage in rural and urban Vietnam and the effect of alcohol consumption. *PLoS One* 9 (3):e91999.

Dasaraju, P. V. and C. Liu. 1996. Infections of the respiratory system. *Medical Microbiology* 4.

Davis, K. A., J. J. Stewart, H. K. Crouch, C. E. Florez, and D. R. Hospenthal. 2004. Methicillin-resistant *Staphylococcus aureus* (MRSA) nares colonization at hospital admission and its effect on subsequent MRSA infection. *Clinical Infectious Diseases* 39 (6):776–782.

Dunyach-Remy, C., C. Ngba Essebe, A. Sotto, and J. P. Lavigne. 2016. *Staphylococcus aureus* toxins and diabetic foot ulcers: Role in pathogenesis and interest in diagnosis. *Toxins (Basel)* 8 (7).

Felmingham, D. and R. N. Gruneberg. 2000. The Alexander Project 1996–1997: Latest susceptibility data from this international study of bacterial pathogens from community-acquired lower respiratory tract infections. *Journal of Antimicrobial Chemotherapy* 45 (2):191–203.

Foster, T. J. 2004. The *Staphylococcus aureus* "superbug". *Journal of Clinical Investigation* 114 (12):1693–1696.

Foster, T. J. 2005. Immune evasion by *staphylococci. Nature Reviews Microbiology* 3 (12):948–958.

Friedman, J. N., M. J. Rieder, and J. M. Walton. 2014. Acute Care Committee Drug Therapy Canadian Paediatric Society, and Committee Hazardous Substances. "Bronchiolitis: recommendations for diagnosis, monitoring and management of children one to 24 months of age." *Paediatrics & Child Health* 19 (9):485–491.

Gagneux, S., K. DeRiemer, T. Van, M. Kato-Maeda, B. C. de Jong, S. Narayanan, M. Nicol et al. 2006. Variable host-pathogen compatibility in *Mycobacterium tuberculosis. Proceedings of the National Academy of Sciences of the United States of America* 103 (8):2869–2873.

Garcia, L. G., S. Lemaire, B. C. Kahl, K. Becker, R. A. Proctor, O. Denis, P. M. Tulkens, and F. Van Bambeke. 2013. Antibiotic activity against small-colony variants of *Staphylococcus aureus*: Review of *in vitro*, animal and clinical data. *Journal of Antimicrobial Chemotheraphy* 68 (7):1455–1464.

Global, regional, and national incidence, prevalence, and years lived with disability for 310 diseases and injuries, 1990–2015: A systematic analysis for the Global Burden of Disease Study 2015. *Lancet* 388 (10053):1545–1602.

Henriques-Normark, B. and E. I. Tuomanen. 2013. The *pneumococcus:* Epidemiology, microbiology, and pathogenesis. *Cold Spring Harbor Perspectives in Medicine* 3 (7):a010215.

Henriqus Normark, B., B. Christensson, A. Sandgren, B. Noreen, S. Sylvan, L. G. Burman, and B. Olsson-Liljequist. 2003. Clonal analysis of *Streptococcus pneumoniae* nonsusceptible to penicillin at day-care centers with index cases, in a region with low incidence of resistance: Emergence of an invasive type 35B clone among carriers. *Microbial Drug Resistance* 9 (4):337–344.

Howden, B. P., J. K. Davies, P. D. Johnson, T. P. Stinear, and M. L. Grayson. 2010. Reduced vancomycin susceptibility in *Staphylococcus aureus*, including vancomycin-intermediate and heterogeneous vancomycin-intermediate strains: resistance mechanisms, laboratory detection, and clinical implications. *Clinical Microbiology Reviews* 23 (1):99–139.

Howlin, R. P., M. J. Brayford, J. S. Webb, J. J. Cooper, S. S. Aiken, and P. Stoodley. 2015. Antibiotic-loaded synthetic calcium sulfate beads for prevention of bacterial colonization and biofilm formation in periprosthetic infections. *Antimicrobial Agents and Chemotherapy* 59 (1):111–120.

Jimenez-Truque, N., E. J. Saye, N. Soper, B. R. Saville, I. Thomsen, K. M. Edwards, and C. B. Creech. 2017. Association between contact sports and colonization with *Staphylococcus aureus* in a prospective cohort of collegiate athletes. *Sports Medicine* 47 (5):1011–1019.

Jones, R., N. Britten, L. Culpepper, D. Gass, D. Mant, and R. Grol. 2005. *Oxford Textbook of Primary Medical Care*. Cham, Switzerland: Springer.

Kikuchi, R., N. Watabe, T. Konno, N. Mishina, K. Sekizawa, and H. Sasaki. 1994. High incidence of silent aspiration in elderly patients with community-acquired pneumonia. *American Journal of Respiratory and Critical Care Medicine* 150 (1):251–253.

Kitti, T., K. Boonyonying, and S. Sitthisak. 2011. Prevalence of methicillin-resistant *Staphylococcus aureus* among university students in Thailand. *Southeast Asian Journal of Tropical Medicine and Public Health* 42 (6):1498–1504.

Kohno, S., Y. Imamura, Y. Shindo, M. Seki, T. Ishida, S. Teramoto, J. Kadota, K. Tomono, and A. Watanabe. 2013. Clinical practice guidelines for nursing-and healthcare-associated pneumonia (NHCAP) [complete translation]. *Respiratory Investigation* 51 (2):103–126.

Liu, C., A. Bayer, S. E. Cosgrove, R. S. Daum, S. K. Fridkin, R. J. Gorwitz, S. L. Kaplan et al. 2011. Clinical practice guidelines by the infectious diseases society of America for the treatment of methicillin-resistant *Staphylococcus aureus* infections in adults and children. *Clinical Infectious Diseases* 52 (3):e18–e55.

Lodha, R., S. K. Kabra, and R. M. Pandey. 2013. Antibiotics for community-acquired pneumonia in children. *The Cochrane Library: Cochrane Reviews*. CD004874.

Merino, S., S. Camprubi, S. Alberti, V. J. Benedi, and J. M. Tomas. 1992. Mechanisms of *Klebsiella pneumoniae* resistance to complement-mediated killing. *Infection and Immunity* 60 (6):2529–2535.

Moffitt, K. L., T. M. Gierahn, Y. J. Lu, P. Gouveia, M. Alderson, J. B. Flechtner, D. E. Higgins, and R. Malley. 2011. T(H)17-based vaccine design for prevention of Streptococcus pneumoniae colonization. *Cell Host and Microbe* 9 (2):158–165.

Murphy, T. F., H. Faden, L. O. Bakaletz, J. M. Kyd, A. Forsgren, J. Campos, M. Virji, and S. I. Pelton. 2009. Nontypeable *Haemophilus influenzae* as a pathogen in children. *Pediatric Infectious Disease Journal* 28 (1):43–48.

Nunes, S., R. Sa-Leao, J. Carrico, C. R. Alves, R. Mato, A. B. Avo, J. Saldanha, J. S. Almeida, I. S. Sanches, and H. de Lencastre. 2005. Trends in drug resistance, serotypes, and molecular types of *Streptococcus pneumoniae* colonizing preschool-age children attending day care centers in Lisbon, Portugal: A summary of 4 years of annual surveillance. *Journal of Clinical Microbiology* 43 (3):1285–1293.

Nunes-Alves, C., M. G. Booty, S. M. Carpenter, P. Jayaraman, A. C. Rothchild, and S. M. Behar. 2014. In search of a new paradigm for protective immunity to TB. *Nature Reviews Microbiology* 12 (4):289–299.

Olson, M. E., S. R. Slater, M. E. Rupp, and P. D. Fey. 2010. Rifampicin enhances activity of daptomycin and vancomycin against both a polysaccharide intercellular adhesin (PIA)-dependent and -independent *Staphylococcus epidermidis* biofilm. *Journal of Antimicrobial Chemotherapy* 65 (10):2164–2171.

Orihuela, C. J., J. Mahdavi, J. Thornton, B. Mann, K. G. Wooldridge, N. Abouseada, N. J. Oldfield, T. Self, D. A. Ala'Aldeen, and E. I. Tuomanen. 2009. Laminin receptor initiates bacterial contact with the blood brain barrier in experimental meningitis models. *Journal of Clinical Investigation* 119 (6):1638–1646.

Osawa, A., S. Maeshima, and N. Tanahashi. 2013. Water-swallowing test: screening for aspiration in stroke patients. *Cerebrovascular Diseases* 35 (3):276–281.

Paczosa, M. K. and J. Mecsas. 2016. *Klebsiella pneumoniae*: Going on the offense with a strong defense. *Microbiology and Molecular Biology Reviews* 80 (3):629–661.

Parker, D. and A. Prince. 2012. Immunopathogenesis of *Staphylococcus aureus* pulmonary infection. *Seminars in Immunopathology* 34 (2):281–297.

Proctor, R. A. and G. Peters. 1998. Small colony variants in staphylococcal infections: Diagnostic and therapeutic implications. *Clinical Infectious Diseases* 27 (3):419–422.

Proctor, R. A., C. von Eiff, B. C. Kahl, K. Becker, P. McNamara, M. Herrmann, and G. Peters. 2006. Small colony variants: A pathogenic form of bacteria that facilitates persistent and recurrent infections. *Nature Reviews Microbiology* 4 (4):295–305.

Ragle, B. E., V. A. Karginov, and J. Bubeck Wardenburg. 2010. Prevention and treatment of *Staphylococcus aureus* pneumonia with a beta-cyclodextrin derivative. *Antimicrobial Agents and Chemotherapy* 54 (1):298–304.

Resman, F., M. Ristovski, J. Ahl, A. Forsgren, J. R. Gilsdorf, A. Jasir, B. Kaijser, G. Kronvall, and K. Riesbeck. 2011. Invasive disease caused by *Haemophilus influenzae* in Sweden 1997–2009; evidence of increasing incidence and clinical burden of non-type b strains. *Clinical Microbiology and Infection* 17 (11):1638–1645.

Richter, J., C. Panayiotou, C. Tryfonos, D. Koptides, M. Koliou, N. Kalogirou, E. Georgiou, and C. Christodoulou. 2016. Aetiology of acute respiratory tract infections in hospitalised children in cyprus. *PLoS One* 11 (1):e0147041.

Rigottier-Gois, L. 2003. Dysbiosis in inflammatory bowel diseases: The oxygen hypothesis. *ISME Journal* 7 (7):1256–1261.

Rock, C., K. A. Thom, M. Masnick, J. K. Johnson, A. D. Harris, and D. J. Morgan. 2014. Frequency of *Klebsiella pneumoniae* carbapenemase (KPC)-producing and non-KPC-producing *Klebsiella* species contamination of healthcare workers and the environment. *Infection Control and Hospital Epidemiology* 35 (4):426–429.

Rosenfeld, R. M., J. F. Piccirillo, S. S. Chandrasekhar, I. Brook, K. Ashok Kumar, M. Kramper, R. R. Orlandi, J. N. Palmer, Z. M. Patel, and A. Peters. 2015. Clinical practice guideline (update): Adult sinusitis. *Otolaryngology—Head and Neck Surgery* 152 (2_suppl):S1–S39.

Rubach, M. P., J. M. Bender, S. Mottice, K. Hanson, H. Y. Weng, K. Korgenski, J. A. Daly, and A. T. Pavia. 2011. Increasing incidence of invasive *Haemophilus influenzae* disease in adults, Utah, USA. *Emerging Infectious Diseases* 17 (9):1645–1650.

Ruuskanen, O., E. Lahti, L. C. Jennings, and D. R. Murdoch. 2011. Viral pneumonia. *Lancet* 377 (9773):1264–1275.

Sahin-Yilmaz, A. and R. M. Naclerio. 2011. Anatomy and physiology of the upper airway. *Proceedings of the American Thoracic Society* 8 (1):31–39.

Salem, A. H., W. F. Elkhatib, and A. M. Noreddin. 2011. Pharmacodynamic assessment of vancomycin-rifampicin combination against methicillin resistant *Staphylococcus aureus* biofilm: A parametric response surface analysis. *Journal of Pharmacy and Pharmacology* 63 (1):73–79.

Schaechter, M. 2009. *Encyclopedia of Microbiology.* Amsterdam, the Netherlands: Academic Press.

Schroeder, A. R. and J. M. Mansbach. 2014. Recent evidence on the management of bronchiolitis. *Current Opinion in Pediatrics* 26 (3):328–333.

Shaw, A. G., T. J. Vento, K. Mende, R. E. Kreft, G. D. Ehrlich, J. C. Wenke, T. Spirk et al. 2013. Detection of methicillin-resistant and methicillin-susceptible *Staphylococcus aureus* colonization of healthy military personnel by traditional culture, PCR, and mass spectrometry. *Scandinavian Journal of Infectious Diseases* 45 (10):752–759.

Shimada, M., S. Teramoto, H. Matsui, A. Tamura, S. Akagawa, K. Ohta, and A. Hebisawa. 2014. Nine pulmonary aspiration syndrome cases of atypical clinical presentation, in which the final diagnosis was obtained by histological examinations. *Respiratory Investigation* 52 (1):14–20.

Siegel, S. J. and J. N. Weiser. 2015. Mechanisms of bacterial colonization of the respiratory tract. *Annual of Review of Microbiology* 69:425–444.

Singh, B., T. Al-Jubair, M. Morgelin, M. M. Thunnissen, and K. Riesbeck. 2013. The unique structure of *Haemophilus influenzae* protein E reveals multiple binding sites for host factors. *Infection and Immunity* 81 (3):801–814.

Smith, I. 2003. *Mycobacterium tuberculosis* pathogenesis and molecular determinants of virulence. *Clinical Microbiology Reviews* 16 (3):463–496.

Teramoto, S. 2009. Novel preventive and therapuetic strategy for post-stroke pneumonia. *Expert Review of Neurotherapeutics* 9 (8):1187–1200.

Teramoto, S., H. Yamamoto, Y. Yamaguchi, Y. Ouchi, and T. Matsuse. 2004. A novel diagnostic test for the risk of aspiration pneumonia in the elderly. *Chest* 125 (2):801–802.

Teramoto, S., M. Kawashima, K. Komiya, and S. Shoji. 2009. Health-care-associated pneumonia is primarily due to aspiration pneumonia. *Chest* 136 (6):1702–1703.

Teramoto, S., T. Matsuse, Y. Fukuchi, and Y. Ouchi. 1999. Simple two-step swallowing provocation test for elderly patients with aspiration pneumonia. *Lancet* 353 (9160):1243.

Teramoto, S., Y. Fukuchi, H. Sasaki, K. Sato, K. Sekizawa, and T. Matsuse. 2008. High incidence of aspiration pneumonia in community- and hospital-acquired pneumonia in hospitalized patients: A multicenter, prospective study in Japan. *Journal of the American Geriatrics Society* 56 (3):577–579.

Tong, S. Y., J. S. Davis, E. Eichenberger, T. L. Holland, and V. G. Fowler, Jr. 2015. *Staphylococcus aureus* infections: Epidemiology, pathophysiology, clinical manifestations, and management. *Clinical Microbiology Reviews* 28 (3):603–661.

Tuchscherr, L., V. Heitmann, M. Hussain, D. Viemann, J. Roth, C. von Eiff, G. Peters, K. Becker, and B. Loffler. 2010. *Staphylococcus aureus* small-colony variants are adapted phenotypes for intracellular persistence. *Journal of Infectious Diseases* 202 (7):1031–1040.

Van Wessel, K., G. D. Rodenburg, R. H. Veenhoven, L. Spanjaard, A. van der Ende, and E. A. Sanders. 2011. Nontypeable *Haemophilus influenzae* invasive disease in The Netherlands: A retrospective surveillance study 2001–2008. *Clinical Infectious Diseases* 53 (1):e1–e7.

Von Eiff, C., K. Becker, D. Metze, G. Lubritz, J. Hockmann, T. Schwarz, and G. Peters. 2001b. Intracellular persistence of *Staphylococcus aureus* small-colony variants within keratinocytes: A cause for antibiotic treatment failure in a patient with Darier's disease. *Clinical Infectious Diseases* 32 (11):1643–1647.

Von Eiff, C., K. Becker, K. Machka, H. Stammer, and G. Peters. 2001a. Nasal carriage as a source of *Staphylococcus aureus* bacteremia: Study group. *New England Journal of Medicine* 344 (1):11–16.

Wenzel, R. P. and A. A. Fowler, 3rd. 2006. Clinical practice: Acute bronchitis. *New England Journal of Medicine* 355 (20):2125–2130.

WHO. 2016. http://www.who.int/mediacentre/factsheets/fs331/en/.

WHO. 2017. http://www.who.int/tb/publications/global_report/Exec_Summary_13Nov2017.pdf?ua=1

WHO. 2018. http://www.who.int/mediacentre/factsheets/fs104/en/.

Wikibooks contributors. Human physiology/the respiratory system. Wikibooks, The Free Textbook Project. https://en.wikibooks.org/w/index.php?title=Human_ Physiology/The_respiratory_system&oldid=3372267

Wikiversity contributors, "WikiJournal of Medicine/Medical gallery of Blausen Medical 2014," 2017. *Wikiversity,* https://en.wikiversity.org/w/index.php?title=WikiJournal_of_Medicine/Medical_gallery_of_Blausen_Medical_2014&oldid=1753045.

Wood, J. M., T. Athanasiadis, and J. Allen. 2014. Laryngitis. *BMJ* 349:g5827.

Zhang, J. R., K. E. Mostov, M. E. Lamm, M. Nanno, S. Shimida, M. Ohwaki, and E. Tuomanen. 2000. The polymeric immunoglobulin receptor translocates pneumococci across human nasopharyngeal epithelial cells. *Cell* 102 (6):827–837.

Zimmerli, W. 2014. Clinical presentation and treatment of orthopaedic implant-associated infection. *Journal of Internal Medicine* 276 (2):111–119.

4

Oral Health

A Delicate Balance between Colonization and Infection

Ana Moura Teles and José Manuel Cabeda

Contents

4.1 Shaping the oral microbiome

The immunological reactions, although essential for microbial control, self-maintenance, and vigilance are a real and severe potential problem. In fact, the destructive nature of most immunological reactions leads the immune system to be the causative agent of disease in chronic inflammation, hypersensitivity, and autoimmunity (Devendra and Eisenbarth 2003; Hennino et al. 2006; Balfour Sartor 2007; Harre and Schett 2017). With this in mind, it is not strange to consider the fact that the primary goal of the immune system is not always to eliminate pathogens, but rather to keep them in check. In fact, in most cases, the microbe itself may be both harmful or symbiotic, depending on the body anatomical location and even the local microbiota (Ramsey et al. 2016). Actually, it is now clear that the human microbiota is indeed a regular provider of genes and biochemical pathways absent from the human genome (Dethlefsen et al. 2007; Turnbaugh et al. 2007), rendering human subjects with a blend of their genome plus the metagenome of microorganisms in the body (He et al. 2015). Thus, a microbe may need to be tolerated in a "safe" anatomical location but must be eradicated from a dangerous one. The obvious "safe" organs seem to be the skin, gut, and aerial and oral cavities because these are continuously subjected to environmental contamination and prone to damaging chronic ineffective immune reactions. Thus, these surfaces are likely to harbor a complex set of interacting microorganisms that influence the host health state in a variable way, depending on age, robustness, nutrition, and other factors.

During the last few years, with the advent of next-generation sequencing (NGS) techniques (Cabeda and Moreno 2014) and the Human Microbiome Project (Turnbaugh et al. 2007; Huttenhower et al. 2012), scientists have been able to better characterize complex environments where many different bacterial species coexist in complex pathogen-pathogen and pathogen-host interactions. These include commensal, mutualistic, agonistic, and pathogenic interactions, leading to a delicate balance of pathogens not only relative to species diversity, but also to their relative numbers (Ebersole et al. 2017). These studies on the human microbiome have led to interesting new concepts and have shed new light on some diseases previously not expected to be related to bacteria such as tumors (Farrell et al. 2012), preterm birth, and low birth weight in infants (Bobetsis et al. 2006; Sacco et al. 2008; Huck et al. 2011; Walia and Saini 2015), diabetes (Mokkala et al. 2017; Knip and Honkanen 2017), hypertension (Raizada et al. 2017), obesity (Roland et al. 2017), cystic fibrosis (de Dios Caballero et al. 2017; Frayman et al. 2017; Boutin et al. 2017), and Alzheimer disease (Itzhaki et al. 2016; Pistollato et al. 2016; Tremlett et al. 2017). The same was true in more expected suspects such as bacteraemia (Bahrani-Mougeot et al. 2008), rheumatic diseases (Zhong et al. 2018), chronic bowel inflammatory disease (Aleksandrova et al. 2017), and asthma (van Meel et al. 2017), among others. These studies have even lead to some strange innovative treatments such as fecal transplant therapy (Pai and Popov 2017; Brandt 2017; Khajah 2017) as well as provided evidence supporting more conventional therapies such as probiotics (Bagarolli et al. 2017).

Microbiome studies have not only been showing that imbalances of the local microbiome contribute to in situ pathology, but has also shown systemic and distant organ-related effects (Looft and Allen 2012; Arimatsu et al. 2014; Hur and Lee 2015; Isolauri 2017). These seem to arise from both metabolic consequences of the microbiome shift and disturbances in the immune cell populations that propagate to other body sites. Thus, health-related consequences can result from both microorganism-derived metabolites and from soluble and cell surface mediators that communicate and shape the immunological repertoire and responses (Looft and Allen 2012; Arimatsu et al. 2014; Hur and Lee 2015; Isolauri 2017).

The NGS studies of the oral microbiome have revealed a complex community with some 700 different species potentially colonizing the mouth, and at least 200 different species present at any time in each individual human mouth (Paster et al. 2006a; Chen et al. 2010; Wade 2013). The complexity is even greater as different regions of the mouth tend to have different bacterial, viral, and fungal communities, in elaborate three-dimensional biofilms that are dynamically influenced by, at least, pH, saliva, immune cell reactions, immune-derived soluble agents, food remains, and oxygen availability. Furthermore, person-to-person contacts contribute to shaping the microbiome with infants showing microbiomes closely related to the ones of their respective mothers (Takahashi et al. 2017). Even sexual behavior may contribute to this diversity with oral sex practitioners, showing sexual transmissible pathogens such as human papillomavirus (HPV) (Tristao et al. 2012; Owczarek et al. 2015).

4.2 Oral anatomy influences the oral microbiome

The oral cavity shows a remarkably diverse array of anatomical niches with different characteristics and susceptibilities. From the highly lubricated and keratin protected tongue, to the thin junctional epithelium in the tooth-gingiva interaction, a diverse set of opportunities and challenges presents to both microorganisms and the immune system. Additionally, availability of food remains is also variable according to the anatomical site (there are significant differences in the proper tooth' anatomy, with smooth or grooved surfaces and if there is no correct alignment of teeth), as is the likelihood of incomplete or inefficient hygiene procedures, rendering microbe populations with different microenvironments. Finally, the likelihood of mechanical tissue damage as a result of not only mastication but also from hygiene procedures or from the occurrence of parafunctional habits, like bruxism, is also singular according to the anatomical site, rendering microbes in different anatomical locations with differential access to inner tissues and differentially subjected to immunological surveillance.

The highly vascularized base of the tongue and mouth is generally considered an intense interaction point between oral antigens and the immune system. Its importance may be highlighted by the presence of lymphoid tissues directly interacting with the oral cavity (the amygdales and the adenoids) and its intense dynamics particularly during early childhood when adaptive immunity is intensively developing. It is also interesting to note that this is the exit point of some of the salivary glands, which are important delivery systems of immunity mediators. Thus, this may be an important anatomical place for both immunological learning and immunological response to the oral microbiota.

Oral tissues, like the tongue mucosa and the palate, that are subjected to frequent mechanical stress from mastication, are protected by a thin layer of keratin (Moutsopoulos and Konkel 2017). This coating provides a higher mechanical protection from tissue damage, thus efficiently preventing pathogens from entering the body. On the other hand, lower pathogen adherence may result from the physical stress of these surfaces constantly removing bacteria by mechanical forces. These surfaces are, at the same time, extensively subjected to a constant saliva flow, which prevents bacterial adhesion and promotes its mechanical wash into the digestive tract, while also promoting some chemical elimination by enzyme mediated bacterial lysis. Similar mechanisms are in action at the surface of teeth. They are subject to extreme mechanical stress but are protected by saliva and specially enamel, the hardest substance in the human body. However, the teeth area closest to the gingiva shows a different picture. This region is generally less involved in mastication and more prone to low efficiency of hygiene procedures, leaving it relatively more likely to bacterial adhesion and biofilm formation. For the most part, the hardness of the teeth surface leaves little room for teeth damage from this microbiota, at least while acid metabolic secretions do not dissolve this protective layer. However, another danger is near the epithelium of the gingival crevice, which lines inside the gingiva, surrounding the teeth. In fact, the gingival epithelium, directed toward the gingival sulcus is a nonkeratinized epithelium that progressively thins toward the base of the sulcus where it meets the teeth and is only three to five cell layers thick. At this point, the epithelium (named junctional epithelium) is linked to teeth by hemi-desmosomes and is particularly vulnerable because it is a passageway of host factors, including cells to the gingival crevice, and thus, a potential gateway for pathogen entrance into the body. It is, however, also a major oral immune system checkpoint because the crevice is filled with the gingival crevicular fluid, constantly renewed through the junctional epithelium and containing a continuous flow of immune system humoral factors and immune system cells such as neutrophils (Moutsopoulos and Konkel 2017).

4.3 Composition of the healthy oral microbiome

Recent estimates indicate that only about 10%–50% of the human body cells are derived from a eukaryotic host (Savage 1977; Wilson 2008; Sender et al. 2016), with the rest composed of prokaryotic unicellular microorganisms, residing in ecological niches of our body. Most of these are gut-resident microbes, followed by the microorganisms present in the oral cavity of the human body (Savage 1977; Wilson 2008; Sender et al. 2016).

The oral microbiome is composed of a large array of viruses, fungi, archaea, and bacteria. Knowledge of its composition has been greatly increased in recent years because of the use of NGS techniques that directly sequence nucleic acid sequences of the genomes present, thus circumventing the difficulty of being able to culture these diverse set of microorganisms (Hasan et al. 2014).

The human microbiome can be cataloged into a "core" microbiome and a "variable" microbiome (Turnbaugh et al. 2007). The essential microbiome comprises the predominant species existing in healthy conditions and is shared among most individuals (Turnbaugh et al. 2007; Zaura et al. 2009; Sonnenburg and Fischbach 2011). The variable one is the product of individual lifestyle, phenotypic, and genotypic determinants, thereby making it exclusive, and it can be as inimitable to the individual as his fingerprint (Dethlefsen et al. 2007). It must however be stressed that microbiomes from the same location on the body are more similar among different individuals than microbiomes from different locations on the same individual (Sonnenburg and Fischbach 2011).

4.3.1 Nonbacterial oral microbiome

Virus: Viruses present in the oral cavity are mostly related to chronic or transient infections. Examples are HIV, HPV, mumps, rabies virus, hepatitis virus, rhinovirus, influenza, and herpes viruses. Additionally, a number of different bacteriophage viruses have been consistently found (Wade 2013), which is not surprising given the large array of bacteria present, but may indicate that these viruses also have a role in shaping the microbiome composition.

Protozoan: Two protozoan species (*Entamoerbagingivalis* and *Trichomonastenax*) were also present in the normal oral microbiome that seem to be associated with poor hygiene and have little or no association with pathology (Wade 2013).

Fungi: As many as 85 different species of fungi have been found in the normal oral microbiota (Ghannoum et al. 2010). The most predominant genera were *Candida*, *Cladosporidium*, *Aureobasidium*, *Saccharomycetales*, *Aspergillus*, *Fusarium*, and *Cryptococcus* (Ghannoum et al. 2010). It is, however, uncertain if all detected species represented active colonization, or if airborne spores were also detected.

Archae: The Archaea component of the oral microbiome seems to be small with only a few methanogens detected in healthy volunteers. However, its presence seems to increase with periodontitis (Wade 2013).

4.3.2 Bacterial oral microbiome

The complexity of the bacterial composition of the oral microbiome has only recently started to be fully appreciated since, at least, 50% of its constituents are noncultivable microorganisms revealed only by recent NGS approaches (Paster et al. 2006a; Wade 2013). The fact that most species identified in the oral cavity have been described at a molecular level (16S rRNA sequencing) has led to the development of an online database describing the characteristics of the bacterial oral microbiome, the Human Oral Microbiome Database (homd.org) (Chen et al. 2010).

NGS studies of different anatomical sites in the mouth and in different individuals have revealed a core oral microbiome where at least 47% of species-specific operational taxonomic units (OTU) are shared between all individuals. This, however, also reveals a considerable heterogeneity because 53% OTUs are divergent (Wade 2013). Furthermore, there are significant differences both between individuals and between anatomical sites. In general terms, we can distinguish three different microbial oral communities: the first one is present in the buccal mucosa, gingivae, and hard palate; the second one is characteristic of the saliva, tongue, tonsils, and throat; and the third one is present in supra and subgingival plaque (Paster et al. 2006a; Wade 2013; Eren et al. 2014).

Bacteria at the Bucal Mucosa: The buccal mucosa has relatively a poor bacterial flora compared to the other buccal anatomical sites. The tongue dorsum of healthy populations is colonized mainly by *Streptococcus salivaris*, *Rothia mucilaginosa*, and uncultivable species of Eubacterium. Despite this relatively low density bacterial population, the presence of bacteria in the tongue is clinically significant because it changes with pathology, and even seems to be associated to a traditional tongue diagnosis in traditional Chinese medicine (Jiang et al. 2012). Also halitosis is associated with an altered bacterial composition in the tongue (Kazor et al. 2003).

Bacteria in the Saliva: At least seven different phyla are present in the saliva (i.e., Actinobacteria, Bacteroides, Firmicutes, Fusobacteria,

Proteobacteria, Spirochetes, and TM7) (Zaura et al. 2009; Huttenhower et al. 2012), making a community of a minimum of 100,000 cells per microliter (He et al. 2015). This microbial community seems stable for a limited time frame in each individual, is not geographically dependent, and is variable between individuals (Lazarevic et al. 2010); it has even been suggested to be used as a disease marker (Relvas et al. 2015).

Bacteria in the Dental Plaque: The dental plaque is a complex three-dimensional array of interacting bacteria, forming a biofilm on the surface of teeth. It contains mainly *Firmicutes* and *Actinobacteria* bacterial groups, but may vary with the plaque type (supragingival or subgingival plaque) and the dental integrity because supragingival plaque shows progressively lower bacterial diversity with the progression of caries; the dominant bacteria may change with disease stage. A similar situation was also found on root surfaces and the subgingival plate in association with periodontal disease. Recently, a host genetic influence has been described for the bacterial composition of healthy teeth plaque, whereas the cariogenic bacteria seemed to be controlled by nongenetic factors (Gomez et al. 2017).

Geographic Influences on Oral Bacteria: Although the geographical diversity of the oral microbiome does not seem to be relevant at the phyla level, significant differences can still be found between individuals at the species level. This seems to indicate that diet and the environment have a much lower influence than the individual per se in shaping the oral microbiome (Wade 2013).

Generally, the oral microbiome includes bacterial representatives from the phyla *Actinobacteria, Bacteroidetes, Firmicutes, Proteobacteria, Spirochetes, Fusobacteria, Synergitetes*, and *Tenericutes* as well as the uncultured GN02, SR1, and TM7 (Wade 2013). Together, the first six phyla represent 96% of the species present in the normal oral microbiota, although, as mentioned before, different anatomical locations may harbor different bacterial populations.

4.4 Shaping oral immunity

The oral immune system may best be described separating the lumen of the oral cavity from its surrounding tissues. In fact, as previously described, the immune system, to some extent, tolerates microorganisms in the lumen of the oral cavity but needs to eradicate them as soon as they gain access into the interior of tissue. As different needs require different actions, different immunological responses arise in these two different anatomical sites.

Nevertheless, it must be noted that there is a continuum and a dependency between these two reactions.

In the lumen of the oral cavity, most immunological components are contained in the saliva, reaching it directly during its production in the various salivary glands, or via the gingival crevicular fluid. Accordingly, saliva may contain immunoglobulins (mostly dimeric IgA), complement' factors, neutrophils, macrophages, lysozyme, histatins, mucins, and even leucocytes (Schenkels et al. 1995).

All these components reach saliva as a response to activation of the supporting oral immunological system that includes the amygdales (palatines, lingual, and pharyngeal) and a network of submucosal lymphoid aggregates resembling the gut Peyer's patches, although not so structurally organized. These lymphoid aggregates are part of a network of lymph capillaries and drain lymph from the mucosa, gingiva, and the dental pulp, thus being capable of detecting antigens present in all these structures.

Now, let's consider the physiology of an immune response to microorganisms that penetrate the mucosal barrier. First of all, we should not forget that, in itself, this may be already considered a major accomplishment on the part of the microorganism. To do that, it had to resist immunoglobulin opsonization and consequent neutrophil or macrophage phagocytosis; it had to survive crevicular complement activation and consequent microorganism lysis; it had to circumvent neutrophil and macrophage direct phagocytosis via pathogen pattern receptors (PPR); if in the masticatory mucosa, it had also to gain access not only through the epithelial cells of the mucosa but also through the external keratin layer; and after breaking through the epithelial layer, it also had to penetrate the basal membrane of the mucosa. However, after accomplishing all this, the microorganism still has to face the mucosal immunity. This starts by the recognition of the microorganism by mucosal resident monocytes and dendritic cells, and as they detect the invader, not only phagocyte it but also migrate to the lymphoid aggregates where they stimulate B and T cells to start responding by presenting to them the antigens of the phagocytized bacteria. Even if resident macrophages fail to phagocyte the microorganism, it will find itself sooner or later dragged by the lymph into the lymphoid aggregates where it will be recognized by B-lymphocytes and start the lymphoid reaction. Similarly, the amygdales can intervene in the process. As fully developed and structured secondary lymphoid organs, their function can be seen to be similar to the lymphoid aggregates, although with a bigger chance of harboring the specific T and B lymphocytes required to respond to the antigenic stimuli. All these stimulated lymphocytes will then divide, producing large numbers of clones of themselves, differentiate from naive into memory or effector cells, and migrate to the tissue where they are most needed. This may be done through the circulatory system or directly into the oral cavity. In the circulatory system, the lymphoid cells can use chemical clues left by the phagocytes that change the circulatory epithelium near the site of infection, making it express the adhesion molecules needed for the lymphocyte to enter the tissue. In turn, the generated memory cells will migrate to other secondary lymphoid organs, leaving with them the memory of the previous encounter and rendering other sites of the body capable of recognizing and responding to the microorganism first found in the oral microenvironment.

All these mechanisms, in turn, influence the set and diversity of immune system cells and soluble mediators continuously being sent to the oral cavity via the saliva and the crevicular flows.

4.4.1 Saliva flow

Saliva (produced at a rate of approximately 500mL per day) has a complex mixture of components, accounting for a diverse set of functions ranging from lubrication to digestion and immunological control of the oral microbiota. Some functions may even be shared, as the lubrication and the mechanical washing of bacteria can be seen as indirect immunological actions as they are efficient ways of preventing oral bacterial adhesion and of directing oral bacteria to the gut where they may help food digestion and absorption. Saliva composition is not only derived from salivary glands, as the crevicular flow (see next section) is also an important contributor of immunological mediators that end up in this physiological fluid. Nevertheless, the salivary glands (primary and accessory) are the major source of saliva components production, including its immunological mediators. In particular, it has been demonstrated that the dimeric IgA component of saliva (including its dimerizing J chain) is mostly derived from plasma B cells of salivary glands. Also, the salivary duct has shown to be responsible not only for the release of saliva, but they also serve as a gateway for antigens to stimulate salivary B lymphocytes to produce the immunoglobulins depending on oral antigen composition (Nair and Schroeder 1983).

Saliva is also rich in lysozyme, peroxidase, lactoferrin, and immune-derived glycoproteins that bind to bacterial adhesins and complement cascade proteins (Schenkels et al. 1995; Fabian et al. 2012), all of which contribute to oral native immunological reactions.

Saliva also harbors a large number of polymorphonuclear cells, mostly reaching saliva via the crevicular flow. Despite the fact that its importance out of the crevicular sulcus is not clear, these cells may react locally to oral antigens by phagocytosis and by releasing active antimicrobial agents such as lysozyme, acid hydrolases, and reactive oxygen species (Moutsopoulos and Konkel 2017).

4.4.2 Crevicular flow

As mentioned previously, the gingival crevice is a particularly sensitive anatomical site, potentially allowing oral bacterial entry into the adjacent tissues. This is even emphasized by the 10-fold increase in the gingival crevicular area in cases of active inflammation, potentially increasing the leakage potential to the inner tissues. However, this is counterbalanced by the fact that this is, in parallel, a major oral immunological active site. Actually, the junctional epithelium of the gingival crevicular is the major source of humoral and cellular immunological components that fill the crevicular sulcus and ends up in the saliva. Contrary to the described composition of the saliva, the crevicular fluid has major influences from the systemic immunological system, as it receives its components not only from local secondary lymphoid tissues, but also from the blood. Thus, the crevicular fluid is not only rich in IgA, but also in IgG and IgM, complement components and its labile fragments produced during complement activation. The presence of these complement activation by-products not only shows that active complement reactions are taking place in the crevicular sulcus, but also indicate that immune cell chemotactic calling and capillary vasodilation with inherent increased permeability is also a result of complement activation in the crevicular sulcus, further potentiating the immune response to bacterial presence in this anatomical area. Native response immune mediators such as neutrophil-derived mediators, among others, have also been described in the crevicular flow and may contribute to the immunological reactions and to differences between deciduous and permanent teeth immunology (Moriya et al. 2017).

Of particular importance to crevicular immunology is the flow of immunological cells. In fact, the leaky nature of the junctional epithelium in this area allows an important flow of immunological cells. Most of these cells are neutrophils (more than 95% of the total leukocytes) migrating from the circulation into the gingival tissue and to the crevicular sulcus at a rate of over 30,000 neutrophils per minute in humans (Schiott and Loe 1970). This is the most important source of immunological cells and renders the oral cavity with a plethora of cells with varying levels of activation and functional states and a potential repertoire of cells for phagocytosis, degranulation, and secretions of immune mediators such as cytokines, anti-microbial peptides, and neutrophil extracellular traps. The importance of these cells has been demonstrated by studies of patients with genetic defects of neutrophils that show severe periodontal immunopathology. Interestingly, it has been shown that both too few or too many neutrophils unbalance the immune reaction toward periodontal immunopathology (Kantarci et al. 2003; Matthews et al. 2007; Dutzan et al. 2016), making the regulation of neutrophils number in the crevice an unknown mechanism of rather critical importance.

4.5 From colonization to infection and to pathology

4.5.1 Dysbiosis—Disease trigger

Once established, the oral microbiome is maintained by host and microbe-derived factors, in a not fully understood process involving pro- and anti-inflammatory bacterial signals (Devine et al. 2015). However, despite the establishment of this flora, infections on the immune-competent oral mucosa are a rather rare event (Zaura et al. 2014). In contrast, patients who are immunosuppressed can often experience life-threatening viral and fungal infections of the mucous membranes (Soga et al. 2011; Petti et al. 2013; Diaz et al. 2013), demonstrating that the pathological potential of this flora exists but is controlled by the host immune system.

The concept of an "exposome" encompassing all nongenetic influences from birth onward promote increased susceptibility or increased protection to a disease has been proposed. The exposome includes the nutritional status, antioxidant status, free radicals generated, and the microbiome (Bogdanos et al. 2013). Inherent to the exposome concept is the notion that maintenance of an oral microbiome healthy state (symbiosis) or a disease state (dysbiosis) results from a complex and dynamic equilibrium between all the resident species in the oral cavity. So, a dysbiotic microbiome is one in which the diversity and relative proportions of taxa within the finely tuned ecosystem is disturbed (Cho and Blaser 2012). Consequently, dysbiosis can be defined as a compositional shift in the normal population of a particular microbial community that promotes development of an inflammatory or disease state. Often it arises through external triggers like poor oral hygiene, dietary habits, and intake of medications which, for instance, change saliva flow or composition, induce gingival inflammation (Marsh et al. 2014; Wu et al. 2016), reduce oxygen tension due to increase in biofilm thickness, altered host defenses, or nutritional, metabolic, or structural stresses within the ecosystem (Cho and Blaser 2012). Dysbiosis can also result from microbes-host relationship shifts, turning from a mutualistic to a parasitic connection, acquisition of virulence factors, or a shift to opportunistic behavior instead of a commensal one (Avila et al. 2009; Parahitiyawa et al. 2010). As a result of the ecological shifts, beneficial bacterial become less capable of inhibiting the growth of the usually well-controlled biofilm pathogenic bacteria (Marsh et al. 2015), resulting in infection (Ruby and Goldner 2007) and disease (Amerongen and Veerman 2002).

4.5.2 From the early colonizers to oral biofilms: A holistic perspective of the microbiome behavior

The oral cavity is a dynamic microbial ecosystem to several allochthonous species (transient visitors) in addition to autochthonous members (stable colonizers) (Krishnan, Chen, and Paster 2017). Furthermore, the ecosystem varies dramatically throughout life from bacteria acquisition at birth (Berkowitz and Jones 1985; Asikainen and Chen 1999) to elderly colonization (Preza et al. 2008). Some of these changes are also associated with modifications in the oral surfaces and microenvironments such as the crevice sulcus with teeth eruption and the availability of synthetic surfaces with the use of prosthesis, for instance. All these changes provide different opportunities for different bacteria depending on the availability of specific adhesion-receptors for bacteria-to-bacteria and bacteria-to-surface contacts (Kolenbrander 2000; Nobbs et al. 2011).

During the first months of life, species of Streptococcus are usually the first pioneering microorganisms to colonize the oral cavity with *Streptococcus salivarius* found mostly on the tongue dorsum and in saliva, *Streptococcus mitis* on the buccal mucosa, and *Streptococcus sanguinis* on the teeth (Socransky and Manganiello 1971; Gibbons and Houte 1975; Smith et al. 1993). The establishment of these herald microorganisms implies local ecological transformations, namely, local redox potential, pH, co-aggregation, and availability of nutrients, thereby enabling more fastidious organisms to colonize after them (Marsh 2000). As a result, we can see the appearance of *Prevotellam elaninogenica*, *Fusobacterium nucleatum*, *Veillonella*, *Neisseria*, and nonpigmented *Prevotella* (Kononen et al. 1992). Latter, the appearance of teeth surfaces (and with it, the gingival crevice) leads to increases of genera such as *Leptotrichia*, *Campylobacter*, *Prevotelladenticola*, and members of the *Fusobacterium* and *Selenomonas genera* (Kononen et al. 1994).

As individuals age, biofilm maturation leads to increases in the numbers of *Streptococcus, Veillonella, Granulicatella, Gamella, Actinomyces, Corynebacterium, Rothia, Fusobacterium, Porphyromonas, Prevotella, Capnocytophaga, Nisseria, Haemophilis, Treponema, Lactobacterium, Eikenella, Leptotrichia, Peptostreptococcus, Staphylococcus, Eubacteria,* and *Propionibacterium* (Aas et al. 2005; Jenkinson and Lamont 2005; Zaura et al. 2009; Bik et al. 2010).

Currently, it is generally accepted that bacteria are present in the oral cavity in the form of biofilms. These are structured and dynamic communities attached to a surface and enwrapped in an extracellular matrix. In this intricate community, different bacteria interact with each other to establish a complex ecosystem where each member contributes in some form to the remaining members of the community (Marsh 2000). These elaborate associations create special microenvironments, modifying the virulence of the resulting biofilm and contributing to the associated pathogenesis (Marsh 2000; Flemming et al. 2016; Koo and Yamada 2016). Thus, microfilm dynamics allows specific low-abundance pathogens to influence disease by altering the "healthy" microflora to a disease state (Hajishengallis et al. 2012), thus changing a healthy community to a disease-inducing one (Krishnan et al. 2017). This has been formally proposed as the "keystone pathogen hypothesis."

4.5.3 From a healthy microbiome to a disease-associated one

The oral microbiome is so deeply linked to the host healthy or diseased state that over the years, several convincing studies have uncovered correlations between its qualitative compositions (Kilian et al. 2016). However, understanding the complexity of microbiome dynamics and its effects on hosts requires a holistic view, including interactions among different residents within the community (bacterial interspecies and interkingdom interactions) as well as their interaction with the host (Baker et al. 2017). In other words, the microbiome must be seen producing commensalism within microorganisms and mutualism with the host (Ruby and Goldner 2007; Zaura et al. 2009; Filoche et al. 2010; Ebersole et al. 2017). Commensal relationships among microbes allow them to flourish at no expense to their cohabitants and, in turn, maintain biodiversity within the oral cavity (Zarco et al. 2012). Research has demonstrated such biodiversity to be crucial to health; children suffering from severe dental caries have less diverse oral microbiome than those who are healthy (Kanasi et al. 2010). Similarly, symptomatic lesions of endodontic origin have a lower level of diversity than asymptomatic ones. It could be suggested that health-associated biodiversity results from the need for specific functions from each species to maintain oral cavity equilibrium and homeostasis (Zarco et al. 2012).

Commensal bacteria not only protect the host simply by niche occupation but also interact with host tissue, promoting the development of proper tissue structure and function; it is now consensual that host-associated polymicrobial communities, such as those found in the oral cavity, co-evolved with us and have become an integral part of who we are (Roberts and Darveau 2015). Also, the residual and commensal bacteria have a critical function called *colonization resistance*, as a result of their proficiency to adapt to a variety of niches, preventing colonization by pathogenic bacteria. In fact, the detrimental effects of commensal bacteria depletion are well demonstrated by the use of broad-spectrum antibiotics (Brook 1999; He et al. 2014). This is also highlighted by the normal commensal microbiota's ability to rapidly adapting to hostile changes in microenvironment, preventing pathogenic bacteria from taking the opportunity. In fact, the normal microbiota rapidly recovers after changes in pH, redox potential, atmospheric conditions, salinity, and water activity from saliva (Badger et al. 2011), resulting from eating, communicating, and oral hygiene (Avila et al. 2009)

Thus, diseases such as caries, gingivitis, and periodontitis may be seen as the result of a failure of the preexisting healthy microbiome to adapt to some trigger that has produced the right conditions for some preexisting controlled bacteria to take over and change the biofilm dynamics into a disease-promoting one. It can thus be expected that further elucidation of these changes will allow us to greatly improve our understanding of oral microbial physiology, pathogenesis, and ecology, as well as our ability to diagnose and treat microbial infections (Baker et al. 2017).

4.5.4 Caries, periodontal disorders, and endodontic lesions: Manifestations of microbiome imbalances

It is well established that the most frequent oral diseases (i.e., caries, periodontal diseases, and endodontic lesions) are caused by microorganisms that, nevertheless, are also present in the healthy human oral microbiota (Socransky and Haffajee 2005; Paster et al. 2006b; Aas et al. 2008; Dewhirst et al. 2010; Johansson et al. 2016). These species are likely living harmlessly in low numbers (often below the limit of detection of conventional microbiology) and, hence, regularly receive the designation of "putative" pathogens or biomarkers of disease (Krishnan et al. 2017). Thus, the presence of the bacteria is not, per se, sufficient to induce the disease.

A remarkable note is that these oral diseases have as a main etiological cause a polymicrobial infection. Another important point to highlight is that those disorders result from an opportunistic infection as a consequence of diet, host immune response, complicating systemic or genetic disorders, pH change, poor oral hygiene, and lifestyle fluctuations (Krishnan et al. 2017). In a simple sentence: when dysbiosis happens.

4.5.4.1 Caries – The most prevalent oral disease, responsible for tooth pain and lost, begins with minor lesions on the enamel and can evolve, without treatment, into the dentin and into the pulp space. This destruction of the mineralized tooth surface results from metabolic actions of opportunistic pathogens. The microorganisms are capable of dietary carbohydrates fermentation with production of acidic by-products, eventually decreasing the micro-environment's pH and destroying the tooth's mineralized surface.

In the early nineteenth century, the nonspecific plaque hypothesis (Rosier et al. 2014) stated that dental infections were the outcome of an overgrowth of plaque nonspecific bacteria, leveraged by its own accumulation and ecological changes in consequence of either ineffective hygiene or diet modification. Also, there was the concept that anyhow, every plaque had the potential to cause disease and as preventive measure, every effort to plaque removal was advised (Rosier et al. 2014).

Later, with classic culture development, isolation and characterization of species were possible, which allowed the observation that kanamycin (Loesche et al. 1977) was particularly effective against caries-associated species, such as streptococci, leading to the emergence of the specific plaque hypothesis. At this point, it was conceivable that only a few species participated in the illness process; therefore, cure or prevention of sickness went through the prescription of antibiotics (Loesche 1979; Rosier et al. 2014). Restrictions of this microbiological identification methods became evident when translation from day-to-day practice with poor long-term clinical benefits was observed (Rosier et al. 2014).

The interaction between the resident oral microbiota and the host environment proposed, in the 1980s, by the ecological plaque hypothesis explained the manifestation of caries and stills. Briefly, an increased frequency of sugar intake, or a reduction in saliva flow, results in plaque biofilms in close proximity to the tooth surface that are exposed for longer and more regular periods to lower pH levels. As a result, there is an increase in the putative pathogens, as they are better adapted to the new acidic conditions, either by producing acids themselves or by being more tolerant to an acidic environment. This occurs at the expense of bacteria that thrive in neutral conditions or contribute to pH neutralization (Liu et al. 2012; Marsh et al. 2015). That is, dysbiosis occurs by the selection of efficient acid-producing and -tolerating bacteria in contrast with daily health, where the biofilm undergoes multiple pH cycles during one day, resulting in enamel de- and re-mineralization. Accordingly, if there is insufficient time at neutral pH during frequent snacking, then demineralization outweighs remineralization which results in mineral loss, and eventually, in enamel lesions (Kilian et al. 2016).

Although a specific microbiome that signals dental caries is yet to be found (Ling et al. 2010), the most common bacteria responsible for dental caries are *Streptococcus mutans*, *Streptococcus sobrinus*, and *Lactobacillus acidophilus* (Selwitz et al. 2007). NGS experiments have demonstrated that in carious lesions with *Streptococcus mutans*, additional species belonging to the genera *Atopobium*, *Propionibacterium*, and *Lactobacillus* were also present at significantly higher levels. In those subjects with no detectable levels of *S. mutans*, *Lactobacillus* species, *Bifidobacterium dentium*, and low-pH non-*S. mutans* streptococci were predominant (Aas et al. 2008). Based on these findings, it was advocated that species of the genera *Veillonella*, *Lactobacillus*, *Scardovia*, and *Propionibacterium*, low-pH non-*S. mutans* streptococci, *Actinomyces* species, and *Atopobium* species may play an important role in caries progression.

Other researchers supported by NGS methodologies on the microbiome of populations with a low and high prevalence of caries found that adolescents in Romania, who had limited access to dental care, were colonized with *S. mutans* and *S. sobrinus*. In contrast, adolescents in Sweden, who benefit from good dental care, were colonized only rarely with those two closely related species of streptococci, but were colonized by more species of *Actinomyces*, *Selenomonas*, *Prevotella*, and *Capnocytophaga* (Johansson et al. 2016).

Among primary and secondary dentitions as well as in root surface caries, there are significant differences in the oral microbiome's composition. In the first one, not surprisingly, *S. mutans* was typically detected at high levels (Becker et al. 2002). Other species like *Actinomycesgerencseriae*, *Scardoviawiggsiae*, *Veillonella*, *Streptococcus salivarius*, *Streptococcus constellatus*, *Streptococcus parasanguinis*, and *Lactobacillus fermentum* were found in the secondary dentition. The root surface, unlike the dental crown, is not covered by enamel. It was found that the predominant taxa included *Actinomyces* species, *Lactobacillus*, *Enterococcus faecalis*, *Mitsuokella* sp. HOT131, *Atopobium* and *Olsenella* species, *Prevotella multisaccharivorax*, *Pseudoramibacter alactolyticus*, and *Propionibacterium acidifaciens*, although *S. mutans* was also present with these kind of caries (Preza et al. 2008).

4.5.4.2 Periodontitis – Periodontal disease, a polymicrobial inflammatory disorder of the periodontium (Pihlstrom et al. 2005), also results from subgingival plaque accumulation that causes remodulation in the microbiota (Horz and Conrads 2007). Its less severe form is gingivitis, a gingiva inflammation caused by pathogenic biofilms that is often reversible (Pihlstrom et al. 2005), if dealt with good oral hygiene in a timely manner (Horz and Conrads 2007).

Periodontitis is an evolution of gingivitis that can be described as an infection that involves all soft tissue and bone that support the periodontium and teeth structures (Horz and Conrads 2007). The pathologic mechanism is similar to the one described for caries: multiple opportunistic pathogens overgrow in dental plaque, become pathogenic (Horz and Conrads 2007), release proteolytic enzymes that break down host tissue, and may result in gingival inflammation, loss of gingival attachment, periodontal pocket formation, alveolar bone ,and even, root resorption. The predominant pathogens involved in periodontitis are *Aggregatibacter actinomycetemcomitans, Porphyromonas gingivalis, Prevotella intermedia, Fusobacterium nucleatum, Tannerella forsythia, Eikenellacorredens*, and *Treponema denticola* (Filoche et al. 2010; Dashiff and Kadouri 2011).

If not treated, periodontal pockets occur, and periodontitis becomes irreversible (Pihlstrom et al. 2005) because the periodontium, once separated, is unable to reattach to the bone. Furthermore, periodontal pathogens develop virulent factors, such as encapsulation factors, that together with the fact that they are deep within the periodontal pockets render them resistant to antibiotics (Horz and Conrads 2007; Van Essche et al. 2011). Also, recolonization is also possible, as the nearby mucous membranes harbor bacterial reserves (Horz and Conrads 2007).

Periodontitis treatment is thus a hard challenge. Mechanical treatment is often insufficient, making antibiotics a much-needed option, despite its aforementioned limitations. Accordingly, the most appropriate and least-invasive form to manage periodontitis consists of regular appointments, acquisition of efficient ambulatory hygiene procedures, and on-time supported use of antibiotics. It is also important that the oral cavity maintains certain gram-positive bacteria that shield pathogens from damaging hard and soft tissues (Van Essche et al. 2011).

4.5.4.3 Endodontic lesions – Infection of the root canal system has been established as the primary cause of apical periodontitis, a condition characterized by the inflammation and destruction of periradicular tissues caused by etiological agents of endodontic origin. The classic pathways between the pulpal space and the outer tissues, in a two-way direction, include the apical foramen, accessory root canals, and the dentinal tubules. Due to ecological factors of the endodontic environment, such as availability of nutrients (namely fermentable carbohydrates), oxygen and pH levels, the surviving microorganisms, belong to restricted group of species which are predominantly anaerobic, with the facultative anaerobes residing in the more coronal part of the root canal system and the obligate anaerobes apically (Alves et al. 2009; Aw 2016). Those species outlast from the breakdown of pulpal tissues and serum proteins (Figdor and Sundqvist 2007), which leads to rise of environment pH (Sundqvist 1992; Figdor and Gulabivala 2008), favoring the proliferation of late-stage bacteria. The different species of the bacterial population also appear to be interdependent for nutrition (Sundqvist and Figdor 2003) because the metabolic products of one species may act as a source of nutrition for others.

In primary infections, predominant taxa detected include species of *Peptostreptococcus*, *Parvimonasmicra*, *Filifactoralocis*, and *P. alactolyticus*, and species of *Dialister*, *F. nucleatum*, *T. denticola*, *P. endodontalis*, *P. gingivalis*, *T. forsythia*, *Prevotella baroniae*, *P. intermedia*, *Prevotella nigrescens*, and *Bacteroidaceae* [G-1] HOT272 (Siqueira and Rocas 2009). *Enterococcus faecalis* was detected, but in lower levels. However, in retreatment cases advocated for secondary or persistent endodontic infections, the predominant taxa include *Enterococcus* species such as *E. faecalis*, *Parvimonas micra*, *Filifactor alocis*, *P. alactolyticus*, *Streptococcus constellatus*, *Streptococcus anginosus*, and *Propionibacterium propionicum* (Aw 2016; Krishnan et al. 2017).

The microbiomes of endodontic-periodontal lesions, which evolve simultaneous infection of pulpal space and of the periodontum, possessed similar profiles including *E. faecalis*, *P. micra*, *Mogibacterium timidum*, *Filifactor alocis*, and *Fretibacterium fastidiosum* (Gomes et al. 2015).

4.6 Oral microbiome as health biomarker

The oral cavity is the primary gateway to the human body; therefore, microorganisms that inhabit that area are capable of spreading to different body sites (Dewhirst et al. 2010).

Pathogens originated in the oral cavity can be often detected in blood cultures as they destroy and pass through oral mucous membranes, periapical lesions, and periodontal pockets (Horz and Conrads 2007). As they get access to the bloodstream, they induce immune responses or produce excessive and deregulated amounts of inflammatory mediators and ultimately can cause disease at different body sites (Williams 2008). Oral bacteria may be used as biomarkers for certain systemic diseases, such as endocarditis (Berbari et al. 1997), ischemic stroke (Joshipura et al. 2003), cardiovascular disease (Beck and Offenbacher 2005; Teles and Wang 2011), pneumonia (Awano et al. 2008), pancreatic cancer (Farrell et al. 2012), diabetes type II (Demmer et al. 2015), pediatric Crohn disease (Docktor et al. 2012), and low weight and preterm birth (Shira Davenport 2010).

Periodontal disease has been shown to predispose individuals to cardiovascular disease through its ability to induce chronic inflammation (Syrjanen 1990).

Also, individuals who are diabetic can experience more often periodontal disease. Similarly, the presence of several anaerobic oral bacterial species, like *A. actinomycetemcomitans* and *S. constellatus* (Shinzato and Saito 1994) has been shown to predispose to bacterial pneumonia. Other renal and cardiovascular pathologies are nowadays a focus of investigation in relation to changes in nitric oxide homeostasis and its effect on vasodilation of vascular smooth muscle. An important role in those disorders maybe played by oral bacteria that reduce dietary nitrates into nitrite (Hezel and Weitzberg 2015), which, in turn, is absorbed by the blood and further reduced to nitric oxide by a variety of mechanisms. In Alzheimer disease, inflammation, a key feature of the disease could be caused in part by peripheral infections, such as periodontal disease (Olsen and Singhrao 2015). Periodontal pathogens such as *A. actinomycetemcomitans* and *P. intermedia* are capable of eliciting systemic inflammation, which results in the release of pro-inflammatory cytokines that traverse the blood–brain barrier. However, it is yet to be established if there is a direct causal relationship between the oral microbiome and these systemic disorders (Krishnan et al. 2017).

All this renders the oral microbiome a complex system capable of influencing health not only locally at the oral cavity, but also influencing the systemic immunological status and capable of influencing distant anatomical status health. Thus, it should be possible to manipulate the microbiome's potential to optimize personal health and identify microbial profiles with potential use to assess disease risk, which is the next phase of the Human Microbiome Project.

Thus, continuous understanding of the oral microbiome dynamics and its influence in both health and disease is a complex endeavor that promises to keep us occupied for many years to come. It may be anticipated that results will change the way we diagnose and also the way we treat many diseases, and not only oral cavity ones.

References

Aas, J. A., A. L. Griffen, S. R. Dardis, A. M. Lee, I. Olsen, F. E. Dewhirst, E. J. Leys, and B. J. Paster. 2008. Bacteria of dental caries in primary and permanent teeth in children and young adults. *Journal of Clinical Microbiology* 46 (4):1407–1417. doi:10.1128/JCM.01410-07.

Aas, J. A., B. J. Paster, L. N. Stokes, I. Olsen, and F. E. Dewhirst. 2005. Defining the normal bacterial flora of the oral cavity. *Journal of Clinical Microbiology* 43 (11):5721–5732. doi:10.1128/JCM.43.11.5721-5732.2005.

Aleksandrova, K., B. Romero-Mosquera, and V. Hernandez. 2017. Diet, gut microbiome and epigenetics: Emerging links with inflammatory bowel diseases and prospects for management and prevention. *Nutrients* 9 (9):962. doi:10.3390/nu9090962.

Alves, F. R. F., J. F. Jr Siqueira, F. L. Carmo, A. L. Santos, R. S. Peixoto, I. N. Rocas, and A. S. Rosado. 2009. Bacterial community profiling of cryogenically ground samples from the apical and coronal root segments of teeth with apical periodontitis. *Journal of Endodontics* 35 (4):486–492. doi:10.1016/j.joen.2008.12.022.

Amerongen, A. V. N., and E. C. I. Veerman. 2002. Saliva—The defender of the oral cavity. *Oral Diseases* 8 (1):12–22.

Arimatsu, K., H. Yamada, H. Miyazawa, T. Minagawa, M. Nakajima, M. I. Ryder, K. Gotoh et al. 2014. Oral pathobiont induces systemic inflammation and metabolic changes associated with alteration of gut microbiota. *Scientific Reports* 4 (May). Article number: 4828. doi:10.1038/srep04828.

Asikainen, S., and C. Chen. 1999. Oral ecology and person-to-person transmission of actinobacillus actinomycetemcomitans and porphyromonas gingivalis. *Periodontology 2000* 20 (June):65–81.

Avila, M., D. M. Ojcius, and O. Yilmaz. 2009. The oral microbiota: Living with a permanent guest. *DNA and Cell Biology* 28 (8):405–411. doi:10.1089/dna.2009.0874.

Aw, V. 2016. Discuss the role of microorganisms in the aetiology and pathogenesis of periapical disease. *Australian Endodontic Journal: The Journal of the Australian Society of Endodontology Inc* 42 (2):53–59. doi:10.1111/aej.12159.

Awano, S., T. Ansai, Y. Takata, I. Soh, S. Akifusa, T. Hamasaki, A. Yoshida, K. Sonoki, K. Fujisawa, and T. Takehara. 2008. Oral health and mortality risk from pneumonia in the elderly. *Journal of Dental Research* 87 (4):334–339. doi:10.1177/154405910808700418.

Badger, J. H., P. C. Ng, and J. C. Venter. 2011. The human genome, microbiomes, and disease. In *Metagenomics of the Human Body,* edited by K. E. Nelson, vol 17, pp. 1–14. Springer Science + Buiness Media, New York.

Bagarolli, R. A., N. Tobar, A. G. Oliveira, T. G. Araújo, B. M. Carvalho, G. Z. Rocha, J. F. Vecina et al. 2017. Probiotics modulate gut microbiota and improve insulin sensitivity in DIO mice. *The Journal of Nutritional Biochemistry* 50 (December): 16–25. doi:10.1016/j.jnutbio.2017.08.006.

Bahrani-Mougeot, F. K., B. J. Paster, S. Coleman, J. Ashar, S. Barbuto, and P. B. Lockhart. 2008. Diverse and novel oral bacterial species in blood following dental procedures. *Journal of Clinical Microbiology* 46 (6):2129–2132. doi:10.1128/JCM.02004-07.

Baker, J. L., B. Bor, M. Agnello, W. Shi, and X. He. 2017. Ecology of the oral microbiome: Beyond bacteria. *Trends in Microbiology* 25 (5):362–374. doi:10.1016/j.tim.2016.12.012.

Balfour, S. R. 2007. Bacteria in crohn's disease. *Journal of Clinical Gastroenterology* 41 (Supplement 1):S37–S43. doi:10.1097/MCG.0b013e31802db364.

Beck, J. D., and S. Offenbacher. 2005. Systemic effects of periodontitis: Epidemiology of periodontal disease and cardiovascular disease. *Journal of Periodontology* 76 (11 Suppl):2089–2100. doi:10.1902/jop.2005.76.11-S.2089.

Becker, M. R., B. J. Paster, E. J. Leys, M. L. Moeschberger, S. G. Kenyon, J. L. Galvin, S. K. Boches, F. E. Dewhirst, and A. L. Griffen. 2002. Molecular analysis of bacterial species associated with childhood caries. *Journal of Clinical Microbiology* 40 (3):1001–1009.

Berbari, E. F., F. R. 3rd Cockerill, and J. M. Steckelberg. 1997. Infective endocarditis due to unusual or fastidious microorganisms. *Mayo Clinic Proceedings* 72 (6):532–542. doi:10.1016/S0025-6196(11)63302-8.

Berkowitz, R. J., and P. Jones. 1985. Mouth-to-mouth transmission of the bacterium streptococcus mutans between mother and child. *Archives of Oral Biology* 30 (4):377–379.

Bik, E. M., C. D. Long, G. C. Armitage, P. Loomer, J. Emerson, E. F. Mongodin, K. E. Nelson, S. R. Gill, C. M. Fraser-Liggett, and D. A. Relman. 2010. Bacterial diversity in the oral cavity of 10 healthy individuals. *The ISME Journal* 4 (8):962–974. doi:10.1038/ismej.2010.30.

Bobetsis, Y. A., S. P. Barros, and S. Offenbacher. 2006. Exploring the relationship between periodontal disease and pregnancy complications. *Journal of the American Dental Association* 137 (October):7S–13S.

Bogdanos, D. P., D. S. Smyk, P. Invernizzi, E. I. Rigopoulou, M. Blank, L. Sakkas, S. Pouria, and Y. Shoenfeld. 2013. Tracing environmental markers of autoimmunity: Introducing the infectome. *Immunologic Research* 56 (2–3):220–240. doi:10.1007/s12026-013-8399-6.

Boutin, S., S. Y. Graeber, M. Stahl, A. S. Dittrich, M. A. Mall, and A. H. Dalpke. 2017. Chronic but not intermittent infection with pseudomonas aeruginosa is associated with global changes of the lung microbiome in cystic fibrosis. *The European Respiratory Journal* 50 (4):1701086. doi:10.1183/13993003.01086-2017.

Brandt, L. J. 2017. Fecal microbiota therapy with a focus on C. difficile infection. *Psychosomatic Medicine* 79 (8):868–873. doi:10.1097/PSY.0000000000000511.

Brook, I. 1999. Bacterial interference. *Critical Reviews in Microbiology* 25 (3):155–172. doi:10.1080/10408419991299211.

Cabeda, J. M., and A. C. Moreno. 2014. *Sequenciação de Ácidos Nucleicos Em Biomedicina*. Porto, Portugal: Edições Fernando Pessoa.

Chen, T., W. H. Yu, J. Izard, O. V. Baranova, A. Lakshmanan, and F. E. Dewhirst. 2010. The human oral microbiome database: A web accessible resource for investigating oral microbe taxonomic and genomic information. *Database* 2010. baq013. doi:10.1093/database/baq013.

Cho, I., and M. J. Blaser. 2012. The human microbiome: At the interface of health and disease. *Nature Reviews Genetics* 13 (4):260–270. doi:10.1038/nrg3182.

Dashiff, A., and D. E. Kadouri. 2011. Predation of oral pathogens by bdellovibrio bacteriovorus 109J. *Molecular Oral Microbiology* 26 (1):19–34. doi:10.1111/j.2041-1014.2010.00592.x.

Demmer, R. T., D. R. Jr Jacobs, R. Singh, A. Zuk, M. Rosenbaum, P. N. Papapanou, and M. Desvarieux. 2015. Periodontal bacteria and prediabetes prevalence in ORIGINS: The oral infections, glucose intolerance, and insulin resistance study. *Journal of Dental Research* 94 (9 Suppl):201S–211S. doi:10.1177/0022034515590369.

Dethlefsen, L., M. McFall-Ngai, and D. A. Relman. 2007. An ecological and evolutionary perspective on human-microbe mutualism and disease. *Nature* 449 (7164):811–818. doi:10.1038/nature06245.

Devendra, D., and G. S. Eisenbarth. 2003. 17. immunologic endocrine disorders. *The Journal of Allergy and Clinical Immunology* 111 (2 Suppl):S624–S636. http://www.ncbi.nlm.nih.gov/pubmed/12592308.

Devine, D. A., P. D. Marsh, and J. Meade. 2015. Modulation of host responses by oral commensal bacteria. *Journal of Oral Microbiology* 7:26941.

Dewhirst, F. E., T. Chen, J. Izard, B. J. Paster, A. C. R. Tanner, W. Yu, A. Lakshmanan, and W. G. Wade. 2010. The human oral microbiome. *Journal of Bacteriology* 192 (19):5002–5017. doi:10.1128/JB.00542-10.

Diaz, P. I., B. Y. Hong, J. Frias-Lopez, A. K. Dupuy, M. Angeloni, L. Abusleme, E. Terzi, E. Ioannidou, L. D. Strausbaugh, and A. Dongari-Bagtzoglou. 2013. Transplantation-associated long-term immunosuppression promotes oral colonization by potentially opportunistic pathogens without impacting other members of the salivary bacteriome. *Clinical and Vaccine Immunology: CVI* 20 (6):920–930. doi:10.1128/CVI.00734-12.

Dios, C., J. De, R. Vida, M. Cobo, L. Máiz, L. Suárez, J. Galeano, F. Baquero, R. Cantón, and R. D. Campo. 2017. Individual patterns of complexity in cystic fibrosis lung microbiota, including predator bacteria, over a 1-Year period. Edited by Julian E. Davies. *mBio* 8 (5): e00959–17. doi:10.1128/mBio.00959-17.

Docktor, M. J., B. J. Paster, S. Abramowicz, J. Ingram, Y. E. Wang, M. Correll, H. Jiang, S. L. Cotton, A. S. Kokaras, and A. Bousvaros. 2012. Alterations in diversity of the oral microbiome in pediatric inflammatory bowel disease. *Inflammatory Bowel Diseases* 18 (5):935–942. doi:10.1002/ibd.21874.

Dutzan, N., J. E. Konkel, T. Greenwell-Wild, and N. M. Moutsopoulos. 2016. Characterization of the human immune cell network at the gingival barrier. *Mucosal Immunology* 9 (5):1163–1172. doi:10.1038/mi.2015.136.

Ebersole, J. L., D. III Dawson, P. Emecen-Huja, R. Nagarajan, K. Howard, M. E. Grady, K. Thompson et al. 2017. The periodontal war: Microbes and immunity. *Periodontology 2000* 75 (1):52–115. doi:10.1111/prd.12222.

Eren, A. M., G. G. Borisy, S. M. Huse, and J. L. Mark Welch. 2014. PNAS plus: From the cover: Oligotyping analysis of the human oral microbiome. *Proceedings of the National Academy of Sciences* 111 (28): E2875–E2884. doi:10.1073/pnas.1409644111.

Fabian, T. K., P. Hermann, A. Beck, P. Fejerdy, and G. Fabian. 2012. Salivary defense proteins: Their network and role in innate and acquired oral immunity. *International Journal of Molecular Sciences* 13 (4):4295–4320. doi:10.3390/ijms13044295.

Farrell, J. J., L. Zhang, H. Zhou, D. Chia, D. Elashoff, D. Akin, B. J. Paster, K. Joshipura, and D. T. W. Wong. 2012. Variations of oral microbiota are associated with pancreatic diseases including pancreatic cancer. *Gut* 61 (4):582–588. doi:10.1136/gutjnl-2011-300784.

Figdor, D., and K. Gulabivala. 2008. Survival against the odds: Microbiology of root canals associated with post-treatment disease. *Endodontic Topics* 18 (1):62–77. doi:10.1111/j.1601-1546.2011.00259.x.

Figdor, D., and G. Sundqvist. 2007. A big role for the very small—Understanding the endodontic microbial flora. *Australian Dental Journal* 52 (1 Suppl):S38–S51. Filoche, S., L. Wong, and C. H. Sissons. 2010. Oral biofilms: Emerging concepts in microbial ecology. *Journal of Dental Research* 89 (1):8–18. doi:10.1177/0022034509351812.

Flemming, H., J. Wingender, U. Szewzyk, P. Steinberg, S. A. Rice, and S. Kjelleberg. 2016. Biofilms: An emergent form of bacterial life. *Nature Reviews Microbiology* 14 (9):563–575. doi:10.1038/nrmicro.2016.94.

Frayman, K. B., D. S. Armstrong, K. Grimwood, and S. C. Ranganathan. 2017. The airway microbiota in early cystic fibrosis lung disease. *Pediatric Pulmonology* 52 (11):1384–1404. doi:10.1002/ppul.23782.

Ghannoum, M. A., R. J. Jurevic, P. K. Mukherjee, F. Cui, M. Sikaroodi, A. Naqvi, and P. M. Gillevet. 2010. Characterization of the oral fungal microbiome (Mycobiome) in healthy individuals. *PLoS Pathogens* 6 (1):e1000713. doi:10.1371/journal.ppat.1000713.

Gibbons, R. J., and J. V. Houte. 1975. Bacterial adherence in oral microbial ecology. *Annual Review of Microbiology* 29:19–44. doi:10.1146/annurev.mi.29.100175.000315.

Gomes, B. P. F. A., V. B. Berber, A. S. Kokaras, T. Chen, and B. J. Paster. 2015. Microbiomes of endodontic-periodontal lesions before and after chemomechanical preparation. *Journal of Endodontics* 41 (12):1975–1984. doi:10.1016/j.joen.2015.08.022.

Gomez, A., J. L. Espinoza, D. M. Harkins, P. Leong, R. Saffery, M. Bockmann, M. Torralba et al. 2017. Host genetic control of the oral microbiome in health and disease. *Cell Host & Microbe* 22 (3):269–278.e3. doi:10.1016/j.chom.2017.08.013.

Hajishengallis, G., R. P. Darveau, and M. A. Curtis. 2012. The keystone-pathogen hypothesis. *Nature Reviews Microbiology* 10 (10):717–725. doi:10.1038/nrmicro2873.

Harre, U., and G. Schett. 2017. Cellular and molecular pathways of structural damage in rheumatoid arthritis. *Seminars in Immunopathology* 39 (4): 355–363. doi:10.1007/s00281-017-0634-0.

Hasan, N. A., B. A. Young, A. T. Minard-Smith, K. Saeed, H. Li, E. M. Heizer, N. J. McMillan et al. 2014. Microbial community profiling of human saliva using shotgun metagenomic sequencing. *PLoS One* 9 (5):e97699. doi:10.1371/journal.pone.0097699.

He, J., Y. Li, Y. Cao, J. Xue, and X. Zhou. 2015. The oral microbiome diversity and its relation to human diseases. *Folia Microbiologica* 60 (1): 69–80. doi:10.1007/s12223-014-0342-2.

He, X., J. S. McLean, L. Guo, R. Lux, and W. Shi. 2014. The social structure of microbial community involved in colonization resistance. *The ISME Journal* 8 (3):564–574. doi:10.1038/ismej.2013.172.

Hennino, A., M. Vocanson, F. Berard, D. Kaiserlian, and J. Nicolas. 2006. [Epidemiology and pathophysiology of eczemas]. *La Revue Du Praticien* 56 (3):249–257. http://www.ncbi.nlm.nih.gov/pubmed/16583949.

Hezel, M. P., and E. Weitzberg. 2015. The oral microbiome and nitric oxide homoeostasis. *Oral Diseases* 21 (1):7–16. doi:10.1111/odi.12157.

Horz, H., and G. Conrads. 2007. Diagnosis and anti-infective therapy of -periodontitis. *Expert Review of Anti-Infective Therapy* 5 (4):703–715. doi:10.1586/14787210.5.4.703.

Huck, O., H. Tenenbaum, and J. L. Davideau. 2011. Relationship between periodontal diseases and preterm birth: Recent epidemiological and biological data. *Journal of Pregnancy* 2011:164654. doi:10.1155/2011/164654.

Hur, K. Y., and M. S. Lee. 2015. Gut microbiota and metabolic disorders. *Diabetes & Metabolism Journal* 39 (3):198–203. doi:10.4093/dmj.2015.39.3.198.

Huttenhower, C., D. Gevers, R. Knight, S. Abubucker, J. H. Badger, A. T. Chinwalla, H. C. Heather et al. 2012. Structure, function and diversity of the healthy human microbiome. *Nature* 486 (7402): 207–214. doi:10.1038/nature11234.

Isolauri, E. 2017. Microbiota and obesity. *Nestle Nutrition Institute Workshop Series* 88. 95–105. doi:10.1159/000455217.

Itzhaki, R. F., R. Lathe, B. J. Balin, M. J. Ball, E. L. Bearer, H. Braak, M. J. Bullido et al. 2016. Microbes and alzheimer's disease. *Journal of Alzheimer's Disease: JAD* 51 (4):979–984. doi:10.3233/JAD-160152.

Jenkinson, H. F., and R. J. Lamont. 2005. Oral microbial communities in sickness and in health. *Trends in Microbiology* 13 (12):589–595. doi:10.1016/j.tim.2005.09.006.

Jiang, B., X. Liang, Y. Chen, T. Ma, L. Liu, J. Li, R. Jiang, T. Chen, X. Zhang, and S. Li. 2012. Integrating next-generation sequencing and traditional tongue diagnosis to determine tongue coating microbiome. *Scientific Reports* 2:936. doi:10.1038/srep00936.

Johansson, I., E. Witkowska, B. Kaveh, P. Lif Holgerson, and A. C. R. Tanner. 2016. The microbiome in populations with a low and high prevalence of caries. *Journal of Dental Research* 95 (1):80–86. doi:10.1177/0022034515609554.

Joshipura, K. J., H. Hung, E. B. Rimm, W. C. Willett, and A. Ascherio. 2003. Periodontal disease, tooth loss, and incidence of ischemic stroke. *Stroke* 34 (1):47–52.

Kanasi, E., F. E. Dewhirst, N. I. Chalmers, R. Jr Kent, A. Moore, C. V. Hughes, N. Pradhan, C. Y. Loo, and A. C. R. Tanner. 2010. Clonal analysis of the microbiota of severe early childhood caries. *Caries Research* 44 (5):485–497. doi:10.1159/000320158.

Kantarci, A., K. Oyaizu, and T. E. V. Dyke. 2003. Neutrophil-mediated tissue injury in periodontal disease pathogenesis: Findings from localized aggressive periodontitis. *Journal of Periodontology* 74 (1):66–75. doi:10.1902/jop.2003.74.1.66.

Kazor, C. E., P. M. Mitchell, A. M. Lee, L. N. Stokes, W. J. Loesche, F. E. Dewhirst, and B. J. Paster. 2003. Diversity of bacterial populations on the tongue dorsa of patients with halitosis and healthy patients. *Journal of Clinical Microbiology* 41 (2)558–563.

Khajah, M. A. 2017. The potential role of fecal microbiota transplantation in the treatment of inflammatory bowel disease. *Scandinavian Journal of Gastroenterology* 52 (11):1172–1184. doi:10.1080/00365521.2017.1347812.

Kilian, M., I. L. C. Chapple, M. Hannig, P. D. Marsh, V. Meuric, A. M. L. Pedersen, M. S. Tonetti, W. G. Wade, and E. Zaura. 2016. The oral microbiome—An update for oral healthcare professionals. *British Dental Journal* 221 (10):657–666. doi:10.1038/sj.bdj.2016.865.

Knip, M., and J. Honkanen. 2017. Modulation of type 1 diabetes risk by the intestinal microbiome. *Current Diabetes Reports* 17 (11):105. doi:10.1007/s11892-017-0933-9.

Kolenbrander, P. E. 2000. Oral microbial communities: Biofilms, interactions, and genetic systems. *Annual Review of Microbiology* 54:413–437. doi:10.1146/annurev.micro.54.1.413.

Kononen, E., S. Asikainen, and H. Jousimies-Somer. 1992. The early colonization of gram-negative anaerobic bacteria in edentulous infants. *Oral Microbiology and Immunology* 7 (1):28–31.

Kononen, E., S. Asikainen, M. Saarela, J. Karjalainen, and H. Jousimies-Somer. 1994. The oral gram-negative anaerobic microflora in young children: Longitudinal changes from edentulous to dentate mouth. *Oral Microbiology and Immunology* 9 (3):136–141.

Koo, H., and K. M. Yamada. 2016. Dynamic cell-matrix interactions modulate microbial biofilm and tissue 3D microenvironments. *Current Opinion in Cell Biology* 42 (October):102–112. doi:10.1016/j.ceb.2016.05.005.

Krishnan, K., T. Chen, and B. J. Paster. 2017. A practical guide to the oral microbiome and its relation to health and disease. *Oral Diseases* 23 (3):276–286. doi:10.1111/odi.12509.

Lazarevic, V., K. Whiteson, D. Hernandez, P. Francois, and J. Schrenzel. 2010. Study of inter- and intra-individual variations in the salivary microbiota. *BMC Genomics* 11 (September):523. doi:10.1186/1471-2164-11-523.

Ling, Z., J. Kong, P. Jia, C. Wei, Y. Wang, Z. Pan, W. Huang, L. Li, H. Chen, and C. Xiang. 2010. Analysis of oral microbiota in children with dental caries by PCR-DGGE and barcoded pyrosequencing. *Microbial Ecology* 60 (3):677–690. doi:10.1007/s00248-010-9712-8.

Liu, Y., M. Nascimento, and R. A. Burne. 2012. Progress toward understanding the contribution of alkali generation in dental biofilms to inhibition of dental caries. *International Journal of Oral Science* 4 (3):135–140. doi:10.1038/ijos.2012.54.

Loesche, W. J. 1979. Clinical and microbiological aspects of chemotherapeutic agents used according to the specific plaque hypothesis. *Journal of Dental Research* 58 (12):2404–2412. doi:10.1177/002203 45790580120905.

Loesche, W. J., D. R. Bradbury, and M. P. Woolfolk. 1977. Reduction of dental decay in rampant caries individuals following short-term kanamycin treatment. *Journal of Dental Research* 56 (3):254–265. doi:10.1 177/00220345770560031101.

Looft, T., and H. K. Allen. 2012. Collateral effects of antibiotics on mammalian gut microbiomes. *Gut Microbes* 3 (5):463–467. doi:10.4161/gmic.21288.

Marsh, P. D. 2000. Oral ecology and its impact on oral microbial diversity. In *Oral Bacterial Ecology: The Molecular Basis*, pp. 11–66. Edited by H. K. Kuramitsu and R. P. Hellen, Wymondham, UK: Horizon Scientific Press.

Marsh, P. D., D. A. Head, and D. A. Devine. 2015. Ecological approaches to oral biofilms: Control without killing. *Caries Research* 49 (Suppl 1):46–54. doi:10.1159/000377732.

Marsh, P. D., D. A. Head, and D. A. Devine. 2014. Prospects of oral disease control in the future—An opinion. *Journal of Oral Microbiology* 6:26176.

Matthews, J. B., H. J. Wright, A. Roberts, P. R. Cooper, and I. L. C. Chapple. 2007. Hyperactivity and reactivity of peripheral blood neutrophils in chronic periodontitis. *Clinical and Experimental Immunology* 147 (2):255–264. doi:10.1111/j.1365-2249.2006.03276.x.

Mokkala, K., N. Houttu, T. Vahlberg, E. Munukka, T. Rönnemaa, and K. Laitinen. 2017. Gut microbiota aberrations precede diagnosis of gestational diabetes mellitus. *Acta Diabetologica*, October. doi:10.1007/s00592-017-1056-0.

Moriya, Y., T. Obama, T. Aiuchi, T. Sugiyama, Y. Endo, Y. Koide, E. Noguchi et al. 2017. Quantitative proteomic analysis of gingival crevicular fluids from deciduous and permanent teeth. *Journal of Clinical Periodontology* 44 (4):353–362. doi:10.1111/jcpe.12696.

Moutsopoulos, N. M., and J. E. Konkel. 2017. Tissue-specific immunity at the oral mucosal barrier. *Trends in Immunology* 1–12. doi:10.1016/j.it.2017.08.005.

Nair, P. N., and H. E. Schroeder. 1983. Retrograde access of antigens to the minor salivary glands in the monkey macaca fascicularis. *Archives of Oral Biology* 28 (2):145–152.

Nobbs, A. H., H. F. Jenkinson, and N. S. Jakubovics. 2011. Stick to your gums: mechanisms of oral microbial adherence. *Journal of Dental Research* 90 (11):1271–1278. doi:10.1177/0022034511399096.

Olsen, I., and S. K. Singhrao. 2015. Can oral infection be a risk factor for alzheimer's disease? *Journal of Oral Microbiology* 7:29143.

Owczarek, K., A. Jalocha-Kaczka, M. Bielinska, J. Urbaniak, and J. Olszewski. 2015. Evaluation of risk factors for oral cavity and oropharynx cancers in patients under the week activity program of head and neck cancers prevention in lodz. *Otolaryngologia Polska* 69 (6): 20–25. doi:10.5604/00306657.1182713.

Pai, N., and J. Popov. 2017. Protocol for a randomised, placebo-controlled pilot study for assessing feasibility and efficacy of faecal microbiota transplantation in a paediatric ulcerative colitis population: PediFETCh trial. *BMJ Open* 7 (8): e016698. doi:10.1136/bmjopen-2017-016698.

Parahitiyawa, N. B., C. Scully, W. K. Leung, W. C. Yam, L. J. Jin, and L. P. Samaranayake. 2010. Exploring the oral bacterial flora: Current status and future directions. *Oral Diseases* 16 (2):136–145. doi:10.1111/j.1601-0825.2009.01607.x.

Paster, B. J., I. Olsen, J. A. Aas, and F. E. Dewhirst. 2006a. The breadth of bacterial diversity in the human periodontal pocket and other oral sites. *Periodontology 2000* 42 (1):80–87. doi:10.1111/j.1600-0757.2006.00174.x.

Paster, B. J., I. Olsen, J. A. Aas, and F. E. Dewhirst. 2006b. The breadth of bacterial diversity in the human periodontal pocket and other oral sites. *Periodontology 2000* 42:80–87. doi:10.1111/j.1600-0757.2006.00174.x.

Petti, S., A. Polimeni, P. B. Berloco, and C. Scully. 2013. Orofacial diseases in solid organ and hematopoietic stem cell transplant pecipients. *Oral Diseases* 19 (1):18–36. doi:10.1111/j.1601-0825.2012.01925.x.

Pihlstrom, B. L., B. S. Michalowicz, and N. W. Johnson. 2005. Periodontal diseases. *Lancet (London, England)* 366 (9499):1809–1820. doi:10.1016/S0140-6736(05)67728-8.

Pistollato, F., S. S. Cano, I. Elio, M. M. Vergara, F. Giampieri, and M. Battino. 2016. Role of gut microbiota and nutrients in amyloid formation and pathogenesis of alzheimer disease. *Nutrition Reviews* 74 (10):624–634. doi:10.1093/nutrit/nuw023.

Preza, D., I. Olsen, J. A. Aas, T. Willumsen, B. Grinde, and B. J. Paster. 2008. Bacterial profiles of root caries in elderly patients. *Journal of Clinical Microbiology* 46 (6):2015–2021. doi:10.1128/JCM.02411-07.

Raizada, M. K., B. Joe, N. S. Bryan, E. B. Chang, F. E. Dewhirst, G. G. Borisy, Z. S. Galis et al. 2017. Report of the national heart, lung, and blood institute working group on the role of microbiota in blood pressure regulation: Current status and future directions. *Hypertension (Dallas, Tex.: 1979)* 70 (3): 479–485. doi:10.1161/HYPERTENSIONAHA.117.09699.

Ramsey, M. M., M. O. Freire, R. A. Gabrilska, K. P. Rumbaugh, and K. P. Lemon. 2016. Staphylococcus aureus shifts toward commensalism in response to corynebacterium species. *Frontiers in Microbiology* 7 (August): 1230. doi:10.3389/fmicb.2016.01230.

Relvas, M., I. Tomas, M. de Los Angeles Casares-De-Cal, and C. Velazco. 2015. Evaluation of a new oral health scale of infectious potential based on the salivary microbiota. *Clinical Oral Investigations* 19 (3):717–28. doi:10.1007/s00784-014-1286-2.

Roberts, F. A., and R. P. Darveau. 2015. Microbial protection and virulence in periodontal tissue as a function of polymicrobial communities: Symbiosis and dysbiosis. *Periodontology 2000* 69 (1):18–27. doi:10.1111/prd.12087.

Roland, B. C., D. Lee, L. S. Miller, A. Vegesna, R. Yolken, E. Severance, E. Prandovszky, X. E. Zheng, and G. E. Mullin. 2017. Obesity increases the risk of small intestinal bacterial overgrowth (SIBO). *Neurogastroenterology & Motility The Official Journal of the European Gastrointestinal Motility Society* 30(3): e13199. doi:10.1111/nmo.13199.

Rosier, B. T., M. D. Jager, E. Zaura, and B. P. Krom. 2014. Historical and contemporary hypotheses on the development of oral diseases: Are we there yet? *Frontiers in Cellular and Infection Microbiology* 4:92. doi:10.3389/fcimb.2014.00092.

Ruby, J., and M. Goldner. 2007. Nature of symbiosis in oral disease. *Journal of Dental Research* 86 (1):8–11. doi:10.1177/154405910708600102.

Sacco, G., D. Carmagnola, S. Abati, P. F. Luglio, L. Ottolenghi, A. Villa, C. Maida, and G. Campus. 2008. Periodontal disease and preterm birth relationship: A review of the literature. *Minerva Stomatologica* 57 (5):233-246-250.

Savage, D. C. 1977. Microbial ecology of the gastrointestinal tract. *Annual Review of Microbiology* 31: 107–133.

Schenkels, L. C., E. C. Veerman, and A. V. Nieuw Amerongen. 1995. Biochemical composition of human saliva in relation to other mucosal fluids. *Critical Reviews in Oral Biology and Medicine: An Official Publication of the American Association of Oral Biologists* 6 (2):161–175.

Schiott, C. R., and H. Loe. 1970. The origin and variation in number of leukocytes in the human saliva. *Journal of Periodontal Research* 5 (1):36–41.

Selwitz, R. H., A. I. Ismail and N. B. Pitts. 2007. Dental caries. *Lancet (London, England)* 369 (9555):51–59. doi:10.1016/S0140-6736(07)60031-2.

Sender, R., S. Fuchs, and R. Milo. 2016. Revised estimates for the number of human and bacteria cells in the body. *PLoS Biology* 14 (8):e1002533. doi:10.1371/journal.pbio.1002533.

Shinzato, T., and A. Saito. 1994. A mechanism of pathogenicity of 'streptococcus milleri group' in pulmonary infection: Synergy with an anaerobe. *Journal of Medical Microbiology* 40 (2):118–123. doi:10.1099/00222615-40-2-118.

Shira Davenport, E. 2010. Preterm low birthweight and the role of oral bacteria. *Journal of Oral Microbiology* 2(1):5779. doi:10.3402/jom.v2i0.5779.

Siqueira, J. F. Jr, and I. N. Rocas. 2009. Diversity of endodontic microbiota revisited. *Journal of Dental Research* 88 (11):969–981. doi:10.1177/0022034509346549.

Smith, D. J., J. M. Anderson, W. F. King, J. van Houte, and M. A. Taubman. 1993. Oral streptococcal colonization of infants. *Oral Microbiology and Immunology* 8 (1):1–4.

Socransky, S. S., and A. D. Haffajee. 2005. Periodontal microbial ecology. *Periodontology 2000* 38:135–187. doi:10.1111/j.1600-0757.2005.00107.x.

Socransky, S. S., and S. D. Manganiello. 1971. The oral microbiota of man from birth to senility. *Journal of Periodontology* 42 (8):485–496. doi:10.1902/jop.1971.42.8.485.

Soga, Y., Y. Maeda, F. Ishimaru, M. Tanimoto, H. Maeda, F. Nishimura, and S. Takashiba. 2011. Bacterial substitution of coagulase-negative staphylococci for streptococci on the oral mucosa after hematopoietic cell transplantation. *Supportive Care in Cancer: Official Journal of the Multinational Association of Supportive Care in Cancer* 19 (7):995–1000. doi:10.1007/s00520-010-0923-9.

Sonnenburg, J. L., and M. A. Fischbach. 2011. Community health care: Therapeutic opportunities in the human microbiome. *Science Translational Medicine* 3 (78):78ps12. doi:10.1126/scitranslmed.3001626.

Sundqvist, G. 1992. Associations between microbial species in dental root canal infections. *Oral Microbiology and Immunology* 7 (5):257–262.

Sundqvist, G., and D. Figdor. 2003. Life as an endodontic pathogen. Ecological differences between the untreated and root-filled root canals. *Endodontic Topics* 6 (1):3–28. doi:10.1111/j.1601-1546.2003.00054.x.

Syrjanen, J. 1990. Vascular diseases and oral infections. *Journal of Clinical Periodontology* 17 (7 (Pt 2)):497–500.

Takahashi, K., R. F. Cunha, and E. G. J. Junior. 2017. Periodontal pathogen colonization in young children by PCR quantification—A longitudinal survey. *Journal of Clinical Pediatric Dentistry* 41 (6):456–461. doi:10.17796/1053-4628-41.6.7.

Teles, R., and C. Y. Wang. 2011. Mechanisms involved in the association between periodontal diseases and cardiovascular disease. *Oral Diseases* 17 (5):450–461. doi:10.1111/j.1601-0825.2010.01784.x.

Tremlett, H., K. C. Bauer, S. Appel-Cresswell, B. B. Finlay, and E. Waubant. 2017. The gut microbiome in human neurological disease: A review. *Annals of Neurology* 81 (3): 369–382. doi:10.1002/ana.24901.

Tristao, W., R. M. P. Ribeiro, C. A. D. Oliveira, J. C. Betiol, and J. D. S. R. Bettini. 2012. Epidemiological study of HPV in oral mucosa through PCR. *Brazilian Journal of Otorhinolaryngology* 78 (4):66–70.

Turnbaugh, P. J., R. E. Ley, M. Hamady, C. M. Fraser-Liggett, R. Knight, and J. I. Gordon. 2007. The human microbiome project. *Nature* 449 (7164):804–810. doi:10.1038/nature06244.

Van Essche, M., M. Quirynen, I. Sliepen, G. Loozen, N. Boon, J. Van Eldere, and W. Teughels. 2011. Killing of anaerobic pathogens by predatory bacteria. *Molecular Oral Microbiology* 26 (1):52–61. doi:10.1111/j.2041-1014.2010.00595.x.

Van Meel, E. R., V. W. V. Jaddoe, K. Bønnelykke, J. C. de Jongste, and L. Duijts. 2017. The role of respiratory tract infections and the microbiome in the development of asthma: A narrative review. *Pediatric Pulmonology* 52 (10): 1363–1370. doi:10.1002/ppul.23795.

Wade, W. G. 2013. The oral microbiome in health and disease. *Pharmacological Research* 69:137–43. doi:10.1016/j.phrs.2012.11.006.

Walia, M., and N. Saini. 2015. Relationship between periodontal diseases and preterm birth: Recent epidemiological and biological data. *International Journal of Applied & Basic Medical Research* 5 (1):2–6. doi:10.4103/2229-516X.149217.

Williams, R. C. 2008. Understanding and managing periodontal diseases: A notable past, a promising future. *Journal of Periodontology* 79 (8 Suppl): 1552–1559. doi:10.1902/jop.2008.080182.

Wilson, M. 2008. *Bacteriology of Humans: An Ecology Perspective*. Malden, MA: Blackwell.

Wu, J., B. A. Peters, C. Dominianni, Y. Zhang, Z. Pei, L. Yang, Y. Ma et al. 2016. Cigarette smoking and the oral microbiome in a large study of american adults. *The ISME Journal* 10 (10):2435–2446. doi:10.1038/ismej.2016.37.

Zarco, M. F., T. J. Vess, and G. S. Ginsburg. 2012. The oral microbiome in health and disease and the potential impact on personalized dental medicine. *Oral Diseases* 18 (2):109–120. doi:10.1111/j.1601-0825.2011.01851.x.

Zaura, E., B. J. F. Keijser, S. M. Huse, and W. Crielaard. 2009. Defining the healthy 'core microbiome' of oral microbial communities. *BMC Microbiology* 9:259. doi:10.1186/1471-2180-9-259.

Zaura, E., E. A. Nicu, B. P. Krom, and B. J. F. Keijser. 2014. Acquiring and maintaining a normal oral microbiome: Current perspective. *Frontiers in Cellular and Infection Microbiology* 4:85. doi:10.3389/fcimb.2014.00085.

Zhong, D., C. Wu, X. Zeng, and Q. Wang. 2018. The role of gut microbiota in the pathogenesis of rheumatic diseases. *Clinical Rheumatology* 37 (1):25–34. doi:10.1007/s10067-017-3821-4.

5

Bacterial Infections in Atherosclerosis

Atherosclerosis Microbiome

Emil Kozarov and Ann Progulske-Fox

Contents

5.1 Introduction

Atherothrombotic vascular disease (AVD), to which atherosclerosis is the main underlying pathology, is the leading cause of morbidity and mortality in adults and the most critical area of medical sciences.

Significant advances in AVD area are crucial, specifically due to increased prevalence with age. Myocardial infarction (MI) and stroke continue to be major causes of morbidity and mortality.

Atherosclerosis is a chronic inflammatory disease. Despite meaningful progress in the identification of risk factors and development of clinical tools, such as statins and PCSK9 inhibitors, AVD-related deaths continue to increase worldwide, due in part to the moderate efficacy of the available drugs. For example, after 104 weeks of maximal-dose therapy, atorvastatin and rosuvastatin led to reductions of only 0.99% (95% confidence interval [CI], −1.19 to −0.63) and 1.22% (95% CI, −1.52 to −0.90), respectively ($P = 0.17$), in percentage of atheroma volume, the primary efficacy end-point (Nicholls et al., 2011).

At the same time, the current treatment modalities targeting hypertension, hyperlipidemia, and controlling hemostasis do not directly address the inflammatory origin of atherosclerosis (Weber and Noels, 2011). Genetic research has also not led to major breakthroughs. A large genome-wide association study (GWAS) of 63,746 coronary artery disease (CAD) cases and 130,681 controls brought the number of identified human genetic variants that are associated with increased coronary disease risk to 46, which still accounts only for about 10% of CAD heritability (Deloukas et al., 2013). The prospect of addressing hundreds of single nucleotide polymorphism (SNP) loci immensely complicates genetic targeting of the predisposition to vascular inflammation.

A large network analysis identified lipid metabolism and inflammation as the two key biological pathways involved in the genetic predisposition to CAD (Deloukas et al., 2013). Nevertheless, many of the individuals with multiple classic risk factors for AVD do not go on to experience such events. Moreover, MI and stroke continue to occur in up to two-thirds of all patients (Libby, 2005). Because many cardiovascular events have not been explained by genetics or other risk factors, and multiple epidemiologic studies have consistently suggested an infectious component, entirely novel approaches for diagnostics and treatment are acutely needed.

These novel approaches should be based on the concept of personalized medicine and addressing additional manageable risk factors to control AVD to achieve longevity, while also increasing the quality of life. There are a variety of avenues that could enable a novel approach to AVD. These are based on the discovery, characterization, and focusing on its infectious component.

5.2 Inflammation as the core process in atherogenesis: Could infection of arterial wall be the culprit?

The inflammatory nature of the atherosclerotic lesion from initiation (fatty streak) to culmination into an acute ischemic event (plaque rupture) has been established. The degree of inflammation correlates with the severity and clinical outcome. Inflammation influences plaque progression and plaque's vulnerability to rupture (Libby, 2002, 2012).

Markers of inflammation may predict development of disease in asymptomatic individuals. C-reactive protein (CRP), an acute-phase reactant elevated in all inflammatory diseases, has been the most widely studied biomarker of inflammation. Indeed, the acceptance of the high-sensitivity C-reactive protein (hsCRP) test as a biomarker for cardiovascular disease (CVD) risk assessment in primary prevention (Ridker, 2016) naturally suggests considering infections as an underlying cause of the inflammation. The increased risk for heart attack associated with elevated CRP has been reported to have prognostic value even among patients with negative cardiac troponin and no evidence of myocyte necrosis (Heeschen et al., 2000), again suggesting an alternative origin of the inflammation.

Defective resolution of inflammation is the basis of most prevalent chronic inflammations, including AVD. Specialized pro-resolving mediators include arachidonic acid-derived lipoxins, omega-3 fatty acid, eicosapentaenoic acid-derived resolvins, docosahexaenoic acid-derived resolvins, protectins, and maresins (Fredman and Tabas, 2017). For in-depth reviews on inflammation in AVD, see also Pant et al. (2014).

Importantly, the abundant evidence accumulated so far points at infections (often of periodontium) as a contributing factor for CVD. Periodontitis provides an "open gate" to the circulation for entry of oral bacteria into the bloodstream, allowing them to activate the host inflammatory response in the direction of atherogenesis (i.e., atheroma formation, maturation, and exacerbation) (Reyes et al., 2013). Similarly, the "leaky gut" syndrome can deliver intestinal species to a new, systemic location, inducing inflammation (Figure 5.1 shows the multiple way the infections contribute to chronic inflammation, including atherogenesis).

Atherosclerosis is a disease of the vasculature characterized by inflammatory lesions in the arterial bifurcations, which, when destabilized, can lead to MI and stroke. AVD does have many characteristics of a chronic infectious disease. For example, clinical studies have indicated that infections with multiple pathogens result in chronic persistent inflammation (Epstein et al., 2009; Campbell and Rosenfeld, 2015). In addition, internalization of many types of bacteria can produce a "privileged niche," where they persist in a dormant, nonreplicating state, sheltered from humoral and cellular immune responses (Tufariello et al., 2003). Lastly, DNA data suggest that various and specific pathogens are associated with atherosclerotic tissue (Kozarov et al., 2006; Ott et al., 2006).

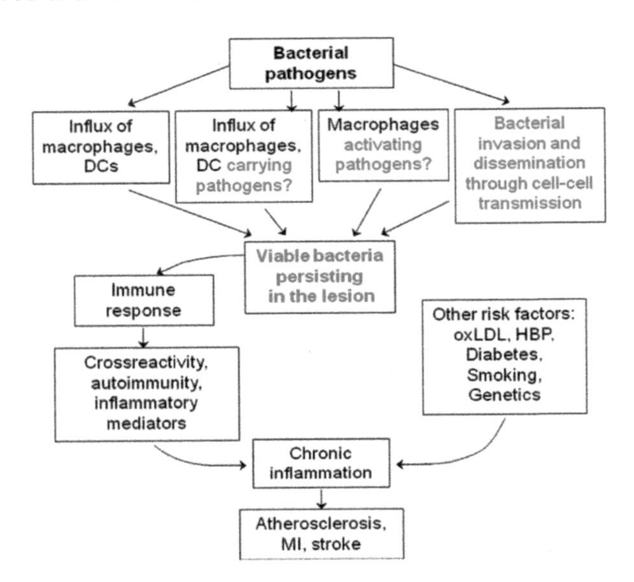

Figure 5.1 Contribution of bacterial pathogens to the "response to injury" model of atherogenesis. Mentioned in red font is the bacterial component of vascular inflammations as suggested in recent publications.

5.3 Epidemiology of AVD

The presently accumulated epidemiological data support the hypothesis that infections contribute to AVD (Amar et al., 2009; Epstein et al., 2009). In particular, periodontal inflammatory components were recognized as contributors to or triggers for systemic inflammatory responses and the risk of developing heart disease was shown to increase by 168% in patients with periodontitis (Genco et al., 2002).

As another example, in the Northern Manhattan Study (NOMAS), a prospective cohort study of stroke incidence and prognosis, it was shown that infectious burden is associated with risk of stroke, a carotid plaque thickness (Elkind, 2010), and also with cognition (Katan et al., 2013). The Atherosclerosis Risk In Communities (ARIC) study found systemic antibody response to periodontal bacteria associated with coronary heart disease in ever and never smokers (Beck et al., 2005). Similarly, the data from the Oral Infections and Vascular Disease Epidemiology Study (INVEST) demonstrated a direct relationship between tooth loss and carotid plaque prevalence (Desvarieux et al., 2003), colonization with pathogenic periodontal pathogens to be associated with subclinical vascular disease (i.e., carotid artery intima-media thickness) (Desvarieux et al., 2005), and showed that severe periodontal bone loss is associated independently with carotid atherosclerosis (odds ratio 3.64, $P < 0.05$) (Engebretson et al., 2005).

5.4 Seroepidemiology

Supporting these data, seroepidemiological investigations showed that infections caused by major periodontal pathogens, *Porphyromonas gingivalis* and *Aggregatibacter actinomycetemcomitans* (in seropositive subjects) is associated with future stroke (Pussinen et al., 2004). Seroepidemiology also demonstrated that anti-*P. gingivalis* antibody is associated with atrial fibrillation as well as carotid artery atherosclerosis. In addition, anti-*Prevotella intermedia* antibody may be associated with stroke through its association with carotid atherosclerosis (Hosomi et al., 2012). Finally, murine and larger animal models of infection corroborated the results from in vitro and studies, demonstrating the exacerbation of the vascular inflammation upon administration of infectious agents (Jain et al., 2003; Brodala et al., 2005).

Most often, direct systemic exposure to infectious agents, such as bacteremia with an oral origin, is communicated. In one study, 80.9% of the patients presented positive cultures after dental cleaning procedure, and 19% of the patients still had microorganisms in the bloodstream after 30 minutes (Lafaurie et al., 2007).

5.5 DNA evidence

Microbial pathogens associated with atheromatous tissues have been identified at DNA level (Haraszthy et al., 2000; Kozarov et al., 2006).Using genomic (16S rDNA signatures) and bioinformatic (molecular phylogenies) tools, the diversity of bacteria in atherosclerotic lesions in patients with AVD was systematically explored. Polymerase chain reaction (PCR), clone libraries, gradient gel analyses, and fluorescence in-situ hybridization (FISH) were used to demonstrate the presence of bacterial DNA in patient tissues. Mean bacterial diversity in atheromas was high, with a score of 12.33 ± 3.81 (range, 5–22), suggesting that diverse bacterial colonization may be important (Ott et al., 2006). Further, another survey of bacterial DNA signatures in atherosclerotic plaques suggested that both oral and intestinal bacteria may correlate with disease markers of atherosclerosis (Koren et al., 2011). We have also shown, using molecular, genetic, and immunological approaches, the association of variety of bacterial species with atheromatous tissues (Kozarov, 2005, 2012), creating for the first time the concept of atherosclerosis microbiome. Interestingly, in a recent study, *P. gingivalis* was the most abundant species detected in coronary and femoral arteries (Mougeot et al., 2017).

5.6 Clinical isolates from atheromas. Atherosclerosis microbiome

However, DNA detection does not fulfill Koch's postulates for association of infectious agents with disease. In a significant advancement, we were able to demonstrate, at several levels, that microbial infection of arterial plaques may be the missing link, a major contributing factor to acute ischemic events. We designed and implemented a breakthrough technology, immune-mediated resuscitation (IMR), for cultivation and identification of a variety of live bacteria in plaques from patients (Rafferty et al., 2011a, 2011b), yielding clinical isolates, which are the basis for establishment of the atherosclerosis microbiome and for development of new diagnostics. Using immunology, cellular microbiology, genomics, and fluorescence microscopy, we also demonstrated vascular cell transmission of bacterial pathogens (Li et al., 2008), association of dormant invasive bacteria with atheromatous tissue (Kozarov, 2005), and most importantly, developed a technology to isolate and identify previously uncultivable bacterial pathogens from such tissue from patients (Rafferty et al., 2011a, 2011b). Patient-specific drug targets were essentially identified, opening the possibility to introduce personalized medicine in the critical area of AVDs.

It is known that periodontal infection accelerates both lipid deposition in arteries and atherosclerosis in animal models (Jain et al., 2003; Li et al., 2008), suggesting that identification of the infectious agents in the atheromatous tissue, followed by a treatment regimen specific for the infecting bacteria, would open an additional avenue to address not only infections, but also hyperlipidemias as a causative factor in atherosclerosis.

The reemergence of bacterial pathogens as potential initiators or exacerbators of AVD poses new challenges to medical research and to an overwhelmed healthcare budget; however, there is a silver lining to it. The possibility of a new approach to CVD creates a perfect opportunity to develop entirely novel theranostics and to introduce personalized medicine in the most critical area of human health.

5.7 Clinical trials with antibiotics

Importantly, the antibiotic clinical trials for the secondary prevention of late-stage coronary heart disease a decade ago failed because all patients were treated with the same antibiotic (O'Connor et al., 2003; Cannon et al., 2005; Grayston et al., 2005). We now know that the negative results were actually expected due to (i) the fact that the trials were not designed to demonstrate causation and (ii) to the fact that we isolated (years later) different pathogens from the plaques from different patients. However, we realized this had not evaded the attention of the clinicians in the field. When presented with our investigational data at meetings and seminars, clinicians asked for further advancements. Identifying the patient-specific pathogens in the vasculature would for the first time provide clinicians with a tool to adequately inform patient-specific treatment, which is the basic tenet of personalized medicine.

5.8 Atherosclerosis microbiome—The most critical segment of human microbiome

The discovery of the atherosclerosis microbiome highlights a rare possibility for chronic inflammation-related biomarker discovery and for individualized diagnosis and treatment of vascular inflammations that lead to MI, stroke, and peripheral vascular disease. From the perspective of public health, currently there is no more critical inflammatory disease than AVD (Figure 5.2 illustrates the variety of avenues exploited by bacterial pathogens for systemic dissemination and internalization in vascular walls, contributing or triggering chronic inflammation). Besides being the number-one killer globally, AVD is estimated to cost the European Union (EU) €169 billion and causes nearly half of all deaths in Europe (48%) and in the EU (42%), making it the main source of morbidity and mortality in the EU (Leal et al., 2006). Similarly, AVD accounted for 32.8% (811,940) of all 2,471,984 deaths in the United States (Roger and O'Donnell 2012), costing the United States $448.5 billion in 2008, greater than any other medical condition (www.americanheart.org). Thus, immediate development of new approaches to treating AVD is needed. Accomplishing this would have tremendous outcomes, in addition, to the economic aspects. For example, successful treatment of MI is estimated to increase the human life span by 16.6 years (Go et al., 2013).

5.9 Conclusion. Infectious component of vascular inflammations

Atherogenesis involves both the innate and adaptive arms of the immune system, similar to bacterial infections. The "response to injury" hypothesis of atherogenesis is presented here as a pathogen-accelerated inflammatory process, leading to endothelial activation, growth factors release, monocytes' adhesion and transmigration (extravasation), while maturing into macrophages, followed by foam cell formation and smooth muscle cell proliferation. This progressive process leads to arterial thickening and eventually to gradual occlusion of the vascular lumen and inflammatory atheromatous lesion formation (Figure 5.2 depicts the bacterial component of atherogenesis).

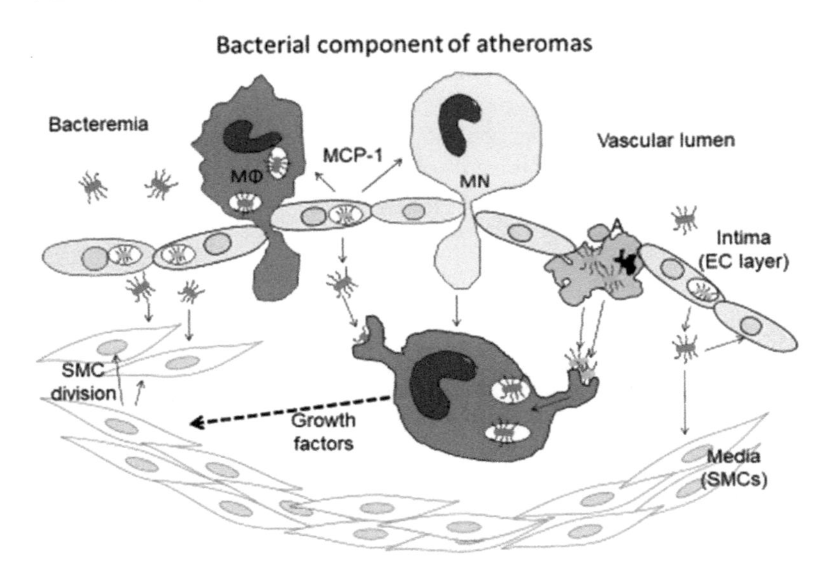

Figure 5.2 Model representing what is now known about the microbial component of atherogenesis. Both bacteremic- and phagocyte-mediated avenues of bacterial delivery to the site of inflammation are proposed. Bacteremia-related bacteria (left) are shown invading the endothelial layer and further spreading into deeper tissue. Activation of infected endothelia are represented with the release of pro-inflammatory chemokines (such as MCP-1) in the lumen, activating blood monocytes (MN) and macrophages (MΦ) and promoting their adhesion and diapedesis. Transmigrating leucocytes (in the center) can harbor internalized viable bacteria, which represents the second avenue for systemic bacterial dissemination to distant sites. The internalization of bacteria can switch them into uncultivable state (red into green), while their internalization by phagocytes can reactivate them (from green to red). Atheromas can grow because of macrophage-secreted growth factors-mediated smooth muscle cell proliferation. Bacteria are also released upon host cell death (depicted at right). Apoptotic EC, apoptotic endothelial cell releasing intracellular bacteria; EC, endothelial cell; MN, monocyte; MΦ, macrophage with internalized bacteria; SMC, smooth muscle cell.

Thus, the thrombosis and the consecutive acute ischemic events that often follow, possibly leading to a fatal outcome, can also be pathogen related.

5.10 Next steps

We strongly suggest closer interaction between interdisciplinary teams of biomedical investigators and better-trained clinicians, including those with both medical and dental clinicians to efficiently address this plague on today's society. To translate the data from the often complicated laboratory and clinical investigations into tangible results, clinicians are needed who will be able to:

1. Address the most critical conditions of AVD in an innovative, efficient manner;
2. Adequately update their core competencies during continuous medical education courses;
3. Place specific emphasis on infections as a key risk factor, thus introducing personalized medicine;
4. Pay critical attention to infections as *modifiable* risk factors for AVD, independent of lifestyle; and
5. Communicate the importance of control of infections, such as periodontitis, to patients.

Creating synergy of research and education, clinicians are those individuals who are empowered to lead the shift in healthcare expenditures from the costly tertiary care toward early diagnostics and prevention, thus reducing the healthcare budgets and their burden on governments, and ushering in the era of personalized (i.e., precision) medicine.

References

Amar, S., S. C. Wu, and M. Madan, M. 2009. Is Porphyromonas gingivalis cell invasion required for atherogenesis? Pharmacotherapeutic implications. *Journal of Immunology* 182:1584–1592.

Beck, J. D., P. Eke, G. Heiss, P. Madianos, D. Couper, D. Lin, K. Moss, J. Elter, and S. Offenbacher. 2005. Periodontal disease and coronary heart disease: A reappraisal of the exposure. *Circulation* 112:19–24.

Brodala, N., E. P. Merricks, D. A. Bellinger, D. Damrongsri, S. Offenbacher, J. Beck, P. Madianos et al. 2005. *Porphyromonas gingivalis* bacteremia induces coronary and aortic atherosclerosis in normocholesterolemic and hypercholesterolemic pigs. *Arteriosclerosis Thrombosis, and Vascular Biology* 25:1446–1451.

Campbell, L. A. and M. E. Rosenfeld. 2015. Infection and atherosclerosis development. *Archives of Medical Research* 46:339–350.

Cannon, C. P., E. Braunwald, C. H. McCabe, J. T. Grayston, B. Muhlestein, R. P. Giugliano, R. Cairns, and A. M. Skene. 2005. Antibiotic treatment of Chlamydia pneumoniae after acute coronary syndrome. *New England Journal of Medicine* 352:1646–1654.

Deloukas, P., S. Kanoni, C. Willenborg, M. Farrall, T. L. Assimes, J. R. Thompson, E. Ingelsson et al. 2013. Large-scale association analysis identifies new risk loci for coronary artery disease. *Nature Genetics* 45:25–33.

Desvarieux, M., R. T. Demmer, T. Rundek, B. Boden-Albala, D. R. Jacobs Jr., P. N. Papapanou, and R. L. Sacco. (2003). Relationship between periodontal disease, tooth loss, and carotid artery plaque: The Oral Infections and Vascular Disease Epidemiology Study (INVEST). *Stroke* 34:2120–2125.

Desvarieux, M., R. T. Demmer, T. Rundek, B. Boden-Albala, D. R. Jacobs Jr., R. L. Sacco, and P. N. Papapanou. (2005). Periodontal microbiota and carotid intima-media thickness: The Oral Infections and Vascular Disease Epidemiology Study (INVEST). *Circulation* 111:576–582.

Elkind, M. S. V., P. Ramakrishnan, Y. P. Moon, B. Boden-Albala, K. M. Liu, S. L. Spitalnik, T. Rundek, R. L. Sacco, and M. C. Paik. 2010. Infectious burden and risk of stroke: The Northern Manhattan Study. *Archives of Neurology* 67:33–38.

Engebretson, S. P., I. B. Lamster, M. S. Elkind, T. Rundek, N. J. Serman, R. T. Demmer, R. L. Sacco, P. N. Papapanou, and M. Desvarieux. 2005. Radiographic measures of chronic periodontitis and carotid artery plaque. *Stroke* 36:561–566.

Epstein, S. E., J. Zhu, A. H. Najafi, and M. S. Burnett. 2009. Insights into the role of infection in atherogenesis and in plaque rupture. *Circulation* 119:3133–3141.

Fredman, G. and I. Tabas. 2017. Boosting inflammation resolution in atherosclerosis: The next frontier for therapy. *American Journal of Pathology* 187:1211–1221.

Genco, R., S. Offenbacher, and J. Beck. 2002. Periodontal disease and cardiovascular disease: Epidemiology and possible mechanisms. *Journal of the American Dental Association* 133(Suppl):14S–22S.

Grayston, J. T., R. A. Kronmal, L. A. Jackson, A. F. Parisi, J. B. Muhlestein, J. D. Cohen, W. J. Rogers et al. 2005. Azithromycin for the secondary prevention of coronary events. *New England Journal of Medicine* 352:1637–1645.

Go, A. S., D. Mozaffarian, V. L. Roger, E. J. Benjamin, J. D. Berry, W. B. Borden, D. M. Bravata et al. 2013. Heart disease and stroke statistics – 2013 update: A report from the American Heart Association. *Circulation* 127 (1):e6–e245.

Haraszthy, V. I., J. J. Zamboni, M. Trevisan, M. Zeid, R. J. Genco. 2000. Identification of periodontal pathogens in atheromatous plaques. *Journal of Clinical Periodontology* 71 (10):1554 –1560.

Heeschen, C., C. W. Hamm, J. Bruemmer, and M. L. Simoons. 2000. Predictive value of C-reactive protein and troponin T in patients with unstable angina: A comparative analysis. CAPTURE investigators. Chimeric c7E3 antiplatelet therapy in unstable angina Refractory to standard treatment trial. *Journal of the American College of Cardiology* 35:1535–1542.

Hosomi, N., S. Aoki, K. Matsuo, K. Deguchi, H. Masugata, K. Murao, N. Ichihara et al. 2012. Association of serum anti-periodontal pathogen antibody with ischemic stroke. *Cerebrovascular Diseases* 34:385–392.

Jain, A., E. L. Batista Jr., C. Serhan, G. L. Stahl, and T. E. Van Dyke. 2003. Role for periodontitis in the progression of lipid deposition in an animal model. *Infection and Immunity* 71:6012–6018.

Katan, M., Y. P. Moon, M. C. Paik, R. L. Sacco, C. B. Wright, and M. S. Elkind. 2013. Infectious burden and cognitive function: The Northern Manhattan Study. *Neurology* 80:1209–1215.

Koren, O., A. Spor, J. Felin, F. Fak, J. Stombaugh, V. Tremaroli, C. J. Behre et al. 2011. Human oral, gut, and plaque microbiota in patients with atherosclerosis. *Proceedings of the National Academy of Sciences USA* 108(Suppl 1):4592–4598.

Kozarov, E. 2012. Bacterial invasion of vascular cell types: Vascular infectology and atherogenesis. *Future Cardiology* 8:123–138.

Kozarov, E., B. Dorn, C. Shelburne, W. Dunn, A. Progulske-Fox. 2005. Human atherosclerotic plaque contains viable invasive *Porphyromonas gingivalis* and *Actinobacillus actinomycetemcomitans. Arteriosclerosis, Thrombosis, and Vascular Biology* 25:e17–e18.

Kozarov, E., D. Sweier, C. Shelburne, A. Progulske-Fox, and D. Lopatin. 2006. Detection of bacterial DNA in atheromatous plaques by quantitative PCR. *Microbes and Infection* 8:687–693.

Lafaurie, G. I., I. Mayorga-Fayad, M. F. Torres, D. M. Castillo, M. R. Aya, A. Baron, and P. A. Hurtado. 2007. Periodontopathic microorganisms in peripheric blood after scaling and root planing. *Journal of Clinical Periodontology* 34:873–879.

Leal, J., R. Luengo-Fernandez, A. Gray, S. Petersen, and M. Rayner. 2006. Economic burden of cardiovascular diseases in the enlarged European Union. *European Heart Journal* 27:1610–1619.

Li, L., R. Michel, J. Cohen, A. DeCarlo, and E. Kozarov. 2008. Intracellular survival and vascular cell-to-cell transmission of *Porphyromonas gingivalis. BMC Microbiology* 8:26–36.

Libby, P. 2002. Inflammation in atherosclerosis. *Nature* 420:868–874.

Libby, P. 2005. The forgotten majority: Unfinished business in cardiovascular risk reduction. *Journal of the American College of Cardiology* 46:1225–1228.

Libby, P. (2012). Inflammation in atherosclerosis. *Arteriosclerosis, Thrombosis, and Vascular Biology* 32:2045–2051.

Mougeot, J. C., C. B. Stevens, B. J. Paster, M. T. Brennan, P. B. Lockhart, and F. K. Mougeot. 2017. *Porphyromonas gingivalis* is the most abundant species detected in coronary and femoral arteries. *Journal of Oral Microbiology* 9:1281562.

Nicholls, S. J., C. M. Ballantyne, P. J. Barter, M. J. Chapman, R. M. Erbel, P. Libby, J. S. Raichlen et al. 2011. Effect of two intensive statin regimens on progression of coronary disease. *New England Journal of Medicine* 365:2078–2087.

O'Connor, C. M., M. W. Dunne, M. A. Pfeffer, J. B. Muhlestein, L. Yao, S. Gupta, R. J. Benner, M. R. Fisher, and T. D. Cook. 2003. Azithromycin for the secondary prevention of coronary heart disease events: The WIZARD study: A randomized controlled trial. *JAMA* 290:1459–1466.

Ott, S. J., El N. E. Mokhtari, M. Musfeldt, S. Hellmig, S. Freitag, A. Rehman, T. Kuhbacher et al. 2006. Detection of diverse bacterial signatures in atherosclerotic lesions of patients with coronary heart disease. *Circulation* 113:929–937.

Pant, S., A. Deshmukh, G. S. Gurumurthy, N. V. Pothineni, T. E. Watts, F. Romeo, and J. L. Mehta. 2014. Inflammation and atherosclerosis—Revisited. *Journal of Cardiovascular Pharmacology and Therapeutics* 19:170–178.

Pussinen, P. J., G. Alfthan, H. Rissanen, A. Reunanen, S. Asikainen, and P. Knekt. 2004. Antibodies to periodontal pathogens and stroke risk. *Stroke* 35:2020–2023.

Rafferty, B., S. Dolgilevich, S. Kalachikov, I. Morozova, J. Ju, S. Whittier, R. Nowygrod, and E. Kozarov. 2011a. Cultivation of *Enterobacter hormaechei* from human atherosclerotic tissue. *Journal of Atherosclerosis and Thrombosis* 18:72–81.

Rafferty, B., D. Jönsson, S. Kalachikov, R. T. Demmer, R. Nowygrod, M. S. Elkind, H. Bush Jr., and E. Kozarov. 2011b. Impact of monocytic cells on recovery of uncultivable bacteria from atherosclerotic lesions. *Journal of Internal Medicine* 270:273–280.

Reyes, L., D. Herrera, E. Kozarov, S. Roldan, and A. Progulske-Fox. 2013. Periodontal bacterial invasion and infection: Contribution to atherosclerotic pathology. *Journal of Clinical Periodontology* 40(Suppl 14):S30–S50.

Ridker, P. 2016. From C-reactive protein to interleukin-6 to interleukin-1: moving upstream to identify novel targets for atheroprotection. *Circulation Research* 118:145–156.

Roger, V. L. and C. J. O'Donnell. 2012. Population health, outcomes research, and prevention. Example of the American Heart Association 2020 goals. *Circulation Cardiovascular Qual Outcomes* 5 (1):6–8.

Tufariello, J. M., J. Chan, and J. L. Flynn. 2003. Latent tuberculosis: Mechanisms of host and bacillus that contribute to persistent infection. *Lancet Infectious Diseases* 3:578–590.

Weber, C. and H. Noels. 2011. Atherosclerosis: Current pathogenesis and therapeutic options. *Nature Medicine* 17:1410–1422.

6

Neonatal Bacterial Infection

Insights into Pathogenic Strategy and Onset of Meningitis and Sepsis

Koilmani Emmanuvel Rajan and Christopher Karen

Contents

6.1 Introduction

Bacterial invasion in neonates can cause life-threatening diseases like meningitis and sepsis. Accordingly, it accounts a major cause of morbidity and mortality in global scenario (Park et al. 2001; Neher and Brown 2007; Tzialla 2015). Despite several preventive measures, including the improvement of socioeconomic status, appropriate antimicrobial therapy, and modern vaccination strategies, the episodes of meningitis and sepsis remains alarming (Harvey 1999; Park et al. 2001; Agrawal 2011; Van Sorge 2012; Miller et al. 2013; Camacho-Gonzalez 2013).

Meningitis is a neurologic emergency characterized by inflammation of the meninges in response to microbial infection (Heckenberg 2014). The hallmark feature of meningitis is the inflammation in pia matter, arachnoids, and subarachnoid space. Moreover, some forms of neurological sequelae in neonatal meningitis have been associated with deficit in learning and memory due to prominent neuronal injury or damage in the two brain structures, namely the cortex and the hippocampus (Bifrare et al. 2003; Leib et al. 2003; Blaser 2011; Barichello et al. 2013a, 2013b). Sepsis is defined as the systemic response to infection. The infection manifests two or more clinical symptoms, including changes in body temperature (>38°C or <36°C), heart rate (>90 beats/min), respiratory rate (>20 breaths/min) or partial CO_2 (<32 mm Hg), and white blood cell (>12,000/cu mm or <4,000/cu mm). Severe sepsis condition causes hypotension and multiple organ dysfunctions. Septic shock is the sepsis-induced hypotension, which is unresponsive to adequate fluid resuscitation, associated with abnormalities (Bone et al. 1992; Markiewski 2008). As soon as the infection occurs, it deteriorates the neonatal health, and the neonate develops septic shock even before the identification of the cause (Segura-Cervantes 2016).

Generally, the immune system (i.e., humoral and cellular immunity and phagocytic function) of neonates are immature; thus, they are prone to bacterial infections. Due to immature immune system functioning, neonates are particularly susceptible to infection and can lead to chronic sequelae such as deafness, blindness, cerebral palsy, and other neurodevelopment disorders. Newborn babies, especially who are premature and have very-low birth weight are vulnerable for any infection. The rapid change in epidemiology of meningitis and sepsis due to immunization practices makes it more challenging to develop new antibiotics. Also preterm infants will be lacking enough immunoglobulins transplacentally derived from their mother (Markiewski et al. 2008; Agrawal and Nadel 2011). The introduction of conjugate vaccines has virtually controlled the epidemiology of meningitis among neonates. Yet, average rates of 0.1–0.4 incidence have been reported per 1000 live births. The prevalence of the disease is higher in preterm with very-low birth weight and chronically hospitalized infants. In diseased victims, it causes severe injury and even causes death in one-fifth of infected population (Bogaert et al. 2004; Gordan 2017). Globally, one million deaths have been occurring every year due to neonatal sepsis; this accounts for 26% of all neonatal deaths (Miller et al. 2013; Tzialla et al. 2015). The possible routes of infection of sepsis are respiratory, abdominal, and urogenital tracks (Markiewski et al. 2008).

In this chapter, we focus on the causative bacteria and its pathophysiology in neonatal meningitis and sepsis.

6.2 Causative organisms of meningitis and sepsis

The common bacterial pathogens that cause meningitis are Group B Streptococcus (GBS), which is also called *Streptococcus agalactiae* (50%), *Escherichia coli* (20%), *Listeria monocytogenes* (5%–10%), *Streptococcus pneumonia* (6%), *Staphylococcus aureus, Neisseria meningitidis, Haemophilus influenza* type b (Hib), other Enterobacteriaceae (e.g., *Klebsiella* spp., *Enterobacter* spp., *Citrobacter* spp., *Serratia* spp., and *Proteus mirabilis* (Harvey et al. 1999; Bonacorsi 2005; Agrawal and Nadel 2011; Van Sorge and Doran 2012; Barichello et al. 2013a, 2013b; Ramakrishnan 2013) (Table 6.1).

Bacteria involved in sepsis infection are GBS (43%), *E. coli* (15.5%–29%), coagulase-negative *Staphylococcus* (CoNS; 42%–45%), *S. aureus* (10%–13%), *S. pneumonia, Klebsiella* spp., *Pseudomonas aeruginosa, L. monocytogenes* and other Enterobacteriaceae like *Cirtobacter, Serratia*, and *Enterobacter* (Markiewski 2008; Heath and Okike 2010; Miller 2013; Camacho-Gonzalez et al. 2013; Shah and Padbury 2014; Tzialla et al. 2015) (Table 6.1).

6.3 Etiology

The etiology of meningitis and sepsis includes several factors like infectious pathogens, pathophysiology, and virulence factors used by pathogen, host-pathogen interaction, host immune response, and evading strategies adapted by pathogen/factors and age of the host. Not to mention that exposure to invasive procedures during delivery, neonates are often handled with medical facilities (e.g., iatrogenic devices, tracheal cuff balloons in ventilation); contaminated neonatal intensive care environment, parenteral

Table 6.1 Organisms Associated with Early and Late Onset of Meningitis and Sepsis

Disease Condition	Type	Organism	Occurrence
Meningitis	Gram positive	Group B *Streptococcus*/ *Streptococcus agalactiae*	Both Early- & Late-onset meningitis
		Listeria monocytogenes	Both Early- & Late-onset meningitis
		Streptococcus pneumoniae	Late-onset meningitis
		Staphylococcus aureus	Late-onset meningitis
	Gram negative	*Escherichia coli* K1	Both Early- & Late-onset meningitis
		Neisseria meningitidis	Late-onset meningitis
		Haemophilus influenza type b	Late-onset meningitis
		Other enterobacteriaceae e.g., *Klebsiella* spp, *Enterobacter* spp, *Citrobacter* spp, *Serratia* spp, *Proteus mirabilis*	Late-onset meningitis
Sepsis	Gram positive	Group B *Streptococcus*/ *Streptococcus agalactiae*	Early-onset sepsis
		Coagulase-negative *Staphylococcus*	Late-onset sepsis
		Staphylococcus aureus	Late-onset sepsis
		Streptococcus pneumoniae	Late-onset sepsis
		Listeria monocytogenes	Late-onset sepsis
	Gram negative	*Escherichia coli* K1	Both Early- & Late-onset sepsis
		Pseudomonas aeruginosa	Late-onset sepsis
		Klebsiella spp	Late-onset sepsis
		Other enterobacteriaceae Cirtobacter spp, *Serratia* spp *and Enterobacter* spp	Late-onset sepsis

Source: Tzialla, C. et al., *Clin. Chim. Acta*, 451, 71–77, 2015; Barichello, T. et al., *J. Med. Microbiol.*, 62, 1781–1789, 2013a; Bonacorsi, S. and Bingen, E., *Int. J. Med. Microbiol.*, 295, 373–381, 2005.

nutrition, and powdered infant formula are the major causes. The sequential steps of bacterial infection to the progress of meningitis are invasion, proliferation, and colonization in central nervous system (CNS), inflammation, and acute brain damage (Cahill 2008; Grandgirard 2010; Ramakrishnan 2013; Pai et al. 2015).

6.4 Pathophysiology of bacterial meningitis

Generally, the bacterial pathogens colonize in the mucosal epithelium take hematogenous route from nasopharynx (especially in neonates and children) and infect the subarachnoid cavity. Later in further stages, the infection reaches the meninges. Mostly, neonatal meningitis is associated with sepsis (Koedel et al. 2010).

6.4.1 Infection

6.4.1.1 Mode of transmission – Neonates are thought to acquire meningitis in two possible ways, such as vertical and horizontal mode of transmission. Vertical transmission that occurs during the first week of life is the early onset of the disease, and it gets transmitted from mother to her newborn either in utero or during the passage through birth canal. Some studies show that the disease presents itself within 24 hours to first 7 days of life. On the other hand, horizontal transmission is also called the late onset of disease that occurs after the first week of life; it can be acquired by nosocomial infections. It can manifest from 7th day of life to the 89th day of life (Grandgirard and Leib 2010; Puopolo and Baker 2012; Barichello et al. 2013a). Frequently, the former are more prone to sepsis, pneumonia, respiratory disease, and meningitis and the latter to fever and meningitis (Zaidi et al. 2009; Heath and Okike 2010; Camacho-Gonzalez 2013; Shah and Padbury 2014; Tzialla et al. 2015) (Table 6.2).

Table 6.2 Risk Factors for Neonatal Meningitis/Sepsis

<table>
<tr><th colspan="2">Risk Factors for Neonatal Meningitis/Sepsis</th></tr>
<tr><th>Early-Onset</th><th>Late-Onset</th></tr>
<tr><td>Maternal colonization</td><td>Damage in primary defense (skin and mucosa)</td></tr>
<tr><td>Chrorioamnionitis</td><td>Invasive procedures</td></tr>
<tr><td>Premature rupture of membranes</td><td>Contaminated catheters</td></tr>
<tr><td>Maternal urinary tract infection</td><td>Infant powder formula</td></tr>
<tr><td>Preterm delivery <37 weeks</td><td>Very low birth weight </= 2500 g</td></tr>
<tr><td>High gravidity</td><td>Prolonged hospital stay</td></tr>
<tr><td colspan="2">Lack of prenatal care</td></tr>
<tr><td colspan="2">Urinary tract infection</td></tr>
<tr><td colspan="2">Intrapartum fever</td></tr>
</table>

Source: Camacho-Gonzalez, A. et al., *Pediatr. Clin. North Am.*, 60, 367–389, 2013; Wynn, J.L. and Wong, H.R. *Clin. Perinatol.*, 37, 439–479, 2010.

6.4.1.2 Colonization and invasion – The colonization of nasopharynx, oropharynx, or sinuses is the critical prerequisite for the development of bacterial meningitis. The virulent type of the pathogen determines the mechanisms behind invasion and colonization. Adhesion to the host tissue is the essential and initial step in bacterial infection. Bacterial adhesions extend to attach, colonize, and evade most of the defensive mechanisms of the host. The appendage or cell surface components that facilitate bacterial adhesion are called *adhesins*. Among gram-negative pathogens, pili or fimbriae are the major bacterial adhesive surface structures. They are mostly filamentous and can pass through or between capsular layers. Type 1, P-pili, type IV, and curli are the well-characterized pilus structures in gram-negative bacteria. On the other hand, in gram-positive bacteria, surface proteins, polysaccharides, or lipids displayed the adhesive functions (Kline 2009; Grandgirard and Leib 2010). Adhesins helps meningeal pathogens to successfully colonize the mucosal epithelial surfaces (i.e., respiratory, intestinal, and genitourinary tract) of the host (Kc et al. 2017). *E. coli* K1, GBS, and *N. meningitidis* use pili or fibrils to initiate binding in brain microvasculature endothelial cells (BMEC). Occasionally, the toxins released by the bacteria can lower the blood–brain barrier and facilitate penetration. Penetration of the mucosal barrier by the bacteria through or between the epithelial cells is very specific to most of the bacteria (Van Sorge and Doran 2012).

6.4.2 Bacteremia

Bacteremia is defined as the presence of bacteria in the bloodstream. It may be transient, intermittent, or continuous. When bacteria cross the mucosal barrier, it gains entry into the systemic circulation. The polysaccharide capsule helps in the survival of the pathogen in bloodstream of host by rendering resistance against host responses like complement-mediated lysis and phagocytosis by polymorphonuclear leukocytes and macrophages. Moreover, the survival of the internalized bacteria can also be enhanced by the inactivation of secretory IgA by the specific bacteria endopeptidase (Grandgirard and Leib 2010; Pai et al. 2015).

6.4.3 CNS invasion

6.4.3.1 Bacterial entry into CNS – Bacteria enters the CNS either transcellularly or paracellularly, through cerebral vasculature (i.e., blood–brain or blood–choroid barrier). The blood–brain barrier acts as a structural and functional barrier of brain. It helps to maintain the homeostasis inside the brain. It is composed of specialized BMEC composed of pericytes, astrocytes, and a basal membrane. It helps to prevent and protect the brain from harmful substances found in the bloodstream (i.e., these cells keeps a checkpoint and infilters the blood proteins and cells). The blood–brain barrier can also provide necessary nutrients to the brain and make it an immune-privileged organ (Dejana et al. 1995; Pulzova et al. 2009). Additionally, it also possesses an adherens junction (AJ) and tight junction (TJ). AJs are dependent on the interaction between the cytoplasmic tail of the components and TJs are composed of membrane proteins (Schulze and Firth 1993; Wolburg and Lippoldt 2002). The critical function of AJs and TJs are to regulate the permeability of blood–brain barrier. Breakdown or any disruption in the blood–barrier is the hallmark characteristic in pathophysiology of bacterial meningitis. Interestingly, GBS and *S. pneumoniae* disturb the blood–brain barrier by producing a pore-forming toxin (Van Sorge and Doran 2012).

Bacteria that cross the blood–brain barrier in live condition called *neurotropic pathogens* where they form complex interplay with host receptors on the surface endothelial cells of the brain. For the traversal, they require the use receptors like endoplasmin, CD46, the 37 kDa laminin, and the platelet-activating factor. It has been shown that *L. monocytogenes* cross the blood–brain barrier by transmigration of *L. monocytogenes*-infected monocytes, which is called a Trojan horse mechanism (Kim 2010). When the pathogen enters CNS, it leads to the activation of endothelium, inflammatory, and proinflammatory cytokine. Rather than the pathogen itself, the immune response of the host can cause major neurological damages. For instance, during an infection, the blood–brain barrier becomes more permeable and allows the influx of leukocytes. This influx of a large number of leukocytes leads to edema within the cerebrum and swelling of meningitis. Though, the mechanism of how the pathogen crosses the blood–brain barrier is poorly understood, several studies have reported that the first step could be the destruction of the endothelial cell layers with the help of pneumolysin (PLY), which subsequently disrupts the tight junctions in the blood–brain barrier during traversal and makes a way for pathogen to cross the blood–brain barrier by transcytosis (Kim 2008; Grandgirard and Leib 2010; Barichello et al. 2013a; Lovino 2013).

6.4.3.2 Survival of pathogens in CNS – The CNS compartment serves as favorable environment to pathogens when compared to the bloodstream because they get benefitted and can rapidly multiply inside the CNS. Additionally, they also penetrate brain, spinal cord, and the Virchow-Robin space along the vessels. In the CNS, particularly in the subarachnoid space, there is limited host defense because of the anatomical design of the blood–brain barrier and less access to the cerebrospinal fluid (CSF).

- The subarachnoid space does not have a completely organized drainage by the lymphatic system (Johnston et al. 2004).
- Absence of the complement factors of soluble pattern recognition receptors (PRRs) help in recognition of bacteria and in phagocytosis (Dujardin et al. 1985; Stahel et al. 1997).
- The systemic circulation contains more blood components that are prevented from crossing the blood–brain barrier to enter the CSF (Pachter et al. 2003).
- Generally CSF contains anti-inflammatory and immunosuppressors that actively suppresses immune reactivity (Niederkorn 2006).

6.4.3.3 Bacterial multiplication in CSF – However, the macrophages and dendritic cells are present in the tissues lining the CSF, namely, leptomeninges, choroid plexus, and perivascular spaces that function as sentinel cells. Sentinel cells through PRRs help to recognize the bacteria in CSF. Some PRRs expressed can act as a macrophage scavenger receptor, the mannose receptor, and complement receptors. They bind to the bacteria, thus mediating their internalization by phagocytes. While in brain, toll-like receptors (TLRs) are present in the immunocompetent cells. These make the CSF more susceptible and suppress the immune reactivity. When bacteria invade the CSF, they can easily proliferate and spread throughout the brain without any hindrance (Grandgirard and Leib 2010; Koedel et al. 2010). The bacterial concentration in infected CSF can reach a similar value to the bacterial concentration of the broth-culture in vitro (Small et al. 1986).

6.4.3.4 Host response to invasion – During inflammation, both microglia and astrocytes get activated, and their morphology changes in a process called *astrogliosis*. Microglia interacts with neurons and astrocytes to neutralize infections as quickly as possible and acts as a first immune defense against infections. Additionally, macrophages can phagocytose the pathogens present in the brain and spinal cord and present them to T cells. It can also travel to the site of infection and release proinflammatory cytokines. Astrocytes (astroglia) are star-shaped abundant cells found in the brain (Lovino et al. 2013).

Neurological damage is not only caused by viable bacteria but also by subcapsular components. Moreover, the unsheathed pathogens that are at stationary phase or damaged by the antibiotic treatment undergo autolysis. It concomitantly results in the release of the bacterial components like peptidoglycan, teichoic acid, lipoteichoic acid (LTA), bacterial DNA, PLY, and proinflammatory cytokines such as tumor necrosis factor alpha (TNF-α), interleukin 1 beta (IL-1β) and interleukin 8 (IL-8), free radicals, and excitatory amino acids (Schmidt et al. 1999; Koedel et al. 2010).

Table 6.3 TLR Dependant Activation and Bacterial Components

TLR	Bacterial Components	Neurodegeneration	Reference
2	Peptodoglycan	Neuronal loss and damage, Apoptosis	Hirschfeld et al. (1999), Han et al. (2003), Schröeder et al. (2003)
2	Lipoteichoic acid		Hirschfeld et al. (1999), Han et al. (2003), Schröeder et al. (2003)
4	Pneumolysin		Poltorak et al. (1998), Malley et al. (2003)
1/2	Lipoprotein		Barichello et al. (2013a)
4	Lipopolysaccharide		Poltorak et al. (1998), Malley et al. (2003)
5	Flagellin		Hayashi et al. (2001)

During meningitis, TLRs play the key inflammatory mediators that trigger immune response and promote proinflammatory cytokines production, which finally causes neuronal damage (Amor et al. 2010). On the cell surface, TLRs that have been identified are TLR1, 2, 4, 5, 6, and 10, which facilitate to identify the subcapsular bacterial components; additionally, TLR3, 4, 7, 8, and 9 trigger signaling cascades. It has been reported that TLR2 along with TLR1 and CD14 recognize LTA, lipoproteins, and peptidoglycan (Hirschfeld et al. 1999; Han et al. 2003; Schröeder et al. 2003), wherein lipopolysaccharides (LPS) and PLY can be engaged by TLR4 (Poltorak et al. 1998; Malley et al. 2003) and bacterial flagellin can be recognized by TLR5 (Hayashi et al. 2001) (Table 6.3). TLRs activate mitogen activated protein kinase (MAPK) and nuclear factor κB (NF-κB). These findings suggest that TLRs have to combine to activate inflammatory response. The individuals with meningitis show general characteristics of high levels of proinflammatory cytokines like IL-1β, IL-6, and TNF-α in CSF (Dunne and O'Neill 2003; Mogensen et al. 2006).

6.5 Pathophysiology of neonatal sepsis

Neonatal sepsis is the growth of microorganism in the blood of a newborn with systemic signs of infection and hemodynamic compromise (Ganatra et al. 2010; Tosson and Speer 2011). Neonates with sepsis may progress to septic shock where the progressions of the infection have not been stopped; multiple organ damage can occur and results in death (Wynn and Wong 2010).

6.5.1 Infection

6.5.1.1 Onset of sepsis – Neonatal sepsis is a blood infection that continues to be the common cause of mortality in neonates. Infection during early stages is termed early-onset sepsis (EOS), which is transmitted prenatally or during time of birth from mother (i.e., occurs within 72 hours to first week of life), whereas the infection that occurs after 1 week is termed late-onset sepsis (LOS), which is considered to be acquired from the environment sources (i.e., occurs after 1 week of life) (Miller et al. 2013; Shah and Padbury 2014; Tzialla et al. 2015).

6.5.1.2 Immune system of neonatal sepsis – Among all age groups, neonates are susceptible to infectious pathogens because of the developmental immunodeficiencies, and suboptimally functioning of innate immune system, hyporesponsive mononuclear phagocytes to physiologic and pathologic signaling mechanisms. In addition, preterm neonates have very low IgG levels because maternal IgG passively transfers across the placenta during pregnancy. Low serum levels will hinder opsonization. Deficiency in cellular immunity decreases the production of proinflammatory cytokines like interferons IL-12/IL-23 and IL-18, and this weakens immune system, which in turn fails to fight against the infection (Zaidi 2009).

6.5.1.3 Factors that cause sepsis – During the antenatal, intrapartum, and neonatal period, intrinsic and extrinsic factors include the risk of sepsis faced in developing countries. Intrinsic factors in the developing world are premature births, intrauterine growth restriction, and prolonged rupture of membranes during pregnancy, maternal peripartum or intrapartum infections, rectovaginal colonization with GBS, and chorioamnionitis. Extrinsic factors include no care taken during pregnancy and insanitary delivery practices. Based on the World Health Organization (WHO) statistical report on birth practices in developing countries, only 35% of the births are handled by professionals. This results in the unhygienic practices like using unsterile instruments and lack of sterile environment during delivery and postnatal period. Early-onset sepsis also includes factors like socioeconomic and cultural practices and the burden of poverty (Ganatra et al. 2010).

Infections are also transmitted by the multiuse syringes, surgical tools, and failure to follow aseptic techniques, especially during invasive procedures. Further it has been reported that the use of intravenous alimentation and prolonged use of mechanical ventilation/central venous catheters for the neonates has increased the risk of sepsis and septic shock (Wynn and Wong 2010) (Table 6.2).

6.5.1.4 Microbiology of sepsis and septic shock – Sepsis is caused predominantly by bacteria, but viruses also can cause it; in both cases, there is a high mortality. Among neonates who are septic, gram-positive bacteria are prevalent, whereas gram-negative infection accounts for 38% of septic shock and 62.5% sepsis mortality. Sepsis is dominated by GBS and coagulase-negative *Staphylococcus* (Hyde et al. 2002; Kermorvant-Duchemin et al. 2008; Chen et al. 2015).

6.5.2 Molecular and cellular events

6.5.2.1 Signaling of PRRs, PAMPs, and DAMPs – Once the local barrier has been compromised, the pathogen is recognized by body's first line of defense (i.e., local immune sentinel cells). This recognition has been done with the help of activated PRRs and TLRs (Kawai and Akira 2009). TLRs are capable of recognizing extracellular and intracellular pathogens, and they are present in multiple cell types. LPS is one of the products that is termed a pathogen-associated molecular patterns (PAMPs). This is the mediator for systemic inflammation, septic shock, and multiple organ failure (Trinchieri and Sher 2007; Kumagai et al. 2008; Rittirsch et al. 2008). LPS signals through TLR4 in conjugation with cell surface adaptor proteins CD14 and myeloid differentiation MD265, and another PAMP (i.e., LTA) signals through TLR2. Generally, more than one TLR are activated at the same time for the specific host-pathogen interaction. In neonates, during both gram-positive and -negative infection, there is upregulation of TLR2 and TLR4 mRNA in leukocytes (Zhang et al. 2007).

Along with PAMPs, TLRs is also activated by damage or danger associated molecular patterns (DAMPs). High-mobility group box-1 (HMGB-1) is a DAMP produced by endothelial cells and macrophages. It gets activated with LPS/TNF-α and signals through TLR2/TLR4/receptor for advanced glycation end products (RAGE). HMGB-1 helps in cytokine production, coagulation activation and recruitment of neutrophil, and involves the disruption of epithelial junctions in the gut. This HMGB-1 advances the disease progression of sepsis to septic shock (Mullins et al. 2004; Lotze et al. 2005; Zhang et al. 2007; Rittirsch et al. 2008; Van Zoelen et al. 2009); in addition, heat shock proteins (Hsps) and uric acid are DAMPs that are also involved in the physiological process of septic shock. In pediatric septic shock, there were elevated levels of Hsp60 and Hsp70 and are associated to death. Although uric acid can increase cytokine production and dendritic cell stimulation, it acts as an anti-oxidant. It is found in lower levels in septic neonates (Batra et al. 2000; Pack et al. 2005; Wheeler et al. 2005, 2007; Kapoor et al. 2006; Kono and Rock 2008).

Nucleotide binding oligomerization domain (NOD), a NLR (i.e., NOD-like receptor) is an intercellular non-TLR PRRs. Peptidoglycan of gram-positive bacteria in the cytosol is detected by NLR. After the pathogens are engaged, immune response is initiated by PRRs and through MAPKNF-κB, proinflammatory cytokines are produced (Trinchieri and Sher 2007; Wynn and Wong 2010).

6.5.2.2 Proinflammatory response – In response to infection, innate inflammatory response is amplified following the stimulation of PRR and cytokine production. It has been reported that during sepsis and septic shock, proinflammatory cytokines such as IL-1β, IL-6, IL-8, IL-2, IL-18, INF-γ, and TNF-α are elevated (Ng 2004). Septic neonates have decreased cytokine production compared to adults who are septic, which is another reason why the neonates are at risk (less IL-1β, IL-12, INF-γ, and TNF-α). These proinflammatory cytokines activate endothelial cells, which result in increased expression of cell adhesion molecules (CAMs) and chemokines. CAMs include soluble ICAM, VCAM, L-selectin, P-selectin, and E-selectin, which helps in leukocyte recruitment and diapedesis (Dollneer et al. 2001; Kourtis et al. 2003; Turunen et al. 2005; Figueras-Aloy 2007). Leukocyte recruitment is an important key inflammatory response step because it prevents the propagation of systemic infection (Sullivan et al. 2002). Generally, in sepsis condition, chemokines (IL-18 and interferon gamma-induced protein 10 [IP-10] can be used as a sensitive marker during infection), and other chemo-attractive molecules like complement proteins (C3a and C5a), host defense proteins (i.e., cathelicidins and defensins) and invaded bacterial components are increased (Ng 2004; Kingsmore 2008; Rittirsch 2008; Zaidi 2009; Wynn and Wong 2010).

6.5.2.3 Interplay between anti-inflammatory and proinflammatory responses – Sometimes in the immune mechanism, there is inappropriate response that results in systemic inflammatory response syndrome (SIRS). When there is no homeostasis in the inflammatory system, it ends up either in excessive immunity or meager immunity. The stage when former immunity occurs is termed as *cytokine storm*, and latter immunity is called *immune paralysis*. Apoptosis occur in both of these stages. To prevent this during infection, anti-inflammatory products (i.e., IL-4, IL-10, IL-11, IL-13, and TGF-β) are produced simultaneously to counter the activity of proinflammatory cytokines. Using receptor antagonists like TNFR2, sIL-6R, sIL2, and IL-1ra, the activity of proinflammatory mediators can be monitored along with anti-inflammatory cytokines (Sriskandan and Altmann 2008; Wynn and Wong 2010).

6.5.2.4 Complement system – The complement system is like an alarm system, which gets activated when pathogens get invaded or also during tissue damage. Complement activation pathway is the one that is the central constituent of immune system. It orchestrates and connects the inflammatory responses of innate and adaptive immunity. They help in coagulation, cytokine production, and activation of lymphocytes. Three properties of complement are opsonization, phagocytosis, and lysis of pathogens, which are events associated with immune response to infection. Three pathways are involved: classical, lectin, and alternative (Notarangelo et al. 1984; Wolach et al. 1997).

Generally, neonates have lower complement proteins and poor response in opsonization mediated by complement. The contribution of the complement system to the invasion of pathogens appears to be contradictory in some cases. The C3 component helps in the dysfunctioning of the effector which leads to an increase in septic infection (Quezado et al. 1994). It has been studied that C3a and C5a are up-regulated during sepsis infection; therefore, inhibition of C5a signaling can prevent from further infection (Ward 2008). Apparently, during early stages, the proinflammatory substance of complement pathway helps in the widespread of sepsis, but in the later stages, C5a along with sepsis leads to systemic infection and finally multiple organ failure.

During sepsis, there is up-regulation of complement-mediated activation of leukocytes along with cell surface receptors (i.e., CR1 and CR3). The receptor for C3b is CR1 (i.e., CR1–C3b) and helps in opsonization. C5a-C5aR facilitates the redistribution of blood flow, inflammation, aggregation of platelets, and release of reactive oxygen intermediates (ROI) (Snyderman and Goetzl 1981; Vogt 1986). CR3 is involved in adhesion of leukocytes, phagocytosis, and the recognition of pathogen and its products. Unfortunately, in neonates, the activation of CR3 on neutrophils is less compared to adults, which is the another reason infants get infected. Similarly, in neonates, C5aR is deficient, which eliminates the possibility to respond to C5a. Thus, it leads to increase in infection (Markiewski et al. 2008; Wynn and Wong 2010).

6.6 Concluding remarks

Despite advancements in treatment, sepsis and meningitis remain the most infectious diseases. Neonatal systemic bacterial infection (i.e., meningitis and sepsis) causes burden of mortality in some cases, and in survivors, it lasts as long-term morbidity in newborns. The pathophysiology of these diseases is complex and multifactorial. A key rhetorical question to neonatologists, neurobiologists, and obstetricians is how much the fetus or a neonate can withstand the inflammatory stress during infection and whether there is time until the infection has been identified to start medication.

To address these queries, there is a need to recognize authentic biomarkers for meningitis and sepsis. This may contribute to control several neonatal incidences of meningitis and sepsis.

References

Agrawal, S. and S. Nadel. 2011. Acute bacterial meningitis in infants and children: Epidemiology and management. *Paediatric Drugs* 13 (6):385–400.

Amor, S., F. Puentes, D. Baker, and P. Van der Valk. 2010. Inflammation in neurodegenerative diseases. *Immunology* 129 (2):154–169.

Barichello, T., G. D. Fagundes, J. S. Generoso, S. G. Elias, L. R. Simões and A. L. Teixeira. 2013a. Pathophysiology of neonatal acute bacterial meningitis. *Journal of Medical Microbiology* 62 (Pt 12):1781–1789.

Barichello, T., J. C. Lemos, J. S. Generoso, M. M. Carradore, A. P. Moreira, J. R. Zanatta, S. S. Valvassori and J. Quevedo. 2013b. Evaluation of the brain-derived neurotropic factor, nerve growth factor and memory in adult rats survivors of the neonatal meningitis by *Streptococcus agalactiae. Brain Reseasrch Bulletin* 92:56–59.

Batra, S., R. Kumar, A. K. Seema, Kapoor, and G. Ray. 2000. Alterations in antioxidant status during neonatal sepsis. *Annals of Tropical Paediatrics* 20 (1):27–33.

Bifrare, Y. D., C. Gianinazzi, H. Imboden, S. L. Leib and M. G. Täuber. 2003. Bacterial meningitis causes two distinct forms of cellular damage in the hippocampal dentate gyrus in infant rats. *Hippocampus* 13 (4):481–488.

Blaser, C., M. Wittwer, D. Grandgirard and S. L. Leib. 2011. Adjunctive dexamethasone affects the expression of genes related to inflammation, neurogenesis and apoptosis in infant rat pneumococcal meningitis. *PLoS One* 6 (3):e17840.

Bogaert, D., R. De Groot, and P. W. Hermans. 2004. Streptococcus pneumonia colonisation: The key to pneumococcal disease. *The Lancet Infectious Diseases* 4 (3):144–154.

Bonacorsi, S. and E. Bingen. 2005. Molecular epidemiology of *Escherichia coli* causing neonatal meningitis. *International Journal of Medical Microbiology* 295 (6–7):373–381.

Bone, R. C., R. A. Balk, F. B. Cerra, R. P. Dellinger, A. M. Felin, W. A. Knaus, R. M. Schein and W. J. Sibbald. 1992. Definitions for sepsis and organ failure and guidelines for the use of innovative therapies in sepsis. The ACCP/SCCM consensus conference committee. American college of chest physicians/Society of critical care medicine. *Chest* 101 (6):1644–1655.

Cahill, S. M., I. K. Wachsmuth, L. Costarrica Mde, and P. K. Ben Embarek. 2008. Powdered infant formula as a source of *Salmonella* infection in infants. *Clinical Infectious Diseases* 46 (2):268–273.

Camacho-Gonzalez, A., P. W. Spearman, and B. J. Stoll. 2013. Neonatal infectious diseases: Evaluation of neonatal sepsis. *Pediatric Clinics of North America* 60 (2):367–389.

Chen, X. C., Y. F. Yang, R. Wang, H. F. Gou, and X. Z. Chen. 2015. Epidemiology and microbiology of sepsis in mainland China in the first decade of the 21st century. *International Journal of Infectious Diseases* 31:9–14.

Dejana, E., M. Corada, and M. G. 1995. Lampugnani. Endothelial cell-to-cell junctions. *FASEB* 9 (10):910–918.

Dollneer, H., L. Vatter, and R. Austgulen. 2001. Early diagnostic markers for neonatal sepsis: Comparing C-reactive protein, interleukin-6, soluble tumour necrosis factor receptors and soluble adhesion molecules. *Journal of Clinical Epidemiology* 54 (12):1251–1257.

Dujardin, B. C., P. C. Driedijk, A. F. Riojers and T. A. Out. 1985. The determination of the complement components C1q, C4, and C3 in serum and cerebrospinal fluid by radioimmunoassay. *Journal of Immunological Methods* 80:227–237.

Dunne, A. and L. A. O'Neill. 2003. The interleukin-1 receptor. Toll-like receptor superfamily: Signal transduction during inflammation and host defense. *Science STKE*, 171:re3.

Figueras-Aloy, J., L. Gomez-Lopez, J. M. Rodriguez-Miguelez et al. 2007. Srum soluble ICAM-I, VCAM-1, L-selectin, and P-selectin levels as markers of infection and their relation to clinical severity in neonatal sepsis. *American Journal of Perinatology* 24 (6):331–338.

Ganatra, H. A., B. J. Stoll, and A. K. M. Zaidi. 2010. International perspective on early-onset neonatal sepsis. *Clinics in Perinatology* 37:501–523.

Gordan, S. M., L. Srinivasan, and M. C. Harris. 2017. Neonatal meningitis: Overcoming challenges in diagnosis, prognosis, and treatment with omics. *Frontiers in Pediatrics*. 16:5–139.

Grandgirard, D. and S. L. 2010. Leib. Meningitis in neonanates: Bench to bedside. *Clinics Perinatology* 37 (3):655–676.

Han S. H., J. H. Kim, M. Martin, S. M. Mickalek, and M. H. Nahm. 2003. Pneumococcal lipoteichoic acid (LTA) is not as potent as staphylococcal LTA in stimulating Toll-like receptor 2. *Infection and Immunity* 71 (10):5541–5548.

Harvey, D., D. E. Holt, and H. Bedford. 1999. Bacterial meningitis in the newborn: A prospective study of mortality and morbidity. *Seminars in Perinatology* 23 (3):218–225.

Hayashi, F., K. D. Smith, A. Ozinsky, T. R. Hawn, E. C. Yi, D. R. Goodlett, J. K. Eng, S. Akira, D. M. Underhill, and A. Aderem. 2001. The innate immune response to bacterial flagellin is mediated by Toll-like receptor 5. *Nature* 410:1099–1103.

Heath, P. T. and I. O. 2010. Okike. Neonatal bacterial meningitis: An update. *Paediatrics Child Health* 20 (11):526–530.

Heckenberg, S. G. B., M. C. Brouwer, and D. Van De Beek. 2014. *Handbook of Clinical Neurology* Vol. 121, 3rd series. Neurological aspect of systemic diseases (Part III). pp. 1361–1375. Philadelphia, PA: Elsevier.

Hirschfeld, M., C. J. Kirschning, R. Schwandner, H. Wesche, J. H. Weis, R. M. Wooten, and J. J. Weis. 1999. Cutting edge: Inflammatory signalling by Borrelia burgdorferi lipoproteins is mediated by Toll-like receptor 2. *Journal of Immunology* 163:2382–2386.

Hyde, T. B., T. M. Hilger, A. Reingold, M. M. Farley, K. L. O'Brien, and A. Schuchat. 2002. Trends in incidence and antimicrobial resistance of early-onset sepsis: Population-based surveillance in San Francisco and Atlanta. *Pediatrics* 110 (4):690–695.

Johnston, M., A. Zakharov, C. Papaiconomou, G. Salmasi and D. 2004. Armstrong. Evidence of connections between cerebrospinal fluid and nasal lymphatic vessels in humans, non-human primates and other mammalian species. *Cerebrospinal Fluid Research* 1 (1):2.

Kapoor, K., S. Basu, B. K. Das, and B. D. Bhatia. 2006. Lipid peroxidation and antioxidants in neonatal septicaemia. *Journal of Tropical Pediatrics* 52 (5):372–375.

Kawai, T. and S. Akira. 2009. The roles of TLRs, RLRs and NLRs in pathogen recognition. *International Immunology* 21 (4):317–337.

Kc, R., S. D. Shukla, E. H. Walters, and R. F. O'Toole. 2017. Temporal upregulation of host surface receptors provides a window of opportunity for bacterial adhesion and disease. *Microbiology* 163 (4):421–430.

Kermorvant-Duchemin, E., S. Laborie, M. Rabilloud, A. Lapillonne, and O. Claris. 2008. Outcome and prognostic factors in neonates with septic shock. *Pediatric Critical Care Medicine* 9 (2):186–191.

Kim, K. S. 2008. Mechanisms of microbial traversal of the blood-brain barrier. *Nature Reviews Microbiology* 6 (8):625–634.

Kim, K. S. 2010. Acute bacterial meningitis in infants and children. *The Lancet Infectious Diseases* 10 (1):32–42.

Kingsmore, S. F., N. Kennedy, H. L. Halliday et al. 2008. Identification of diagnostic biomarkers for infection in premature neonates. *Molecular & Cellular Proteomics* 7 (10):1863–1875.

Kline, K. A., S. Fälker, S. Dahlberg, S. Normark, and B. Henriques-Normark. 2009. Bacterial adhesins in host-microbe interactions. *Cell Host Microbe* 5 (6):580–592.

Koedel, U., M. Klein, and H. W. Pfister. 2010. New understandings on the pathophysiology of bacterial meningitis. *Current Opinion in Infectious Diseases* 23 (3):217–223.

Kono, H. and K. L. Rock. 2008. How dying cells alert the immune system to danger. *Nature Reviews Immunology* 8 (4):279–289.

Kourtis, A. P., F. K. Lee, and B. J. Stoll. 2003 Soluble L-selectin, a marker of immune activation, in neonatal infection. *Clinical Immunology* 109 (2):22408.

Kumagai, Y., O. Takeuchi, and S. Akira. 2008. Pathogen recognition by innate receptors. *Journal of Infection and Chemotherapy* 14 (2):86–92.

Leib, S. L., C. Heimgartner, Y. D. Bifrare, J. M. Loeffler and M. G. Täuber. 2003. Dexamethasone aggravates hippocampal apoptosis and learning deficiency in pneumococcal meningitis in infant rats. *Pediatric Research* 54 (3):353–357.

Lotze, M. T. and K. J. Tracey. 2005. High-morbidity group, box 1 protein (HMGB1): Nuclear weapon in the immune arsenal. *Nature Reviews Immunology* 5 (4):331–342.

Lovino, F., C. J. Orihuela, H. E. Moorlag, G. Molema, and J. E. E. Bijlsma. 2013. Interactions between blood-borne *Streptococcus pneumoniae* and the blood-brain barrier preceding meningitis. *PLoS One* 8 (7):e68408.

Malley, R., P. Henneke, S. C. Morse, M. J. Cieslewicz, M. Lipsitch, C. M. Thompson, E. Kurt-Jones, J. C. Paton, M. R. Wessels, and D. T. Golenbock. 2003. Recognition of pneumolysin by Toll-like receptor 4 confers resistance to pneumococcal infection. *Proceedings of the National Academy of Sciences of the United States of America* 100 (4):1966–1971.

Markiewski, M. M., R. A. DeAngelis, and J. D. Lambris. 2008. Complexity of complement activation in sepsis. *Journal of Cellular and Molecular Medicine* 12 (6A):2245–2254.

Miller, A. E., C. Morgan, and J. Vyankandondera. 2013. Causes of puerperal and neonatal sepsis in resource-constrained settings and advocacy for an integrated community-based postnatal approach. *International Journal of Gynaecology & Obstetrics* 123 (1):10–15.

Mogensen, T. H., S. R. Paludan, M. Kilian, and L. Ostergaard. 2006. Live *Streptococcus pneumoniae, Haemophilus influenzae,* and *Neisseria meningitides* activate the inflammatory response through Toll-like receptors 2, 4, and 9 in species-specific patterns. *Journal of Leukocyte Biology* 80:267–277.

Mullins, G. E., J. Sunden-Cullberg, A. S. Johansson, A. Rouhiainen, H. Erlandsson-Harris, H. Yang, K. J. Tracey et al. 2004. Activation of human umbilical vein endothelial cells leads to relocation and release of high-morbidity group box chromosomal protein 1, *Scandinavian Journal of Immunology* 60 (6):566–573.

Neher, J. J. and G. C. Brown. 2007. Neurodegenaration in models of gram-positive bacterial infections of the central nervous system. *Biochemical Society Transactions* 35 (Pt 5):1166–1167.

Ng, P. C. 2004. Diagnostic markers of infection in neonates. *Archives of Disease in Childhood—Fetal and Neonatal Edition* 89 (3):F229–F235.

Niederkorn, J. Y. 2006. See no evil, hear no evil, do no evil: The lessons of immune privilege. *Nature Immunology* 7:354–359.

Notarangelo, L. D., G. Chirico, A. Chiara et al. 1984. Activity of classical and alternative pathways of complement in preterm and small for gestational age infants. *Pediatric Research* 18 (3):281–285.

Pachter, J. S., H. E. De Vries, and Z. Fabry. 2003. The blood-brain barrier and its role in immune privilege in the central nervous system. *Journal of Neuropathology & Experimental Neurology* 62:593–604.

Pack, C. D., U. Kumaraguru, S. Suvas, and B. T. Rouse. 2005. Hear-shock protein 70 acts as an effective adjuvant in neonatal mice and confers protection against challenge with herpes simplex virus. *Vaccine* 23 (27):3526–3534.

Pai, S., D. A. Enoch, and S. H. 2015. Aliyu. Bacteremia in children: Epidemiology, clinical diagnosis and antibiotic treatment. *Expert Review of Anti-infective Therapy*13 (9):1073–1088.

Park, W. S., Y. S. Chang, and M. Lee. 2001. 7-Nitroindazole, but not aminoguanidine, attenuates the acute inflammatory responses and brain injury during the early phase of Escharichia coli meningitis in the newborn piglet. *Biology of the Neonate* 80 (1):53–59.

Poltorak, A., X. He, I. Smirnova, M. Y. Liu, C. Van Huffel, X. Du, D. Birdwell et al. 1998. Defective LPS signalling in C3H/HeJ and C57BL/10ScCr mice: Mutations in Tlr4 gene. *Science* 282:2085–2088.

Pulzova, L., M. R. Bhide, and K. Andrej. Pathogen translocation across the blood-brain barrier. 2009. *FEMS Immunology & Medical Microbiology* 57 (3):203–213.

Puopolo, K. M. and C. J. Baker. 2012. *Group B Streptococcal* infection in neonates and young infants, In: Edwards, M. S., Weisman, L. E. (Eds.). *UpToDate*, Waltham, MA.

Quezado, Z. M. N., W. D. Hoffman, J. A. Winkelstein, I. Yatsiv, C. A. Koev, L. C. Cork, R. J. Elin, P. Q. Eichacker, and C. Natanson. 1994. The third component of complement protects against *Escherichia coli* endotoxin-induced shock and multiple organ failure. *Journal of Experimental Medicine* 179:569–578.

Ramakrishnan, K. A., M. Levin, and S. N. 2013. Faust. Bacterial meningitis and brain abscess. *Medicine* 41 (12):671–677.

Rittirsch D., M. A. Flierl, and P. A. Ward. 2008. Harmful molecular mechanisms in sepsis. *Nature Reviews Immunology* 8 (10):776–787.

Schmidt, H., K. Stuertz, F. Trostdorf, V. Chen, I. Sadowski, W. Brück and R. Nau. 1999. Streptococcal meningitis: Effect of CSF filtration on inflammation and neuronal damage. *Journal of Neurology* 246:1063–1068.

Schröeder, N. W., S. Morath, C. Alexander, L. Hamann, T. Hartung, U. Zähringer, U. B. Göbel, J. R. Weber and R. R. Schumann. 2003. Lipoteichoic acid (LTA) of *Streptococcus pneumoniae* and *Staphylococcus aureus* activates immune cells via Toll-like receptors (TLR)-2, lipopolysaccharide-binding protein (LBP), and CD14, whereas TLR-4 and MD-2 are not involved. *Journal of Biological Chemistry* 278 (18):15587–15594.

Schulze, C. and J. A. 1993. Firth. Immunohistochemical localization of adherens junction components in blood-brain barrier microvessels of the rat. *Journal of Cell Science* 104 (Pt 3):773–782.

Segura-Cervantes, E., J. Mancilla-Ramírez, J. González-Canudas, E. Alba, R. Santillán-Ballesteros, D. Morales-Barquet, G. Sandoval-Plata, and N. 2016. Galindo-Sevilla. Inflammatory responses in preterm and very preterm newborns with sepsis. *Mediators of Inflammation* 2016:6740827.

Shah, B. A. and J. F. 2014. Padbury. Neonatal sepsis: An old problem with new insights. *Virulence* 5 (1):171–1718.

Small, P. M., M. G. Täuber, C. J. Hackbarth, and M. A. Sande. 1986. Influence of body temperature on bacterial growth rates in experimental pneumococcal meningitis in rabbits. *Infection Immunity* 52 (2):484–487.

Snyderman, R. and E. J. Goetzl. 1981. Molecular and cellular mechanics of leukocyte chemotaxis. *Science* 213 (4510):830–837.

Sriskandan, S. and D. M. Altmann. 2008. The immunology of sepsis. *The Journal of Pathology* 214:211–223.

Stahel, P. F., D. Nadal, H. W. Pfister, P. M. Paradisis and S. R. Barnum. 1997. Complement C3 and factor B cerebrospinal fluid concentrations in bacterial and aseptic meningitis. *Lancet* 349:1886–1887.

Sullivan, S. E., S. L. Staba, J. A. Gersting et al. 2002. Circulating concentrations of chemokines in cord blood, neonates, and adults. *Pediatric Research* 51 (5):653–657

Tosson, A. M. and C. P. Speer. 2011. Microbial pathogens causative of neonatal sepsis in Arabic countries. *The Journal of Maternal-Fetal & Neonatal Medicine* 24 (8):990–994.

Trinchieri, G. and S. Sher. 2007. Cooperation of Toll-like receptor signals in innate immune defence. *Nature Reviews Immunology* 7 (3):179–190.

Turunen, R., S. Andersson, I. Nupponen, H. Kautiainen, S. Siitonen, and H. Repo. 2005. Increased CD11b-density on circulating phagocytes as an early sign of late-onset sepsis in extremely low-birth-weight infants. *Pediatric Research* 57 (2):270–275.

Tzialla, C., A. Borghesi, M. Pozzi and M. Stronati. 2015. Neonatal infections due to multi-resistant strains: Epidemiology, current treatment, emerging therapeutic approaches and prevention. *Clinica Chimica Acta* 451 (Pt A):71–77.

Van Sorge, N. M. and K. S. Doran. 2012. Defense at the border: The blood-brain barrier versus bacterial foreigners. *Future Microbiology* 7 (3):383–394.

Van Zoelen, M. A., H. Yang, S. Florguin, J. C. Meijers, S. Akira, B. Arnold, P. P. Nawroth, A. Bierhaus, K. J. Tracey, and T. van der Poll. 2009. Roll of toll-like receptors 2 and 4, and the receptor for advanced glycation end products in high-mobility group box 1-induced inflammation in vivo. *Shock* 31 (3):280–284.

Vogt, W. 1986. Anaphylatoxins: Possible roles in disease. *Complement* 3 (3):177–188.

Ward, P. A. 2008. Role of the complement in experimental sepsis. *Journal of Leukocyte Biology* 83:467–470.

Wheeler, D. S., L. E. Fisher Jr., J. D. Catravas, B. R. Jacobs, J. A. Carcillo, and H. R. Wong. 2005. Extracellular hsp70 levels in children with septic shock. *Pediatric Critical Care Medicine* 6 (3):308–311.

Wheeler, D. S., P. Lahni, K. Odoms, B. R. Jacobs, J. A. Carcillo, L. A. Doughty, and H. R. Wong. 2007. Extracellular heat shock protein 60 (Hsp60) levels in children with septic shock. *Inflammation Research* 56 (5):216–219.

Wolach, B., T. Dolfin, R. Regev, S. Gilboa, and M. Schlesinger. 1997. The development of the complement system after 28 weeks' gestation. *Acta Paediatrica* 86 (5):523–527.

Wolburg, H. and A. Lippoldt. 2002. Tight junctions of the blood-brain barrier: Development, composition and regulation. *Vascular Pharmacology* 38 (6):323–337.

Wynn, J. L. and H. R. Wong. 2010. Pathophysiology and treatment of septic shock in neonates. *Clinics in Perinatology* 37 (2):439–479.

Zaidi, A. K., D. Thaver, S. A. Ali, and T. A. Khan. 2009. Pathogens associated with sepsis in newborns and young infants in developing countries. *The Pediatric Infectious Disease Journal* 28 (1 Suppl):S10–S18.

Zhang, S. Y., E. Jouanguy, S. Uglini, A. Smahi, G. Elain, P. Romero, D. Segal et al. 2007. TLR3 deficiency in patients with herpes simplex encephalitis. *Science* 317 (5844):1522–1527.

7

Bacterial Infections of the Oral Cavity

Bacterial Profile, Diagnostic Characteristics, and Treatment Strategies

P. S. Manoharan and Praveen Rajesh

Contents

7.1 Introduction

The oral hygiene status, pathogenicity of a microorganisms, and host response influence the general oral health and overall well-being of an individual. There are various commensals present in the mouth and throat. An understanding of the normal commensals in the oral environment, its symbiosis, and interaction is needed for the healthcare provider. It helps to appreciate pathogenesis of a microorganism and understand the nature of a disease (Gendron et al., 2000).

Bacterial infections contribute to a major deal in the dental, oral, and general health of the individual. Dental caries and periodontal disease are the most common diseases of the oral cavity. The disease-causing microorganisms seem to exhibit a definite site specific pathogenicity. *Streptococcus mutans*—a caries-producing microorganism causes lesion only when on the tooth structure. *Lactobacillus acidophilus* is commonly seen in deep carious lesions. *Aggregatibacter actinomycetemcomitans* and *Porphyromonas gingivalis*, which are associated with periodontal infections when present on enamel structure, was not found to be cariogenic. But species particular to caries or periodontal disease are not isolated, although associations were present (Aas et al., 2005). The complexity of the oral environment demands the study of pathogens to be carried out as a consortium of microorganisms. Culture-independent molecular techniques, site, and subject specificity of the microorganisms seem to play a major role in isolating the etiology and understanding the behavior of the bacterial infections (Loesche et al., 1992). Periodontal disease is not a universal phenomenon. It is surprising that severe forms of this disease affect a group of population who are abnormally susceptible (Genco and Borgnakke, 2013).

The oral cavity being the abode to diverse microbiota was found to influence the systemic health of the individual. Lifestyle, diet, habitat, and personal habits have been found to influence the oral microbiome. These complexities have opened new avenues for translatory research. The course of systemic diseases and conditions were also found to be altered by oral infections. This reversal of paradigm has also directed the healthcare provider to consider the oral health status in handling other medical conditions.

This chapter provides an overview on the present trend in understanding the nature and behavior of various pathogens of common dental and oral diseases. Moreover, an insight on the applied aspects for the healthcare provider in handling patients who present with such infections is also discussed.

7.2 Morphological attributes of the structures in the oral cavity

7.2.1 Tooth or the dental apparatus

The uniqueness of the tooth begins with the description of it being partly embedded in the hard connective tissue, the bone, and part of it open to the oral environment (Figure 7.1). The surface is covered by the hardest substance in the human body—the enamel. The surface of the coronal tooth structure is not smooth and well rounded, but present with elevations called *cusps* and depressions called *pits* and *fissures* (Figure 7.2). The second layer from the enamel is the dentin, which is composed of tubular structures that travel down into the living pulpal tissue layer. These also provide channels for microorganisms to percolate if the dentin is exposed. The enamel does not extend to the root structure. The root dentin is covered with cementum all over the surface. The nature of disease and progression

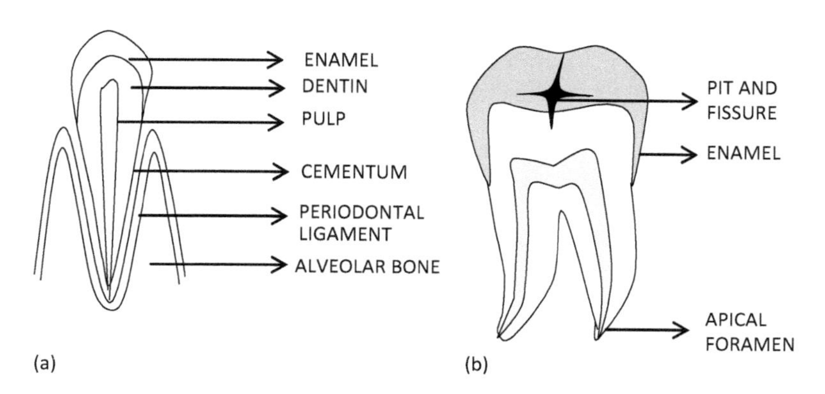

Figure 7.1 Longitudinal section of the tooth and supporting apparatus. (a) Longitudinal section of the tooth showing various layers and the periodontal apparatus; (b) caries in the pits and fissures.

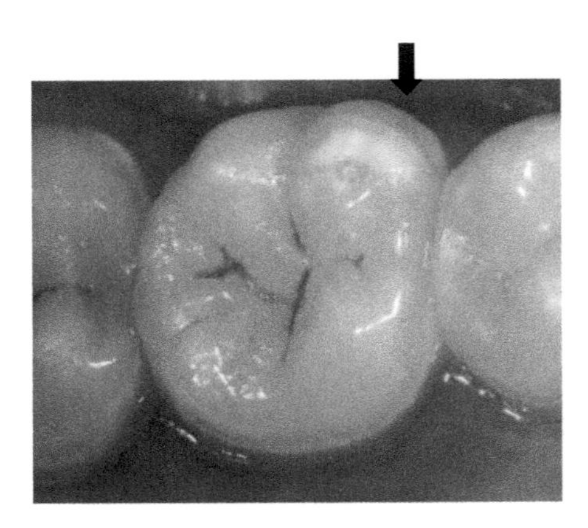

Figure 7.2 Tooth structure. Pits and fissures of a molar tooth—susceptibility to caries. Arrow indicates hypocalificed area on a smooth surface which may breakdown further.

of disease is different in a smooth surface compared to that of pit and fissure of the enamel. The type of bacteria that harbor the tooth also depends on the various locations mentioned previously.

7.2.2 Periodontal apparatus

The root of the tooth is embedded in the alveolar bone of the maxilla and mandible. The cementum is a hard tenacious layer that covers the root surface, which is 15–150 microns thick. The tooth is anchored to the alveolar bone through periodontal ligament, a fibro-epithelial tissue layer that is sensitive to pressure stimuli, but at the same time protective for the tooth by acting as a cushion against microtrauma. The periodontal ligament is seen extending from the cementum of the root to the cortical plate of the alveolar bone (supporting bone surrounding the tooth structure of maxilla and mandible). The thin cortical plate aids in attachment to the periodontal ligaments, which is described as a lamina dura and is a radiopaque line in radiographs.

7.2.3 Junctional epithelium

The attachment of the soft tissue, gingiva, to the tooth is through junctional epithelium, which can be considered a weak zone and is prone for infection (Figure 7.3). The harbored microorganisms and its products proximal to this area can ingress to the supporting periodontal apparatus. Another point of entry is through the apical foramen present at the end of the tooth where the pulpal nerves and blood supply enters. The course of the disease and nature of treatment is completely different, depending on the pathway of infection.

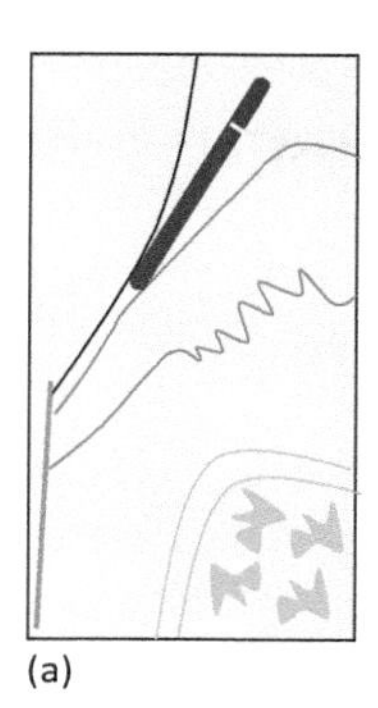

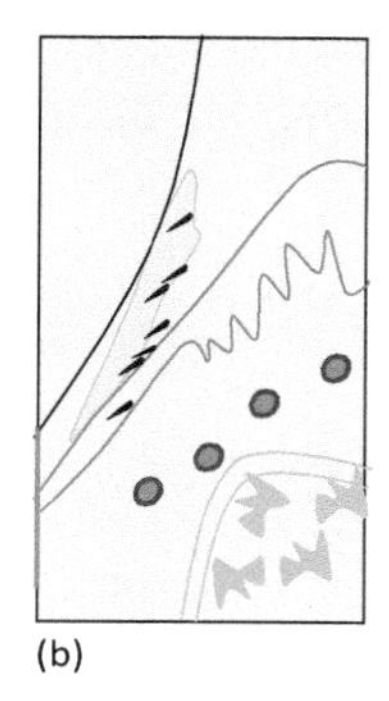

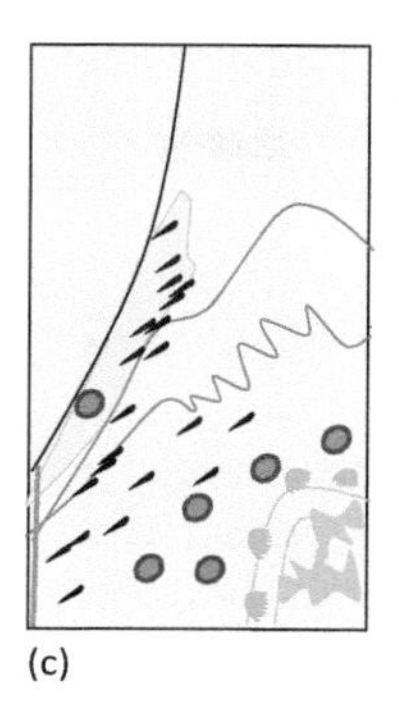

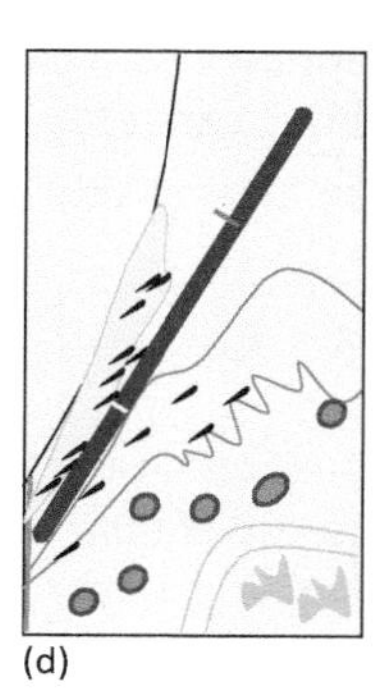

Figure 7.3 Junctional epithelium in health and disease. (a) In health. The instrument probed shows normal depth of sulcus—yellow marking; (b) plaque deposits harboring microorganisms; (c) bacteria and its toxins infliterating the connective tissue, breakdown of junctional epithelium; (d) deepening of sulcus. The instrument probed shows periodontal pocket—red marking.

The epithelial component, which is described as the attachment apparatus, has been extensively studied (Schroeder and Listgarten, 2003). It is the most active yet sensitive area of tooth structure. A high turnover of cells are noticed in this junction and under continuous shedding process as new cells are formed. The junctional epithelium is seen at the base of the sulcus, 15 to 30-cell-layer thickness and is found attached partly to the enamel at the neck region of the tooth and partly over the cementum. Plaque deposits that harbor bacteria at this junction can be potentially harmful. The subgingival aggregation of plaque forms a niche from which the disease process is aggravated by further accumulation of plaque and other debris.

7.2.4 The oral mucosa

The mucosa covering the oral cavity is described as stratified squamous epithelium. Some areas like the gingiva palate are keratinized. The cheek part of the lip and floor of the mouth are covered by nonkeratinized epithelium. The tongue has epithelium modified to specialized mucosa and papilla, which lodge taste buds. *Candida* infections are more common on the tongue than bacterial infections. Reports of bacterial infections are seen with tongue piercing. Autoimmune desquamative diseases, such as lichen planus and pemphigus vulgaris, can occur in relation to mucosa tongue and cheek. Viral infections, like herpetic gingivostomatitis and ulcerative gingivitis, are more common than bacterial in the gingival area. However, bacterial superinfection was found to be associated in most of the cases. Superinfection can also be observed in simple traumatic to recurrent aphthous ulcers. Bacterial colonies are also seen in salivary films adhering to mucosa of lip cheek and tongue. However, their active role in producing illness as an entity by itself is not recognized.

7.3 Bacteria of dental caries and periodontal diseases

The oral environment is flooded with more than 10^8 bacteria, which amount to 400 of the 700 species or phenotypes in the oral cavity. There is a chance to harbor at least 150 of the bacterial species at any point in a lifetime. Many species have not been isolated, but their role in producing oral infections is obscure. This microbiota can be seen adherent to the surface of the tooth mucosa and tongue surface as biofilms. In the context to the oral cavity, the bacterial deposits can be termed *dental plaque* or *bacterial plaque*. The pathogenesis can be attributed to the irritants that form as a result of bacterial colonization, and it follows a carefully choreographed pathway (Socransky et al., 1998). Acids, endotoxins, antigens from the bacteria, and the chemical mediators released in response to these products decide the course of the disease. The outcome is based on the balance between pathogenicity of the bacteria and the response of the host to the irritants. The degeneration of the tooth, epithelium, periodontal ligament, and bone happens, and it becomes progressively severe, if the disease process is not controlled in the initial stage.

Bacterial plaque are colorless and transparent when they form. With time, the presence of chromogenic bacteria imparts color to the plaque. Long-standing plaque gets mineralized to a hard tenacious deposit called the *calculus*, which can be creamish white, dark yellow. or even brown. Heavy deposits are seen adjacent to salivary gland excretory ducts (i.e., lingual aspect of lower anterior teeth and buccal aspect of maxillary molars). Plaque, when calcified from the minerals in the saliva and blood, is called *calculus*. The calculus present above the marginal gingiva on the tooth is called *supragingival calculus* and below the marginal gingival is called *subgingival calculus*. Subgingival calculus are not visible unlike supragingival deposits. They can be found only with tactile exploration with a diagnostic instrument like an explorer and might get noted in radiographs. The time required for a supragingival calculus to form is 2 weeks with 80% of mineralization. Mature deposit formation may take several months to years. The subgingival calculus should be regarded a secondary product following infection and should not be treated as if it was the primary etiology. Calculus deposits can occur in any rough surface at a faster rate. The demineralized rough surface enamel by the acid products of plaque is prone to calculus deposition. Rough surfaces of restorations, fixed or removable dentures, implants, and improper contours may also lead to deposition of plaque and calculus. The tooth or replacement structures lose their surface smoothness. The crystals of calculus penetrate the surface and are virtually locked to the tooth. In the root surfaces the covering layer, cementum, gets an irregular surface, and deposits may be seen that are extremely difficult to remove. The nature of such a plaque was found to be similar to that of the deposit that was noticed on the teeth.

The relationship of microorganism to the disease is not clearly understood because there are number of other criteria related such as association, elimination, response of the host, virulence factors, animal model studies, and assessment of risk. All these criteria put together contribute to the "weight of evidence" and would lead to specific diagnosis. The consensus report from World Workshop in Periodontology have concluded that *A. actinomycetemcomitans* and *Tannerella forsythia* as periodontal pathogens. Classification of all the types of periodontal diseases was also put forth (Armitage, 1999). Table 7.1 shows the weight of evidence for the three common periodontal pathogens mentioned above (Lindhe et al., 2008).

Spirochetes are seen in some wet mount preparations with stains specific to Spirochetes and dark field microscopy in destructive periodontitis. Fusiform bacteria was also found frequently isolated with destructive periodontitis and also in a kind of diphtheria like pseudomembranous infection—acute necrotizing ulcerative gingivitis [Vincent in 1889]. *Streptococci* was also found to be commonly isolated from supragingival plaque.

7.4 Mechanism of colonization and adherence of bacteria

Plaque biofilm is seen as a firmly adherent mass that develops soon after immersion into a fluid media. The mouth is constantly bathed in saliva, and the adsorption of the contents of the saliva to form a matrix takes place in a sequential manner.

The first step is the formation of a conditioning film called *acquired pellicle* (Phase I). The film contains glycoproteins (i.e., mucins) and antibodies from the saliva. This acts as a matrix to harbor the microorganisms. Primary colonization is mainly by facultative anaerobic gram-positive cocci. In the first 24 hours, the plaque consists mainly of *Streptococcus sanguis*. The bacteria along with polymorphonuclear leucocytes and epithelial cells adhere to the tooth surface. Early deposits can be easily removed. The surface energy is altered, which leads to the adhesion of bacteria on to the tooth. Loose bacteria (i.e., planktonic) slowly transform to a consolidated adherence (Gibbons and Houte, 1975). A typical corncob appearance of plaque is because of adherence of cocci on a filamentous bacteria (Listgarten et al., 1973).

Table 7.1 Weight of Evidence Table

Bacteria	Association	Elimination	Host Response	Virulence Factors	Animal studies
A. Actinomycetemcomitans	⬆ Localised aggressive or prepubertal periodontitis adult periodontal lesions ⬇ Healthy, edentulous sites, gingivitis	Therapy successful Recurrent lesions species seen	⬆ Antibody in serum or saliva of Localised aggressive periodontitits, Chronic periodontitis	Endotoxin, Leucotoxin, epitheliotoxin, collagenase, fibroblast inhibiting factor, bone resorption inducing factor, cytokines from macrophages, cytolethal distending toxin[apoptosis], neutrophil function modified, immunoglobulin degradation	Gnotobiotic rats showed positive disease presentation Caused subcutaneous abscess in mice
P. gingivalis	⬆ Periodontitis aggressive forms, attachment loss and bone loss	Therapy successful Reccurrent lesions species seen	⬆ Antibody in serum or saliva of various forms of periodontitis	Collagenase, endotoxin, proteolytic trypsin like activity, fibnrinolysin, hemolysin, proteases including gingipain, phospholipase A, Fibroblast inhibiting factor, hydrogen sulphide, ammonia, fatty acids, neutrophil modifying factors, cytokine production from host cells, migration of Polymorophonuclear lympocytes, immunoglobulin degradation	Gnotobiotic rats, sheep, monkeys and dog experiments showed positive disease presentation Negative impact of immunization on disease in experimental animals
T. forsythia	⬆ Periodontitis Actively progressing lesion, Abscess, refractory periodontitis, periodontal pockets, attachment and bone loss ⬇ health, gingivitis	Therapy successful Reccurrent lesions species seen	⬆ Antibody in serum ⬆ More antibody in refractory periodontitis	Endotoxin, fatty acids, methylglyoxal production, cytokine production, apoptosis and invasion of epithelial cells	Increased ligature induced periodontitis in dogs Gnotobiotic rats showed positive disease presentation

Source: Lindhe, J. et al., *Clinical Periodontology and Implant Dentistry*, Blackwell Munksgaard, Oxford, UK, 2008.

Bacteria adheres by means of fimbria and extracellular polymeric substances. Slowly gram-positive rods dominate, especially the *Actinomyces* species. The gram-negative bacteria (i.e., *Veillonella* and *Fusobacterium*) find difficulty in attaching to the pellicle. The receptors found on the surface of the gram-positive rods and streptococci allow adhesion of the gram-negative bacteria (Phase II). Once attached, they rapidly grow and multiply (Phase III) and synthesize cellular membrane component (Phase IV). As the thickness increases, they present with deeper layers under an anaerobic environment. The bacteria are deprived of the nutrition from the dietary products, so they produce enzymes that can breakdown host macromolecules of the periodontium into peptides and amino acids. The destructive disease onset thus occurs and can cause irreversible damage to the periodontal apparatus. The matrix material seen between the microorganisms amounts to almost 25% of the plaque volume. This intermicrobial matrix can be either filamentous or granular in appearance. The polysaccharides are mainly levans (i.e., fructans from the diet) and glucans (i.e., dextran and mutans). The stability of intercellular matrix is offered by mutans. Uncharacterized lipids also are seen in the matrix, which are mainly the lipopolysaccharide toxins from gram-negative bacteria.

This adherent biofilm becomes complex and resistant as it gains more shear strength against removal. The bacterial deposits become site specific, and a distinct composition are noted on various surfaces (e.g., pit and fissure/smooth surface plaque, shallow/deep crevices of sulcus). The time period of the lesions also determines the type of bacteria. Established lesions are presented with more gram-negative bacteria and anaerobes. Around 10^9 bacteria can be found on a supragingival plaque in a tooth. Millions to billions of bacteria may be found to colonize the subgingival crevice in diseased and even in healthy oral cavity. The inverse relationship between the development of disease and the presence of microorganisms has opened up avenues of research in this field.

Some of the commonly isolated pathogens in relation to periodontal disease are *A. (Actinobacillus)actinomycetemcomitans, T. forsythia, Campylobacter rectus, Eubacterium nodatum, Fusobacterium nucleatum, Peptostreptococcus micros, P. gingivalis, Prevotella intermedia, Prevotella nigrescens, Streptococcus intermedius*, and *Treponema* spp. (Socransky et al., 1998).

The subgingival plaque presents with all the aforementioned characteristics along with a special entity—the cuticle, which is presumed to be secretory product of the epithelial cells (Schroeder and Listgarten, 2003). Apart from gram-positive and -negative anaerobes, Spirochetes and other flagellated bacteria are also seen.

7.5 Microorganisms in dental caries

No single microorganism or group of microorganisms may be held solely responsible for the initiation and progression of caries. Thus, it is less important to remember the specific names of the bacteria than to understand their potential role in the biofilm community in health and disease. However, few bacteria, which have a very strong association with caries pathogenesis, need to be discussed.

Several studies have shown *Streptococcus mutans* to have strong association with initiation of dental caries. (Loesche et al., 1975; Loesche and Straffon, 1979). They are mainly isolated from the surface of teeth. They synthesize a variety of extracellular polysaccharides, including water-soluble and nonwater-soluble glucan and fructan from sucrose. These polysaccharides promote bacterial colonization and are key virulence factors in the formation of dental caries

Lactobacilli were more common from sites with soft and necrotic dentin (Schüpbach et al., 1996). Their role is still important in contributing to the proteolytic and collagenolytic activities associated with breakdown of the tissue. Recently, molecular analyses have been performed on carious dentin to describe more fully the microbial diversity of lesions. A range of lactobacilli, comprising 50% of the species detected, were identified (Nadkarni et al., 2004; Chhour et al., 2005).

Root surface caries have a different environment, and hence, a different flora can be expected. Gram-positive filamentous bacteria, especially *Actinomyces* spp., play a key role in root-surface caries. A close relationship between *A. odontolyticus* and the earliest stages of enamel demineralization and the progression of small caries lesions have been reported. The most important human pathogen is *A. israelii*.

7.5.1 *Streptococcus mutans*

S. mutans is a general term for several closely related species of *Streptococcus* originally described as different serotypes of *S. mutans*. The name *mutans* results from its frequent transition from coccal phase to coccobacillary phase. They are facultative anaerobes. The specific name, *S. mutans*, is now limited to human isolates previously belonging to serotypes c, e, and f. This is the most common species isolated from human dental biofilms. The next most prevalent species is *S. sobrinus* (previously *S. mutans* serotypes d and g), which has the potential to produce more acids from sucrose compared to *S. mutans*.

Mutans streptococci are catalase-negative, gram-positive spherical or oval cocci occurring in pairs and chains; 0.7–0.9 μm in diameter. Colonies of *S. mutans* grown on blood agar after 48-hour anaerobic incubation are either regular and smooth or irregular, hard, and sticky. The diameter of colonies is 0.5–1.0 mm. Zones of α- or γ-hemolysis can be observed around colonies (Xuedong and Yuqing, 2015).

7.5.2 Lactobacilli

Lactobacilli are gram-positive, with cocco-bacillary forms (mostly bacillary), alpha or nonhemolytic, facultative anaerobes. These organisms ferment carbohydrates to form acids (i.e., they are acidogenic) and can survive well in acidic milieu (i.e., they are aciduric); they may be homofermentative or heterofermentative. A special selective medium, tomato juice agar (pH 5.0), promotes the growth of lactobacilli while suppressing other bacteria. The question as to whether they are present in carious lesions because they prefer the acidic environment or whether they generate an acidic milieu and destroy the tooth enamel, has been debated for years—the classic "chicken and egg" argument (Samaranayake, 2006).

7.5.3 *Actinomyces*

Most *Actinomyces* are soil organisms, the potentially pathogenic species are commensals of the mouth in humans and animals. A number of *Actinomyces* species are isolated from the oral cavity. These include *A. israelii*, *A. gerencseriae*, *A. odontolyticus*, *A. naeslundii* (genospecies I and 2), *A. myeri*, and *A. georgiae*. They are gram-positive filamentous branching rods that are non-motile, non-sporing, and non-acid-fast. Clumps of the organisms can be seen as yellowish "sulfur granules" in pus discharging from sinus tracts, or the granules can be squeezed out of the lesions. The colonies resemble breadcrumbs or the surface of "molar" teeth. Sulfur granules in lesions are a clue to their presence. When possible, these granules should be crushed, gram-stained, observed for gram-positive and branchin filaments, and also cultured in preference to pus (Samaranayake, 2006).

7.6 Microorganisms of periodontal infections

Some common bacteria and their associations with diseases are mentioned as discussed in various studies and reports given in the literature. In 1968, Sigmund Socransky classified bacterial species into colored complexes. Bleeding-associated bacteria were named red complex. *P. gingivalis*, *T. forsythiam*, and *T. denticola* are the bacteria in this group, and they are also associated in deep periodontal pockets (Haffajee et al., 2006). The second is the orange complex which constitutes *F. nucelatum, P. intermedia, P. nigrescens, P. micros, S. constellatus, E. nodatum, C. showae, C. gracilus*, and *C. rectus*. *P. intermedia* along with *P. gingivalis* was found to occur in deep pockets. The third yellow complex include bacteria such as *S. sanguis, S. oralis, S. mitis, S. gordonii*, and *S. intermedius*. *Capnocytophaga* spp., *Eikkenella corrodens, Campylobacter concisus*. and *Actinobacillus actinomycetemcomitans* [serotype a] form the fourth green complex. *Veillonella parvula* and *Actinomyces odontolyticus* form the fifth purple complex.

A. actinomycetemcomitans was found to adhere the oral epithelium through a protein adhesion Aae, which binds to a carbohydrate receptors on buccal epithelial cells. It was also found that it migrates to gingival area and to the tooth surface. The bacterial fimbriae with extracellular carbohydrate polymer attaches on hard tooth surface. Sometimes coaggregation to other bacteria can be seen as a means of colonization. Among the six serotypes of *A. actinomycetemcomitans* a, b, c, d, e, and f and sero types a, b, and c are globally dominant (Brigido et al., 2014). Serotype b was found to be associated with localized aggressive periodontitis in American subjects (Zambon, 1985) and serotype a with chronic periodontitis. Sero type c was more commonly associated with periodontally healthy subjects. The isolation of such stereotypes is not similar in other countries (i.e., Korea, Finland, Japan, Taiwan, and Brazil) (Brigido et al., 2014). There is an association of this bacteria in localized aggressive periodontitis; progressive lesions are proved by many researchers in human and animal models. Increased leukotoxin and cytolethal distending toxin was also reported. Previous reports stated that a certain subset of species was not seen in samples of subgingival plaque and did not show enhanced antibody response (Loesche et al., 1992; Moore and Moore, 1994).

P. gingivalis are observed as black colonies in culture on blood agar plates. Burdon in 1928 has grouped such black/brown colony-forming organisms as *Bacteroides melanogenicum*. Deeper pockets have shown presence of serotype I, whereas II and III are isolated commonly from shallow pockets and gingivitis (Dahlen, 1993) In the Bacteroides group, the *P. gingivalis* group are saccharolytic (i.e., breaking down sugar products to produce energy), whereas *P. intermedia* and *P. melanogenica* are sacharolytic. *P. gingivalis* group of microorganisms are found to produce collagenase, hydrogen sulfide, indole, proteases (including those which breakdown immunoglobulins), gingipain, hemolysins, endotoxin, ammonia, and fatty acids. *P. gingivalis* inhibits polymorphonuclear leucocytes to diapedise through epithelium, and it has been demonstrated to influence production and destruction of cytokines by mammalian cells. Its absence in healthy gum or gingivitis and its presence in an advanced and destructive form of periodontal disease where there is increased pocket depth indicate its aggressiveness, behavior, and nutritional source. Immunization using hemagglutinin B, capsular polysaccharide, heat shock protein, gingipain R, and the active sites of RgpA and Kgp proteinases have found to reduce alveolar bone loss in mouse models. And their action was found to raise the level of specific antibodies to *P. gingivalis* antigens.

In 1979, Tanner in 1979 described *T. forsythia* as "fusiform" *Bacteroides*. It requires two weeks to form minute colonies. It is a gram-negative anaerobic, spindle-shaped, highly pleomorphic rod. It is seen commonly in subgingival sites as reported by Socransky et al. (1998). In electron microscopy, this bacteria is found to exhibit a serrated S layer, which was found to mediate hemagglutination, adhesion of epithelial cells, and demonstrate subcutaneous abscess formation in rats. This layer was found to possess two types of glycoproteins of different molecular mass. Along with macrophages and epithelial cells, if these bacteria are cultured, it was found to express proinflammatory cytokines, chemokines, PGE2, and MMP9 as reported by Bodet et al. (2007).

Higher counts and prevalence of this microorganism was found in individuals with different forms of periodontitis than in healthy individuals (Yang et al., 2016; Haffagee et al., 2009; van Winkelhoff et al., 2016). Presence of this microorganism was also demonstrated in progressing active periodontal lesions and dormant lesions (Dzink et al., 1989). Machtei et al. (1992) stated that the subjects with bone loss, attachment loss, and tooth loss have harbored *T. forsythia* compared to subjects who are free of disease. Elimination of this microorganism was found to reduce in counts after scaling and root planning. *T. forsythia* was also found to be the most commonly isolated from periodontal pocket epithelial cells (Dibart et al., 1998).

Spirochetes are gram-negative, highly motile microorganisms commonly seen in destructive periodontal diseases. The etiology of acute necrotizing ulcerative gingivitis was strongly related to spirochetes because large numbers were seen in tissue specimens after biopsy of the infected sites (Listgarten and Socransky, 1964; Listgarten, 1965). Their role is not clear in other forms of periodontitis. However, their occurrence in subgingival plaque, deeper pockets, and severe disease states were demonstrated (Moore and Moore, 1994; Haffagee et al., 2006) Among the virulence factors, dentilisin, coded by prtP gene, affects many protein substrates like fibronectin, fibrinogen, and laminin (Ishihara, 2010).

P. intermedia and *P. nigrescens* are part of the black pigmented *Bacteroides* group of microorganisms that are gram-negative, anaerobic rods. Elevated levels are seen in specimens isolated in acute necrotizing ulcerative gingivitis (Loesche et al., 1982). Some authors have reported elevated levels in some forms of periodontitis (Moore and Moore, 1994; Papapanou 1996).

7.7 Pathogenesis of dental caries and periodontal diseases

7.7.1 Dental caries

Dental caries is the localized destruction of susceptible dental hard tissues by acidic by-products from bacterial fermentation of dietary carbohydrates. Its pathogenesis is best explained by the Keyes tetrad, which states that dental caries can occur on a susceptible host only when both acidogenic bacteria and a carbohydrate-rich diet are present for a definite period of time (Figure 7.4).

The carious process begins with the tenaciously adhered biofilm on the tooth surface expressing acidic metabolites that dissolve the hydroxyapatite of enamel. *S. mutans* are largely responsible for initiating caries due to their exceptional ability to produce extracellular polysaccharides, which are the building blocks for the biofilm. They can also metabolize a number of sugars and glycosides such as glucose fructose sucrose, lactose, galactose, mannose, cellobiose, glucosides, trehalose, maltose group of sugar alcohol. *S. mutans* synthesize intracellular glucose and sucrose polysaccharides and also produce mutacins (bacteriocins) what is considered to be

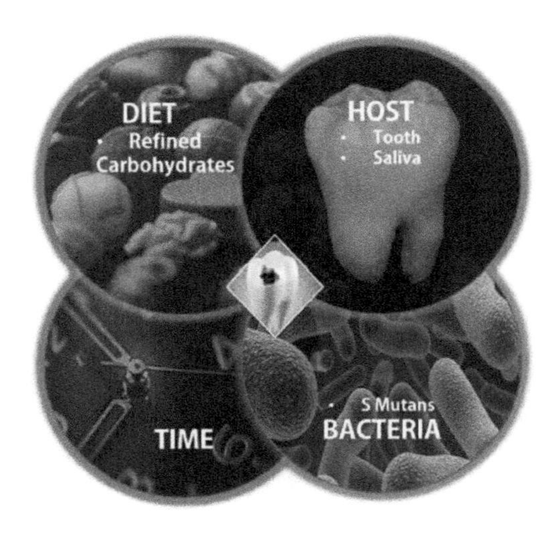

Figure 7.4 Keyes Tetrad—Interplay of four factors for caries pathogenesis.

important factor in the colonization and establishment in the biofilm; these biofilms produce acids from carbohydrates that result in caries (Karpinski and Szkaradkiewicz, 2013).

Gradual demineralization of involved dental hard tissues will be active because of the disturbance in physiologic equilibrium in the biofilm of dental plaque covering the affected site. The bacteria of the genus lactobacillus are the important factor for further progression of dental caries, especially in dentin (Figure 7.5). The caries process consists of three reversible stages. The microflora ions on sound enamel surface contains mainly nonmutants streptococci and actinomyces in which acidification is mild and infrequent; this is compatible with the equilibrium of the demineralization and remineralization balance or shifts the mineral balance toward net mineral gain (dynamic stability stage) when sugar is supplied frequently; acidification becomes moderate and frequent, which may enhance the acidogenicity and acidurance of the non-mutans adaptively. These microbial acids induce the adaptation and selection process and may over time shift the demineralization and remineralization balance toward net mineral loss, leading to progression of dental caries (i.e., acidogenic stage). Under severe and prolonged conditions, more aciduric bacteria becomes dominant through aciduce selection by temporary acid impairment and acid inhibition of growth (i.e., aciduric stage).

The organic components of dentin do not allow acids to cause caries like lesions in dentin, rather the appearance of such lesion requires activation of enzymatic proteolysis by metallo-proteinase (MMPs) in mildly acidic

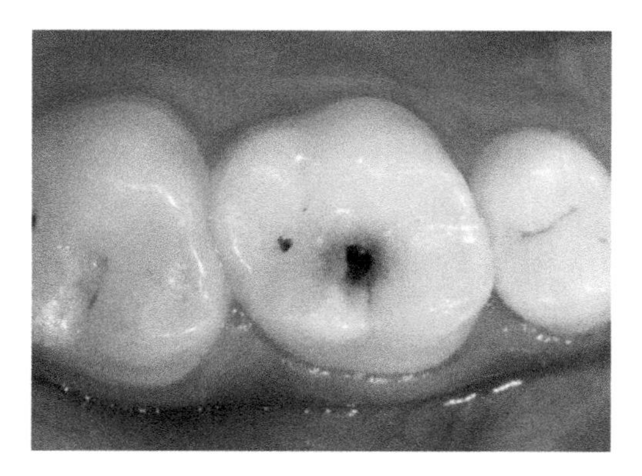

Figure 7.5 Caries progression showing breakdown of tooth structure with discoloration.

SUSCEPTIBLE TOOTH SURFACE

FORMATION OF BIOFILM AND MOLECULAR DEPOSITS

ACID PRODUCTION AND CHANGES IN PH

SHIFT IN DYNAMIC EQUILIBRIUM OF MINERALS

DISSOLUTION OF MINERALS

INITIATION OF DENTAL CARIES

Figure 7.6 Flow chart showing development of dental caries.

condition that leads to cavity formation. It constitutes a large family of calcium-zinc dependent endo-peptidase that contributes extracellular matrix degradation (Femiano et al., 2016) (Figure 7.6).

7.7.1.1 Remineralization–demineralization cycle – The etiological agents of dental caries are *S. mutans* and lactobacillus. These bacteria can generate acids from fermentable carbohydrates. The level of infection that causes damage to the teeth depends on multiple factors. Acids generated by these bacteria diffuse into the subsurface of the enamel and can dissolve calcium and phosphate, which then diffuse out of the tooth; this process is called demineralization.

The presence of healthy saliva will provide buffers so that calcium and phosphate will reverse the early damage caused by demineralization. Due to this repair, the salivary buffers must first neutralize the acid and stop the demineralization. When the calcium and phosphate concentration becomes higher outside the tooth than inside, they will diffuse back into the tooth. This reversal process is called remineralization.

7.7.1.2 Zones of infection – In 1935, Fish attempted to explain the pathogenesis of periradicular inflammation using an animal model (Figure 7.7). He inoculated bacteria into the pulp chamber of virgin teeth and observed the following items.

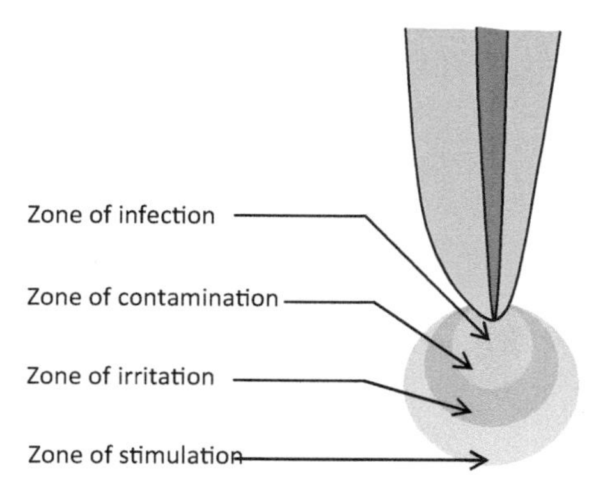

Figure 7.7 Zones of periapical lesion.

A central zone containing pus, polymorphonuclear leukocytes (PMNL) and microorganism are known as a zones of infection.

Zone of contamination are seen around the central zone characterized by the presences of PMNLs and macrophages. The toxins from the bacteria are diluted, and the inflammatory fluid contains antibacterial action. There was an empty lacunae created by autolysis of dead bone cells.

Zones of irritation contain macrophages, lymphocytes, and plasma cells, mediators of inflammation and immune system collagen matrix degraded by the macrophages; the bone is resorbed, leaving a small space. This space is filled with granulomatous tissue, and this prevents the spread of necrosis and initiates repair because it contains new capillaries and fibroblast russell bodies.

Zone of stimulation is the peripheral zone characterized by fibroblast and osteoblast. The fibroblasts lay down collage fibers, creating a wall of defense around the zone of irritation on which osteoblasts reside to deposit new bone in irregular fashion.

7.7.1.3 Root caries – Root caries, which is seen apical to the cemento-enamel junction, is generally characterized by a soft active progressive lesion. Etiology is gingival recession as a result of periodontitis, age, radiation therapy, xero-stomia, abrasion, erosion, abfraction, primary root caries, recurrent caries, or diabetes. Microorganism responsible for root caries are *S. mutants*, Lactobacillus, and Actinobacillus. Root caries is mostly seen when there is periodontal ligament attachment loss.

This exposes the root surface to the oral environment, which leads to infiltration of caries. They appear as a white or discolored, irregular, and progressive lesion (Zaremba et al., 2006).

Histologically, dental caries have been characterized as having two distinct layers: the outer zone (stains with caries detector dyes) where the dentin is highly demineralized and the collagen is denatured and heavily infected with bacteria (often referred to as the infected zone), and the inner zone (does not stain with a caries detector dye) where the dentin is demineralized but the collagen is intact and minimally infected (often referred to as the *caries affected zone*).

7.7.1.4 Intraradicular infections – Endodontic infections have a polymicrobial nature and obligate anaerobic bacteria conspicuously dominate the microbacterial factors in primary infections. The endodontic microbiota presents a high interindividual variation that can significantly vary in species diversity and abundance from individual to individual, indicating that apical periodontitis has a heterogeneous etiology, and multiple bacterial combinations can play role in diverse causation. As the breadth of bacterial diversity in endodontic infection has been unraveled by the molecular biology methods, the list of endodontic pathogens has expanded to include several cultivable and as-yet uncultivated species that had been underrated by culture-dependent methods. Endodontic bacteria are now recognized to belong to 8 of the 12 phyla that have oral representatives namely firmicutes, bacteroides, spirochetes, fusobacterial, actinobacteria, and proteobacteria. Studies have demonstrated that predictable disinfection of the root canal system is only achieved after proper antimicrobial medications are placed in the canals and left therein between appointments.

7.7.1.5 Extraradicular infection – An apical periodontitis lesion is formed in response to intraradicular infection and comprises an effective barrier against the spread of the infection to the alveolar bone and the other body sites. The most common form of extraradicular infection is the acute apical abscess characterized by the purulent inflammation in the periapical tissue in the root canal. Acute abscesses are usually characterized by absence of over symptoms. Extracellular infection can be dependent on or independent of the intraradicular infections. Once the intraradicular infection is properly controlled by root canal treatment or tooth extraction and drainage of pus is achieved, the extraradicular infection is handled by the host defense and usually subsides. There are some situations that permit intraradicular bacteria to reach the periapical tissue and establish an extraradicular infection.

1. The infection may be a result of direct advance of some bacterial species that overcome host defenses concentrated near the apical foramen or that manage to penetrate into the lumen of pocket cyst, which is in direct communication with the apical foramen.
2. The infection may be due to bacterial persistence in the apical periodontitis lesion after remission of acute apical abscess.
3. The infection may be a sequel to apical extrusion of debris during root canal instrumentation after overinstrumentation. The virulence and the quantity of the involved bacteria as well as the host ability to deal with the infection (Figure 7.8).

7.7.2 Periodontal disease

Periodontitis, commonly known as pyorrhea, is a complex disease entity of the oral cavity, affecting the supporting tissues of the tooth. The disease process is likely to initiate at the junction where the tooth emerges out into the oral cavity.

The number of species of pathogenic bacteria present may not be directly related to the periodontal infections. Even microorganism and site specificity is not directly related to the pathology or course of infection (Socransky et al., 1987). It was demonstrated that even an otherwise healthy gingiva or

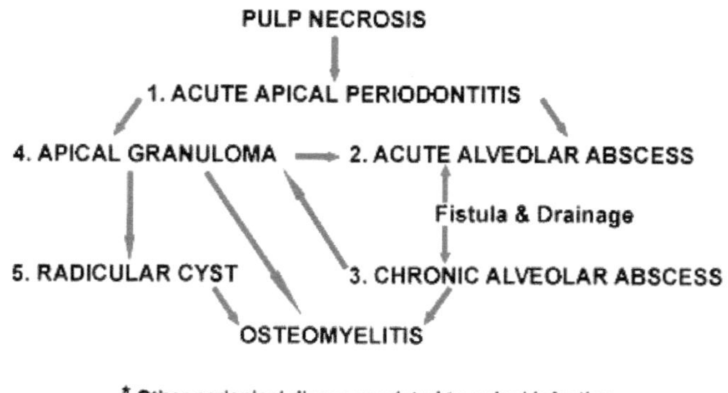

Figure 7.8 Flow chart showing development of pulpal and periapical lesions.

periodontium could harbor pathogens but still could not establish any kind of disease pattern. However, the reduction of periodontal pathogens has been used as a parameter in assessing the positive outcome of periodontal therapy. Fastidious, strict anaerobes may cause initiation of periodontal disease in a shallow periodontal pocket or progression of disease in a deep pocket to some extent. The plaque that is accumulated may be supragingival on the crown surface or subgingival in the sulcus or on the tooth surface root. The nature of microbiota would be different in the aforementioned two situations, which would be based on the partial pressure of oxygen in the environment, depth of the periodontal pocket, redox potential, availability of excretory products from sulcular fluid, and blood products.

7.8 Diagnostic characteristics

7.8.1 Dental caries

Dental caries present as black to black brown discoloration primarily in pit and fissure. It can be either a cavitated or a noncavitated lesion. Noncavitated lesions are not so easy to identify because a physical examination tool like an explorer can push the microorganisms deeper when it is used. Advanced diagnostic tests are made available to disclose the presence of lesions. Smooth surface caries occur on cusp tips and tooth surfaces that may be hypocalcified white spots, which are often marred by the presence of salivary film.

Diagnosis is made possible by drying the tooth surface and examination under good light source. Caries can also be diagnosed by radiographs like intraoral periapical or bitewing radiographs. They present as radiolucent lesion extending from the deepest part of the fissure in a triangular fashion with their base at the dentinoenamel junction. Smooth surface caries break down is also triangular, but their base is at the surface and apex toward the pulp at the dentinoenamel junction. Diagnosis of caries is important for planning the treatment.

Cervical caries and root caries can also be seen in individuals who are geriatric. Nursing bottle caries are common in bottle-fed infants where the milk bottle is left in the mouth. Caries are seen in maxillary central incisors. Rampant caries, yet another entity, presents as caries in the proximal areas of the teeth, and they seem to spread from one tooth to the other, which is common in individuals who neglect their oral hygiene or who have reduced salivary secretion. Reduced salivary secretion is yet another factor for caries. This may be due to any salivary gland pathology, autoimmune conditions of salivary glands like Sjogren syndrome, after radiation, use of antihypertensive, antipsychotic, antineurotic drugs, diabetes, salivary duct obstruction, increased antibiotic usage, and many others.

Caries risk-assessment analysis can be performed using a systematic charting that includes dietary history, genetic factors, and other local presentations (Hillman, 2002). Salivary buffer capacity by analyzing the pH of saliva can tell the risk range from high to low. Such analysis can help the clinician understand the nature of the disease and its severity and to develop preventive measures to preserve, prevent, and restore the tooth. Dental caries can progress to pulp and result in pulpal inflammation, which can be either acute or chronic and is based on the symptoms. It can be classified as reversible or irreversible pulpitis based on the nature of symptoms. Irreversible pulpitis needs a root canal. Irreversible pulpitis can progress to spread the infection beyond the tooth to the surrounding bone where it results in periapical abscess. The abscess can consolidate to a granuloma or spread to the surrounding areas and into potential spaces of the face and oral cavity, resulting in serious infection. Sometimes, the abscess can divert its course to the path of least resistance and drain to the oral cavity in the form of a sinus opening. Granuloma can be a chronic presentation or can progress to a cyst with breakdown of the cells and development of lining of the cavity filled with fluid. The cyst related to the tooth is termed a *radicular cyst* and can either progressively increase in size, expanding to the bone, or can present as a symptomatic painful swelling if it is infected.

All the chronic destructive lesions like granulomas, cysts, or chronic abscesses can be diagnosed by only radiographs. Cysts are characterized by a radiopaque border surrounding a radiolucent lesion, whereas granulomas and abscesses do not have a radiolucent border. The tooth becomes non vital in irreversible pulpitis, abscess, granuloma, and cystic lesions.

7.8.2 Periodontitis

Gingivitis, the inflammation of the gingiva, presents with redness, swelling, bleeding on probing, and altered contour. The inflammation can be either acute or chronic based on the presentation of symptoms.

Periodontal disease presents as bleeding on probing when it affects the gingiva and the superficial soft tissues in an acute phase. Several Indices are used to detect the health of gingiva (i.e., Ramjford's, Sillness and Loe, and Loe and Sillness). Noninflammatory enlargement of gingiva may be as a result of chronic use of certain drugs like phenytoin sodium, nifedipine, and certain immunosuppressants. Severe bleeding may be due to vitamin C deficiency (i.e., scurvy) or acute myeloid leukemia. Linear erythema of marginal gingiva with bleeding is noted in HIV. Gingival inflammation can also be present in acute necrotizing ulcerative gingivitis and herpetic gingivostomatitis. Gingival enlargement may be due to benign or malignant lesions.

Bleeding upon slightest provocation along with clinical signs and symptoms of surface color and texture absence of stippling will denote reactive inflammation. Breakdown of junctional epithelium and deepening of sulcus is known as a periodontal pocket, which is measured using a probing instrument such as a William's periodontal probe or Community Periodontal Index for Treatment Needs (CPITN) probe, which has got a blunt end. Bone loss can be appreciated with radiographs (i.e., orthopantomography or intra-oral periapical).

Bone-loss patterns commonly observed are craters, both horizontal and vertical in nature. Bone sounding with deep probing can be done to perceive the configuration of the defect. Confirmation is done with radiographs. Treatment varies depending on the type of bone defect. Periodontal inflammation can progress to destruction of periodontal cells and accumulation of inflammatory exudate. Periodontal abscess can result in the destruction of the apical area of the tooth or the lateral surface of the tooth. Clear diagnosis can be made if the abscess is in the lateral surface of the tooth and is sensitive to lateral percussion with handle of a mouth mirror.

Apical periodontitis is often confused with periapical abscess. Clinical correlation is of utmost importance to diagnose whether it is a periodontal or an endodontic infection. The periodontal abscess can also take the course similar to periapical abscess. Multiple periodontal abscesses and weak periodontium are characteristic of diabetes mellitus. Other diseases like neutrophil dysfunction (Chedak Higashi Syndrome), palmoplantar keratosis (Papillon Levfre Syndrome), hypophosphatasia can be associated with a typical form of periodontitis called as *aggressive periodontitis*. It can also occur in silos as a separate entity. This periodontitis is formerly known as *juvenile periodontitis* and presents as early loss of teeth and characteristic bone-loss patterns in a young patient in the molar and incisor areas. Refractory periodontitis is another type of periodontitis where the lesions do not respond to conventional modality of treatments. Periodontal diseases are also found to be aggravated by the presence of systemic diseases and vice-versa where it is understood as a reversal of paradigm in understanding lesions accompanied by systemic diseases. The clinician's role is to correlate the findings and treat the patient appropriately.

7.9 Treatment strategies

7.9.1 Dental caries

The treatment is based on the type of lesion and presenting condition.

The use of nonoperative prevention-oriented strategy for caries management is often put aside in favor of restoration placement. One approach is caries management by risk assessment (CAMBRA). The goal is to determine if caries is reduced in high-risk patients receiving treatment with a nonoperative anti-caries agent. The general approach to active caries should be preventive treatment like advice to take less sugar, brushing twice daily with effective fluoride toothpaste, using dental floss, and stimulating saliva by use of sugar-free gums such as xylitol. When active fissure caries have been diagnosed, and fissures are with susceptible morphologic characteristics, sealant with a low-filled resin is indicated.

The operative management of dental caries has traditionally involved removal of all soft demineralized dentin before filling is placed. This tissue is heavily infected with bacteria is removed slowly using a slow speed bur or hand-excavating instruments. There are five basic reasons to place restorations when cavitation occurs as a result of caries:

- To remove infected dentin,
- To protect the pulp and avoid pain,
- To remove the habitat for cariogenic bacteria, and
- To facilitate plaque control.

The primary goal is to regain the lost form, function, and esthetics. In deep carious lesions, direct or indirect pulp capping procedures are recommended to preserve vitality of pulp.

Deep carious lesions can penetrate the root canal system and cause an array of problems for the patient. The complex anatomy of the root canal system forms the ideal playground for microorganisms. They can lodge in the narrow crevices and cause periradicular inflammatory responses. This in turn can cause a sequelae of events

7.9.2 Endodontic treatment

Treatment of a vast majority of endodontic infections include nonsurgical or surgical root canal. The infected root canals are shaped and cleaned until the microorganism load is brought down, and they are filled three dimensionally to achieve a hermetic seal.

7.9.3 Periodontal disease management

Many interesting anecdotes were used previously in the management of periodontal disease. Pierre Fauchard, a French physician known as the father of modern dentistry, advocated techniques to remove deposits using his specially developed instruments. As early as 1929, Rasmussen reported that ultraviolet light was found to reduce the overall bacterial load, thus reducing the oral microbiota. Caustic chemicals have been used to control infections. Some felt that the microorganisms played only a secondary role in the progression of the disease. They attributed the presence and progress of infections to the individual's constitutional defect or a traumatic occlusion. Early researchers also proposed the role of a mixed group of microorganisms causing infections. They found patterns that could cause infections like fuso-spirochetal combination in Vincent disease or Trench mouth. After the mid-1950s, the role of biofilm or plaque and not just a single bacteria was found to be causative for periodontal infections. The treatment strategy also changed accordingly. Plaque was mentioned as a biomass (i.e., nonspecific plaque hypothesis) irrespective of the pathogenicity of some potential microorganisms causing infections (Theilade, 1986). Specific plaque hypothesis stated to tackle only the specific bacteria related to the infections (Loesche and Straffon, 1979). The ability of certain bacteria like *Actinomycosis viscoses* to produce infection in rats free of disease, and experiments such as those strengthened the theory of specificity (Figure 7.9). Drug therapy can be delivered in the form of oral antibiotics or local drug delivery in the form of a miniature wafer or chip using metronidazole and tetracycline. They can be used as adjuvants apart from active mechanical debridement, removal of granulation tissue, and robust maintenance.

The cause of infection due to endogenous or exogenous organisms can also alter the treatment strategy. An opportunistic infection caused by an endogenous organism, which may be a normal commensal in oral cavity, may resolve if the ecologic and immune response is under control, which regulates the bacterial growth and multiplication (Takahashi, 2011). On the other hand, the mere presence of some uncommon microorganisms (i.e., exogenous) can

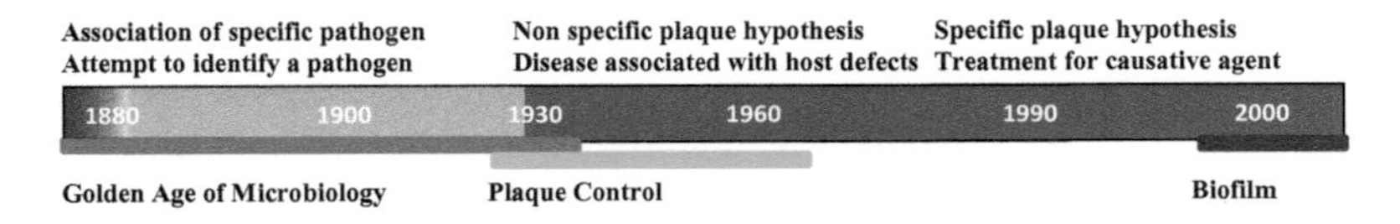

Figure 7.9 Diagramatic representation of various hypothesis.

trigger and complicate the course of infection. In such cases, therapy should be aimed to prevent exposure and elimination of the microorganisms.

It is also vital to understand the risk factors associated with periodontal disease. Some are modifiable such as lifestyle factors (e.g., smoking and alcohol consumption) and some are diseases or conditions like diabetes, obesity, osteoporosis, metabolic syndrome, low calcium in diet, and vitamin D. Managing periodontitis includes control of these modifiable risk factors. Genetics plays a major role in aggressive periodontitis. However, association of a genetic factor with chronic adult periodontitis remains unclear.

The most important phase in periodontal therapy is the maintenance phase where the relapse of the disease can happen after treatment. Motivation to practice and maintain oral hygiene measures are crucial in the prognosis of the disease after treatment. Goal setting, self-monitoring, and planning are found to be effective in improving oral hygiene-related behavior in such patients (Newton and Asimakopoulou, 2015). Various psychological models were proposed to handle periodontal infections after therapy with behavior modification. In a systematic review, all the models were found to play some role in the reduction of plaque and the increased motivation for the initial time period. But no difference was noted at 6 or 12 months follow-up.

The mechanical removal of biofilm is mandatory rather than trying to fight against the microorganisms with antimicrobials and antibiotics. Mechanical debridement can be carried out using hand instruments like scalers. These instruments are modified to various shapes to access the inaccessible areas. Machine-driven instrumentation like the use of an ultrasonic vibration in scaler tips can also effectively remove plaque and tenacious deposits of calculus. The surface of the root is covered by a thin cementum layer. Deposits on this surface along with the granulation tissue should be removed by root planing. The diseased cementum should be removed and the soft tissue approximated back, expecting for healing with long junctional epithelium formation. This procedure was also found to reduce the pocket that resulted because of the disease process.

The instruments mentioned can remove the deposits on the tooth and root surfaces, thus restoring the periodontium to health. The loss of bone and the attached epithelium may not be possible to recover to its original state. With advanced periodontal surgical therapy, guided tissue regeneration and use of graft and other scaffolds can restore the attachment loss to some extent.

Attachment of deposits to the prosthesis intraorally can be mechanically debrided using the mentioned instrumentation. Fabrication of highly glossy surfaces and precision in restorations and replacements can prevent bacterial adhesion. As the adage goes, "Prevention is always better than cure." Early recognition of disease and timely visits to the dentist can prevent complications associated with the disease progression.

7.10 Oral health and systemic diseases and conditions

Several systemic diseases or conditions can increase the severity of gingivitis and periodontitis. Diabetes, HIV, and some drugs like phenytoin, cyclosporine, immunosuppressants, and nifedipine were found to influence the prognosis of bacterial infections and gingival health in individuals. Smoking and mouth breathing may also trigger gingival and periodontal disease and complicate bacterial infections. Prognosis of patients with periodontitis is poor in patients with poorly controlled diabetes.

7.10.1 Reversal of paradigm

Specific pathologic conditions of the oral cavity like caries or periodontal disease can act as foci of infection to trigger systemic conditions elsewhere in the body. It can affect vital systems like the cardiovascular, renal, and endocrine systems. Certain bacteria like *P. gingivalis* is found to increase the formation of atheroma in blood vessels, causing atherosclerosis and thereby increasing the risk of myocardial infarction (Humphrey et al., 2008). It is also shown that increase in bacterial foci of infection can reduce the insulin uptake by the cells and cause poor glycemic control. Its effect was also shown to precipitate obesity, preterm low birth-weight infants, glomerulonephritis, and bacterial endocarditis.

Dental procedures may carry the bacterial load through the bloodstream elsewhere and trigger an inflammatory response in certain targeted tissues. The bacterial endocarditis is an acute condition that can be triggered by an extraction or a simple invasive procedure in patients with artificial valves. The bacterial colonies may aggregate in the artificial valves, causing a local reaction via an inflammatory response and a systemic response (Sen et al., 2017).

Oral periodontopathic bacteria were found to be aspirated by some patients, resulting in aspiration pneumonia. An association between poor oral hygiene and chronic obstructive pulmonary disease was also noted. Periodontal disease-associated enzymes in saliva were found to modify the mucosal surface to promote adhesion and microbial colonization by pathogens lining the respiratory tract (Scannapieco, 1999).

A report of three unrelated patients with interleukin receptor-associated kinase (IRAK) deficiency are resistant to pyogenic infections as opposed to healthy children who developed pyogenic infections (Picard et al., 2008). Toll interleukin receptor (TIR) mediates the recruitment of IRAK complex. This TIR-IRAK signal pathway is found to be important for immunity against some bacteria, but their role is negligible against other microorganisms. In a case control study of 150 patients who were pregnant or who were postpartum revealed that periodontal disease can be significant risk factor of preterm low birth-weight infants (Messinis et al., 2010).

In a meta-analysis, it was found that with active intervention of periodontal therapy one cannot reduce the incidence of preterm low birth-weight infants (Polyzos, 2010). However, pregnant women should undergo periodontal screening and avoid any oral foci of infection. They have also mentioned that quality trials provide strong evidence of no correlation to exist, whereas randomized clinical trials have overestimated the effect of treatment (Offenbacher et al., 1996; Khader and Ta'ani, 2005).

A positive association was seen between bacterial infections and respiratory infection. In a review, there was a mention of possible mechanism of pathogenesis like bacterial aspiration (*P. gingivalis* and *A. actinomycetemcomitans*), modification of respiratory mucosa by enzymes of breakdown products, and cytokines released from periodontal diseases (Scannapieco, 1999). Periodontal and gingival bacterial super infections of individuals affected with HIV was found to present a similar picture to those of individuals who were immune suppressed. In a study done on 31 intact teeth affected by pulp and marginal infections, polymerase chain reaction was used to isolate pathogens like *A. actinomycetemcomitans*, *B. forsythus*, *Eikenella corrodens*, *F. nucleatum*, *P. gingivalis*, *P. intermedia*, and *T. denticola* (Reichart, 2003). The organisms were present in the chronic apical periodontitis and chronic periodontitis. This was suggestive that any refractory course of endo-perio lesions may be attributed to the endodontic and periodontic pathway of infections (Rupf et al., 2000).

Presence of *Helicobacter pylori* in the oral cavity was noted in patients with gastric infection. They have also commented that the oral cavity is the first extragastric reservoir.

Some studies have proven association between bacterial infection and ischemic cerebrovascular disease (Syrjanen et al., 1989). An extremely rare case of brain abscess was found to occur with the causative being *P. gingivalis*, which was identified with culture methods. The bacterial source was found to be from the oral cavity as the patient suffered from recurrent periodontitis (Yoo et al., 2016).

Association of periodontal disease with aortic atheromas was also found. A systematic review and meta-analysis showed that periodontal disease is a strong risk factor for ischemic heart disease, which was found to be independent of other risk factors (Sen et al., 2017). In a systematic review, an association was noted between obesity and periodontal status (Haffajee and Socransky 2009). Although, periodontitis as a risk factor for obesity could not be established because there were other events that could be associated. In clinical practice, a high prevalence of periodontal disease can be noted with individuals who are obese (Chaffee and Weston, 2010).

In another meta-analysis, it was found that Type II diabetes was found to be associated more with destructive periodontitis, than Type I (Chavvary et al., 2009).

In a meta-analysis with 456 patients, a positive association was found in patients with reduction of glycemic index with active periodontal treatment in patients with diabetes. However, none of the reduction was statistically significant (Janket et al., 2005).

7.11 Miscellaneous: Diseases and their possible associations

Tonsillitis and pharyngitis are caused by *beta hemolytic streptococci*. Severe purulent forms are associated with *Staphylococcus*. Scarlet fever caused by streptococci is characterized by reddish diffuse skin rashes of face and neck. Hemolytic property of streptococci can present as petechiae (small hemorrhagic spots) on the palate or strawberry tongue. *F. necrophorum* was also isolated in one study. More studies were recommended to confirm their association (Pallon et al., 2018).

Syphilis caused by *Treponema pallidum* can present itself as various presentations based on the stages of syphilis. Transmitted through the genito-oral route commonly, the primary lesion occurs at the site of inoculation (usually the lip as an ulcer, which is called the *chancre*). Intraoral presentation of the ulcer is painful and covered by greyish white membrane. It is a highly infectious stage and heals usually in 3–4 months. Secondary stage appears 6 weeks after primary lesion if left untreated. Diffuse eruptions of skin and mucous membrane are noted as macules. A thick papule may be noted as a greyish plaque over an ulcerated surface. They are seen commonly in tongue, gingiva, and buccal mucosa. Snail track ulcers are characteristic of the secondary stage. Tertiary stage appears after 6 months. They are not infectious and may present as perforations of mucosa of oral cavity as a punched-out ulcer called a *gumma*. Sloughing of tissues are seen at the base of the ulcer. Active antibiotic therapy is recommended to treat the infection. Gonorrhea, caused by *Neisseria gonorrhea*, is a genitourinary bacterial infection. Transmitted through the genito-oral route, the oral lesions may be seen in mucosa of the tongue as red, dry lesions, which are shiny and painful erosive ulcers.

Tuberculosis caused by *Mycobacterium tuberculae* is characterized by granulomatous inflammation presenting as a painful nonhealing ulcer in the intraoral presentation along with other general characteristics. Chronic active treatment is recommended with antibiotics for a longer period of time (e.g., up to 12–24 months). Systemic spread is dangerous and generalized lymphadenopathy with presentation on the skin as "scrofula" is characteristic of tuberculosis.

Rhinoscleroma, caused by *Klebsiella rhinomatosis* is a rare bacterial infection characterized by reactive proliferative nodular presentation of skin and mucosa of nose lacrimal gland and sinuses. Oral lesions present as ulcer of palate and attached gingiva. It may also present with lymphadenopathy.

Actinomycosis israelii is found to be associated with cervicofacial skin and bone lesions or tongue infections with pus collection. Sulfur granules in the exudate seem to be characteristic of this infection. Long-term high doses of antibiotics are recommended to handle this infection.

Clostridium tetani can cause tetanus, which presents symptoms like lockjaw due to a spasm of the masseter muscle. Risus sardonicus—a condition due to spasm of rhizorius—the grinning muscle is characteristic of this disease.

7.12 Future perspectives

The oral microflora exists as commensal in the oral cavity, which benefits the host as well. There is a symbiotic relationship between the host and the flora that may be identified in the disease process as dysbiosis (Marsh et al., 2014). Identification of factors that can affect the symbiosis, reduction, and elimination of such factors can be the modality of treatment. Oral antimicrobial therapy should be aimed to modulate the activity and growth of oral pathogens. Silico models on biofilms have shown how sublethal concentrations of agents that mildly fluctuate the pH of the oral environment can alter the competitiveness of the bacterial colonies. Buffering capacity of the plaque film can modify to large extent the composition of microorganisms in the biofilm. Elimination of microorganisms to render a disease-free state is almost next to impossible. Maintaining the normal commensals in the patient's mouth, deriving their benefits, maintaining the pH, and buffering capacity of the oral cavity through a holistic approach may prevent dysbiosis. A holistic approach and systemic thinking have also been proposed by many others in handling oral infection (He et al., 2009) Interspecies interaction, microbial community, and microbial interactions in polymicrobial diseases may need to be further researched to obtain insight into the etiology of endodontic and periodontal infections. The role of bacteria was also questioned by many authors, which has been reviewed systematically and have shown that *Streptococcus angiosus*, *Capnocytophaga gingivalis*, *Prevotella melanogenica*, *Bacteroides fragilis*, and *Streptococcus mitis* may have an association with carcinogenesis. Possible mechanisms like inhibiting apoptosis, activating cell proliferation, promoting cellular invasion, and inducting chronic inflammation may set a predisposition for carcinogenesis. However, the author states that in all the studies mentioned, the methodology standardization is questionable because there is a wide variation in tissue sampling, microbial profiling, and DNA extraction (Perera et al., 2016).

The human body has coevolved with the microorganisms and has constantly been modified by lifestyle, diet, habits, and psychological factors. At some point in the state when the balance is lost, the disease process sets in as the immune mechanisms set in an inflammatory pathway, which leads to destruction of the tissues. The human body was mentioned by halobiont or the superorganism where the human being coexist with the commensals by adaptation and functional integration.

With evolution, *S. mutans* has developed a mechanism to compete against other oral species. It has developed defenses against increased oxidative stress and remained resistant to by-products of its own metabolism. Worldwide promotion of toothbrush and flossing have also made considerable modification in oral environment, making resistant biofilms loaded with microorganisms which turn pathogenic. Virulence and pathogenic potential in the recent decades are understood by quorum sensing, which has opened a new avenue for research. The host substrate are the mucosa and the teeth. The mucosa sheds continuously and does not provide a stable home for the microorganisms, whereas the tooth and artificial crown or replacements and restorations can serve as a stable substrate to harbor biofilms, which become more complicated with aforementioned factors.

In the recent age, after the launch of Human Microbiome Database and 16s ribosomal RNA gene community profiling, identification of a bacteria in its original or a variant can be done by simple matching with the available database. Next-generation sequence (NGS) coding tools can allow a high-volume study of genetic material in samples to provide better and specific understanding of oral microbiome.

Treatment should be focused in future in prevention strategies, awareness of the public in the role of diet, lifestyle, and habits on oral health (Kilian et al., 2016). Awareness of effective plaque control measures on tooth surfaces, restorations, replacements, and implants can also minimize the bacterial biofilm from nonshedding surfaces of the oral cavity to an acceptable level, which is compatible to oral health.

References

Aas, J. A., B. J. Paster, L. N. Stokes, I. Olsen, and F. E. Dewhirst. 2005. Defining the normal bacterial flora of the oral cavity. *Journal Clinical Microbiology* 43 (11):5721–5732.

AlJehani, Y. 2014. Risk factors of periodontal disease: Review of the literature. *International Journal of Dentistry* 20:1–9.

Armitage, G. C. 1999. Development of classification system for periodontal diseases and conditions. *Annals of Periodontology* 4(1):1–6.

Bodet, C., E. Andrian, S. Tanabe, and D. Grenier. 2007. Actinobacillus actinomycetemcomitans lipopolysaccharide regulates matrix metalloproteinase, tissue inhibitors of matrix metalloproteinase, and plasminogen activator production by human gingival fibroblasts: A potential role in connective tissue destruction. *Journal of Cellular Physiology* 212 (1):189–194.

Brigido, J. A., V. R. S. Silveira, R. O. Rego, and N. A. P. Nogueira. 2014. Serotypes of aggregatibacter actinomycetemcomitans in relation to periodontal status of geographic origin of individuals—A review of literature. *Medicina Oral Patologia Oral y Cirugia Bucal* 19 (2):184–191.

Chaffee, B. and S. J. Weston. 2010. Association between chronic periodontal disease and obesity: A systematic review and meta-analysis. *Journal of Periodontology* 81 (12):1708–1724.

Chavvary, N. G. M., M. V. Vettore, C. Sansone, and A. Sheiham. 2009. The relationship between diabetes mellitus and destructive periodontal disease: A meta-analysis. *Oral Health Preventive Dentistry* 7:107–127.

Chhour, K.-L., M. A. Nadkarni, R. Byun, F. E. Martin, N. A. Jacques, and N. Hunter. 2005. Molecular analysis of microbial diversity in advanced caries. *Journal of Clinical Microbiology* 43:843–849.

Dahlen, G. G. 1993. Black pigmented gram negative anaerobes in periodontitis. *FEMS Immunology & Medical Microbiology* 6 (2–3):181–192.

Dibart, S., Z. Skobe, K. R. Snapp, S. S. Socransky, C. M. Smith, and R. Kent. 1998. Identification of bacterial species on or in crevicular epithelial cells from healthy and periodontally diseased patients using DNA-DNA hybridization. *Oral Microbiology and Immunology* 13 (1):30–35.

Dzink, J. L., R. J. Gibbons, W. C. Childs 3rd, and S. S. Socransky. 1989. The predominant cultivable microbiota of crevicular epithelial cells. *Oral Microbiology and Immunology* 4 (1):1–5.

Femiano, F., R. Femiano, L. Femiano, A. Jamilian, R. Rullo, and L. Perillo. 2016. Dentin caries progression and the role of metalloproteinases: An update. *European Journal of Paediatric Dentistry* 17(3):243– 247.

Genco, R. J. and W. S. Borgnakke. 2013. Risk factors for periodontal disease. *Periodontology 2000* 62 (1):59–64.

Gendron, R., D. Grenier, and L.-F. Maheu-Robert. 2000. The oral cavity as a reservoir of bacterial pathogens for oral infections. *Microbes and Infection* 2(8):897–906.

Gibbons, R. J., and J. V. Houte. 1975. Bacterial adherence in oral microbial ecoloy. *Annual Reviews in Microbiology* 29:19–44.

Haffajee, A. D., and S. S. Socransky. 2009. Relation of body mass index, periodontitis and Tannerella Forsythia. *Journal of Clinical Periodontology* 36 (2):89–99.

Haffajee, A. D., R. P. Teles, and S. S. Socransky. 2006. Association of Eubacterium nodatum and Treponema denticola with human periodontitis lesions. *Oral Microbiology and Immunology* 21:269–282.

He, X., X. Zhou, and W. Shi. 2009. Oral microbiology: Past, present and future. *International Journal of Oral Science* 1(2):47–58.

Humphrey, L. L., R. Fu, D. I. Buckley, M. Freeman, and M. Helfand. 2008. Periodontal disease and coronary heart disease incidence: A systematic review and meta-analysis. *Journal of General Internal Medicine* 23 (12):2079–2086.

Hillman, J. D. 2002. Genetically modified Streptococcus mutans for the prevention of dental caries. *Antonie Van Leeuwenhoek* 82:361–366.

Ishihara, K. 2010. Virulence factors of *Treponema denticola*. *Periodontology 2000* 54:117–135.

Janket, S. J., A. Wightman, A. E. Baird, T. E. V. Dyke, and J. A. Jones. 2005. Does periodontal treatment improve glycemic control in diabetic patients? A meta-analysis of intervention studies. *Journal of Dental Research* 84 (12):1154–1159.

Karpiński, T. M. and A. K. Szkaradkiewicz. 2013. Microbiology of dental caries. *Journal of Biology and Earth Sciences* 3 (1):21–24.

Khader, Y. S. and Q. Ta'ani. 2005. Periodontal disease and the risk of preterm low birthweight: A meta analysis. *Journal of Periodontology* 76 (2):161–165.

Kilian, M., I. L. C. Chapple, M. Hannig, P. D. Marsh, V. Meuric, A. M. L. Pedersen, M. S. Tonetti, W. G. Wade, and E. Zaura. 2016. The oral microbiome—An update for oral healthcare professionals. *British Dental Journal* 221:657–666.

Loesche, W. J. and L. H. Straffon. 1979. Longitudinal investigation of the role of Streptococcus mutans in human fissure decay. *Infection and Immunity* 26:498–507.

Loesche, W. J., S. A. Syed, B. E. Laughon, and J. Stoll. 1982. The bacteriology of acute necrotizing ulcerative gingivitis. *Journal of Periodontology* 53 (4):223–230.

Loesche, W. J., D. E. Lopatin, J. Stoll, N. van Poperin, P. P. Hujoel. 1992. Comparison of various detection methods for periodontopathic bacteria: Can culture be considered the primary reference standard? *Journal of Clinical Mircobiology* 30:418–426.

Loesche, W. J., J. Rowan, L. H. Straffon, and P. J. Loos. 1975. Association of Streptococcus mutans with human dental decay. *Infection Immunity* 11:1252–1260.

Lindhe, J., N. P. Lang, and T. Karring. 2008. *Clinical Periodontology and Implant Dentistry*. Oxford, UK: Blackwell Munksgaard.

Listgarten, M. A. 1965. Electron microscopic observations on the bacterial flora of acute necrotizing ulcerative gingivitis. *Journal of Periodontology* 36 (4):328–339.

Listgarten, M. A. and S. S. Socransky. 1964. Ultrastructural characteristics of a spirochete in the lesion of acute necrotizing ulcerative gingivostomatitis (Vincents infection). *Archives of Oral Biology* 16:95–96.

Listgarten, M. A., H. Mayo, and M. Amsterdam. 1973. Ultrastructure of the attachment device between coccal filamentous microorganisms in "corn cob" formations of dental plaque. *Archives of Oral Biology* 18 (5):651–656.

Machtei, E. E., L. A. Christersson, S. G. Grossi, R. Dunford, J. J. Zambon, and R. J. Genco. 1992. Clinical criteria for the definition of "established periodontitis." *Journal of Periodontology* 63 (3):206–214.

Marsh, P. D., A. David, and D. A. Devine. 2014. Prospects of oral disease control in the future—An opinion. *Journal of Oral Microbiology* 6:10.

Moore, W. E. C. and W. V. H. Moore. 1994. The bacteria of periodontal diseases. *Periodontolology 2000* 5:66–67.

Nadkarni, M. A., C. E. Caldon, and K. L. Chhour. 2004. Carious dentine provides a habitat for a complex array of novel Prevotella-like bacteria. *Journal of Clinical Microbiology* 42: 5238–5244.

Newton, J. T. and K. Asimakopoulou. 2015. Managing oral hygiene as a risk factor for periodontal disease: A systematic review of psychological approaches to behaviour change for improved plaque control in periodontal management. *Journal of Clinical Periodontology* 42 (16):36–46.

Offenbacher, S., V. Katz, G. Fertik, J. Collins, D. Boyd, G. Maynor, R. McKaig, and J. Beck. 1996. Periodontal Infection as a possible risk factor for preterm low birth weight. *Journal of Periodontology* 67 (10):1103–1113.

Pallon, J., M. Sundquist, and K. Hedin. 2018. A two year follow up study of patients with pharyngotonsillitis. *BMC Infectious Disease Series* 18:3.

Papapanou, P. N. 1996. Periodontal diseases: Epidemiology. *Annals of Periodontology* 1 (1):1–36.

Polyzos, N. P., I. P. Polyzos, A. Zavos, A. Valachis, D. Mauri, E. G. Papanikolaou, S. Tzioras, D. Weber, and I. E. Messinis. 2010. Obstetric outcomes after treatment of periodontal disease during pregnancy: Systematic review and meta-analysis. *British Medical Journal* 341:c7017.

Picard, C., A. Puel, M. Bonnet, C. L. Ku, J. Bustamante, K. Yang, C. Soudais, and S. Dupuis. 2008. Pyogenic bacterial infections in humans with IRAK-4 deficiency. *Science* 299 (5615):2076–2079.

Perera, M., N. N. Al-hebshi, D. J. Speicher, I. Perera, and N. W. Johnson. 2016. Emerging role of bacteria in oral carcinogenesis: A review with special reference to perio-pathogenic bacteria. *Journal of Oral Microbiology* 8:10.

Reichart, P. A. 2003. Oral manifestations in HIV infection: Fungal and bacterial infections, Kaposi's sarcoma. *Medical Microbiology and Immunology* 192 (3):165–169.

Rupf, S., S. Kannengieber, K. Merte, W. Pfister, B. Sigusch, and K. Eschrich. 2000. Comparison of profiles of key periodontal pathogens in periodontium and endodontium. *Dental Traumatology* 16 (6):269–275.

Samaranayake, L. 2006. *Essential Microbiology for Dentistry*, 3rd edn. China: Elsevier, pp. 115–186.

Schroeder, H. E. and M. A. Listgarten. 2003. The junctional epithelium: From strength to defense. *Journal of Dental Research* 82 (3):158–161.

Scannapieco, F. A. 1999. Role of oral bacteria in respiratory infection. *Journal of Periodontology* 70 (7):793–802.

Sen, S., M. Chung, V. Duda, L. Giamberardino, A. HInderliter, and S. Offenbacher. 2017. Periodontal disease associated with aortic arch atheroma in patients with stroke or transient ischemic attack. *Journal of Stroke and Cardiovascular Diseases* 26 (10):2137–2144.

Socransky, S.S., A. D. Haffajee, G. L. Smith, J. L. Dzink. 1987. Difficulties encountered in the search for the etiologic agents of destructive periodontal diseases. *Journal of Clinical Periodontology* 14 (10):588–593.

Socransky, S. S., A. D. Haffajee, M. A. Cugini, C. Smith, and R. L. Kent Jr. 1998. Microbial complexes in subgingival plaque. *Journal of Clinical Periodontology* 25 (2):134–144.

Syrjanen, J., J. Peltola, V. Valtonen, M. Iivanainen, M. Kaste, and J. K. Huttunen. 1989. Dental infections in association with cerebral infarction in young and middle aged men. *Journal of Internal Medicine* 225 (3):179–184.

Takahashi, N. 2011. The role of bacteria in the caries process: Ecological perspectives. *Journal of Dental Research* 90 (3):294–303.

Theilade, E. 1986. The non-specific theory in microbial etiology of inflammatory periodontal diseases. *Journal of Clinical Periodontology* 13 (10):905–911.

van Winkelhoff, A. J., P. Rurenga, G. J. Wekema-Mulder, Z. M. Singadji, and T. E. Rams. 2016. Non-oral gram-negative facultative rods in chronic periodontitis microbiota. Microbial Pathogenesis 94:117–122.

Xuedong, Z. and L. Yuqing. 2015. *Atlas of Oral Microbiology*. Chengdu, China: Elsevier, pp. 41–65.

Yang, X., C. Li, and Y. Pan. 2016. The Influences of Periodontal Status and Periodontal Pathogen Quantity on Salivary 8-Hydroxydeoxyguanosine and Interleukin-17 Levels. *Journal of Periodontology* 87 (5):591–600.

Yoo, J. R., S. T. Heo, M. Kim, C. S. Lee, and Y. R. Kim. 2016. *Porphyromonas gingivalis* causing brain abscess in patient with recurrent periodontitis. *Anaerobe*. 13:165–167.

Zambon, J. J. 1985. *Actinobacillus actinomycetemcomitans* in human periodontal disease. *Journal of Clinical Periodontology* 12:1–20.

Zaremba, M. L., W. Stokowska, A. Klimiuk, T. Daniluk, D. Ro, and D. Waszkiel. 2006. Microorganisms in root carious lesions in adults. *Advance in Medical Science* 51:237–240.

8
Prognosis and Impact of Recurrent Uveitis, the Ophthalmic Infection Caused by *Leptospira* spp.

Charles Solomon Akino Mercy and Kalimuthusamy Natarajaseenivasan

Contents

8.1 Introduction

The eye, a functionally and structurally complex organ, can be infected by the varieties of bacterial, fungal, viral, and parasitic infection. Among this most often ocular infections occur through bacterium. Ophthalmologists are still facing difficulties in managing bacterial eye infections. External and intraocular infections can lead to visual impairments, which is a major public health problem. Many microorganisms have been implicated in the pathogenesis of uveitis in humans, including certain bacteria species of *Salmonella, Campylobacter, Shigella, Klebsiella, Yersinia, Chlamydia, Mycoplasma, Leptospira, Treponema*, or *Borrelia* or viruses such as rubella or HIV. In equine ophthalmology, numerous microorganisms, such as *Brucella abortus, Borrelia burgdorferi, Streptococcus equi, Rhodococcus equi, Onchocer cacervicalis*, and *Toxoplasma gondii*, have also been associated with equine recurrent uveitis (ERU), but the most important pathogen implicated in the pathogenesis of ERU is *Leptospira* (Verma 2012). Leptospirosis, which is caused by *Leptospira*, a spirochete bacterium, is transmitted directly or indirectly from animals to humans and are reported to frequently infect the ocular structure, which occurs worldwide but is most common entity in tropical and subtropical areas. The spectrum of the disease remains extremely wide, ranging from subclinical infection to a severe multiorgan failure with high mortality. Ocular manifestation develops in both acute bacteremia phases and the second immunologic phases but is mostly noted in the second phase of illness, and the severity of the infection ranges from mild self-limiting conditions to those that could be extremely serious and visually threatening. Leptospiral infected persons frequently develop uveitis, with ocular manifestations reported in between 3% and 92% and this often occurring after a year from the initial acute form of leptospirosis. Detection of ocular manifestations remains underdiagnosed because of the prolonged symptom-free period and is frequently misdiagnosed due to its nonspecific symptoms that mimic better-known diseases. Management of patients with ocular infection may involve nothing more than supportive and palliative therapy along with aggressive intervention with antimicrobial and anti-inflammatory agents. This chapter deals with leptospirosis-associated uveitis and its prognosis, impact, diagnosis, and prevention strategies.

8.2 History

Initially leptospires were not cultured but have been observed from an autopsy specimen who died of yellow fever and was named *Spirocheta interrogans* by Stimson (1907). Its contagious nature and microbial origin were proved independently, first in Japan by Inada et al. (1915) (*Spiroch aeta ictero-haemorrhagiae*), and soon after in Germany (*Spirochaeta icterogenes*) by Uhlenhuth and Fromme (1916). Following detailed microscopical and cultural observations Noguchi (1918) proposed the name *Leptospira*, which means thin spirals.

Leptospiral uveitis was first reported by Adolf Weil (1886) in his original article and subsequently, several authors found its varying presentation. The importance of occupation as a risk factor was recognized early during 1917, and rat as a source of human leptospiral infection was discovered subsequently. DukeElder (Duke-Elder and Perkins 1966) reported that among all uveitis cases, 10% contribute to leptospiral uveitis. European authors reported an incidence of 10% to 44%, whereas Brand (Brand and Benmoshe 1963) and Heath (Heath et al. 1965) reported figures of 13% and 2% from Israel and the United States, respectively. In India, leptospirosis was known to occur for many decades in Andaman Islands.

8.3 Bacteriology

8.3.1 Morphological characteristics

Leptospires belong to the order Spirochaetales, family *Leptospiraceae*, genus *Leptospira*. These bacteria are long thin, about 0.1 μm in diameter by 6–20 μm in length. Electron microscopy shows a cylindrical cell body (i.e., protoplasmic cylinder) wound helically around an axistyle (0.01–0.02 μm in diameter), which comprises two axial filaments (a spirochetal form of a modified flagellum) inserted subterminally at the extremities of the cell body, with their free ends directed toward the middle of the cell (Bharti et al. 2003). They are helical bacteria with tight coils and have a typical double-membrane structure in which the cytoplasmic membrane and peptidoglycan cell wall are closely associated and are overlaid by an outer membrane that contains porins that allow solute exchange between the periplasmic space and the environment. Within the outer membrane, the leptospiral lipopolysaccharide (LPS) constitutes the main antigen and has a composition similar to that of other Gram-negative bacteria with lower endotoxic activity. It is characterized by active motility due to the presence of two periplasmic flagella with polar insertions in the periplasmic space and exhibits two distinct forms of movement, translational and non-translational. Either or both the ends of the single organism are blended or hooked. The free living (*L. biflexa*) and parasitic leptospires (*L. interrogans*) are morphologically indistinguishable, although the morphology of individual isolates varies with subculture in vitro and can be restored by passage in hamsters. Both saprophytic and pathogenic leptospires are present in nature. However, pathogenic leptospires mostly present in renal tubules of animals, whereas saprophytic leptospires were found to be present in many types of wet or humid environments.

8.3.2 Growth condition

Leptospires can able to survive in alkaline soil, mud, swamps, streams, rivers, organs, and tissues of live or dead animals and diluted milk. Survival of pathogenic leptospires in the environment is dependent on several factors including pH, temperature, and the presence of inhibitory compound. In general, they are sensitive to dryness, heat, acids, and basic disinfectants (Mohammed et al. 2011). Leptospires are obligative aerobic with an optimal growth temperature of 28°C to 30°C, and an optimal pH 7.2 to 7.6 is required for their growth. Their generation time varies from 7 to 12 hours. They grow in simple media enriched with vitamins B_1, B_{12}, and long-chain fatty acids are the only organic compounds required for their growth. Long-chain fatty acids are used as the sole carbon source and are metabolized by β-oxidation. Free fatty acids cause inherent toxicity and hence must be supplied either bound to albumin or in a nontoxic esterified form because *Leptospira* cannot synthesize fatty acids de novo (Faine et al. 1999). Ammonium salts are an effective source of cellular nitrogen. Several liquid media enriched with rabbit serum were described in the past by Noguchi (1919), Fletcher (1928), Korthof (1932), and Stuart (1946). Currently, the most widely used medium is based on the oleic acid, bovine serum albumin, and polysorbate (Tween) medium EMJH. Some strains require the addition of pyruvate or rabbit serum for initial isolation. In nucleic acid of leptospires purine bases were incorporated but not pyrimidine bases. Hence, it is resistant to the antibacterial activity of the pyrimidine analogue. Therefore, 5-fluorouracil can be used in the leptospiral growth media to eliminate other contamination sources. Other antibiotics have been added to media for culture of veterinary specimens, such as gentamicin, nalidixic acid, or rifampicin, in which contamination is more likely to occur (Adler and De La Pena Moctezuma 2010).

8.3.3 Cultural characteristics

Growth of the *Leptospira* with a pure culture takes 10 to 14 days, whereas primary isolates often grow slow and the cultures are retained for up to 13 weeks. *Leptopsira* can be cultivated in liquid, semi-solid, and solid media. Liquid media are the most commonly used medium for the maintenance of leptospires by periodical subculture for the serological diagnosis of infection and for typing. Semi-solid media can be prepared by incorporating agar at lower concentration (0.1% to 0.2%), and it is used for the long-term maintenance of the strains. Growth in the semi-solid media is easily visualized as one or more rings form from mm to cm below the surface of the medium, which is related to the optimum oxygen tension and is known as a Dinger's ring or disk (Levett 2001; Natarajaseenivasan et al., 1996). It is mainly useful for storing the culture and for long-term storage lyophilization was preferred, and also it can be stored at −70°C for several months. In solid medium, colonial morphology is dependent on agar concentration (1%–1.5%) and serovar; it can be used for cloning the strains, isolating leptospires from contaminated sources, and for the detection of hemolysin production.

8.3.4 Molecular characteristics

The genome of *Leptospira* is composed of two circular chromosomes of the larger replicon, with the sizes various between 3.6 and 4.3 Mb in length; a smaller chromosome, 277- to 350-kb in size contains a plasmid-like origin of replication (Ren et al., 2003). However, the genomes size of *L. interrogans* serovar Copenhageni is approximately 4.6 Mb, whereas the genome of *L. borgpetersenii* serovar Hardjo is only 3.9 Mb in size. The saprophyte *L. biflexa* possesses a third circular replicon of 74 kb, designated p74, not present in the pathogens. The presence of housekeeping genes on p74, which have orthologs located on the large chromosome in pathogenic *Leptospira*, suggests that p74 is essential for the survival of *L. biflexa* (Adler and De La Pena Moctezuma 2010). The guanine plus cytosine (GC) content is between 35% and 41%. *Leptospira* contains two sets of 16S and 23SrRNA genes and one set of 5S rRNA. Several insertion sequence (IS) coding for transposases were found among that IS*1533*; IS*1500* are found in many serovars; *IS1533* has a single open reading frame; and *IS1500* has four. Copy number of IS varies between different serovars and also among same serovars isolated in different resources. Comparative genomics of the two pathogenic and one saprophytic species has identified 2052 genes common to all; the core leptospiral genome is consistent with a common origin for leptospiral saprophytes and pathogens. Genome comparisons allow the identification of pathogen-specific genes (Adler and De La Pena Moctezuma 2010). The relationships between the different *Leptospira* strains provide for horizontal gene transfer and create in the bacterial population the pool of genes important for adaptivity to various conditions (Voronina et al. 2014). The reported genes and their functions are given in Table 8.1.

8.3.5 Classification

The Spirochaetales are an order of bacteria dividing itself into two families: Spirochaetaceae and Leptospiraceae. The Spirochaetaceae family includes Treponema types, Serpulina and Borrelia, whereas the Leptospiraceae

Table 8.1 A Number of Leptospiral Genes Reported with Various Function

Sl. No.	Function	Corresponding Genes
1.	DNA repair	*rec A*
2.	Encoding RNA polymerase	*rpoB*
3.	Encoding rRNA	*Rrs*
4.	Encoding ribosomal proteins	*fus, rplC, rplB, rplE, rplF, rplD, rplS, rplP, rplP, rplN, rplR, rplO, rplQ, rplW, rplV, rplX, rpsC, rpsQ, rpsM, rpsS, rpsS, rpsN, rpsK, rpsJ, rpsH, rpsE, rpsD, rpmD, rpmJ, rpmC, adk, rpoA, infA, secY, tuf*
5.	Amino acid synthesis	*proA, proB, proC, leuA, leuB, leuC, leuD*
6.	Encoding heat shock proteins	*hsp10, hsp58*
7.	Encodes outer membrane proteins	*lipL21, loa22, ligB, ligA, fecA, cirA, ostA, gspD, bamA, cirA, ompA, ompL1,omp85, ompL37, ompL47, ompL54, mcE, fecA, tolC, fadL, lenA, lenF, lp29, lp30, lp49, lipL32, lipL41, lipL36, lipL46, tlyC, hbpA, lepA, lruC, lsA21,*
8.	Encodes flagellar protein	*flaA, flaB, flg B–D, flgF, flgG, flgE, flgK, flgL, fliE, fliD, fliK, flgH, fliF, flgI, flhA, flhB, flhF, fliL, fli Q-S, motB, motA, fliG, fliM, fliN, fliW and fliH, bolA, ftsA, ftsH, ftsI, ftsK, ftsW, ftsZ, gldF, gldG*
9.	Lipopolysaccharide (LPS) synthesis	*rfb loci*

family includes *Leptospira*, which is further classified into *Leptospira* spp. are of importance because different serovars can exhibit different host specificities and may not be associated with a particular clinical form of infection. Classification of *Leptospira* is based on the expression of the surface exposed epitopes in a mosaic of the lipopolysaccharide (LPS) antigens, whereas the specificity of epitopes depends on their sugar composition and orientation (Dikken and Kmety 1978).

8.3.5.1 Serological classification – Basically, *Leptospira* has been divided into pathogenic and nonpathogenic (saprophytic) species. Of that pathogenic species have the potential to cause disease in both animals and humans whereas saprophytic, the free living, are generally considered as nonpathogenic. All pathogenic strains classified under *L. interrogans* (Dikken and Kmety 1978). Saprophytic strains include *L. biflexa*. Serovars that are antigenically related have traditionally been grouped into serogroups; within the species *L. interrogans* more than 300 serovars and with *L. biflexa* more

Table 8.2 Classification of *Leptospira*

Sl. No.	Classification	*Leptospira* Genus
1.	Pathogenic	*L. interrogans, L. borgpetersenii, L. kirschneri, L. noguchii, L. santarosai, L. weilii, L. alexanderi, L. alstonii, L. kmetyi*
2.	Intermediated pathogenic	*L. broomii, L. fainei, L. inadai, L. licerasiae, L. wolffii*
3.	Non pathogenic	*L. biflexa, L. meyeri, L. terpstrae, L. vanthielii, L. wolbachii, L. yanagawae, L. idonii*

than 60 serovars have been recognized. They have no taxonomic standing but is useful for epidemiological understanding.

8.3.5.2 Genotypic classification – The genus is classified into 12 pathogenic and 4 saprophytic species, with more than 300 pathogenic serovars reported. The genomospecies of *Leptospira* do not correspond to the previous two species (*L. interrogans* and *L. biflexa*), and indeed, pathogenic and nonpathogenic serovars occur within the same species (Ramadass et al. 1992; Levett 2001). The new genomic classification system has revealed a pathogenic species, which can contain both pathogenic and nonpathogenic serovars as well as intermediate species such as *L. meyeri*, *L. inadai*, and *L. fainei.* Taxonomy, which is still in progress, classifies *Leptospira* into 22 species. All recognized species have been classified as pathogenic, intermediate, and non-pathogenic (Fouts et al. 2016) (Table 8.2).

8.4 Etiology

It is endemic in tropical countries because of their geoclimatic and social conditions, which influence the epidemiological and geographical distribution of specific entities. The highest prevalence rates remain in tropical, developing countries where leptospirosis cases are on the rise. This rise is associated with urban population growth, urban decay, and flooding. Outbreaks have been related to heavy rainfall in various parts of the world, including Tamil Nadu, India, (Natarajaseenivasan et al. 2011; Prabhakaran et al. 2014) and Salvador, Brazil (Planka and Dean 2000). The incidence of ocular complications were variable, but this probably reflects the long time over which they may occur (Rathinam et al. 1996). In the United States, the incidence was estimated at 3%, whereas in Romania an incidence of 2% was estimated between 1979 and 1985. However, in abattoir workers with evidence of recent leptospirosis, the latter authors reported an incidence of 40%. For example, ERU has various etiologies, with *Leptospira* infection and genetic predisposition being the leading risk factors. In Europe and the United States, *Leptospira* serovars Grippotyphosa, Pomona, and Bratislava have been implicated in pathophysiology of ERU, whereas in the United Kingdom, *Leptospira* infection is not a major factor in the etiology of ERU.

8.5 Epidemiology

Leptospirosis has become an endemic disease in the geographical regions with a mild climate and a high precipitation of flood. It is endemic to areas of the Caribbean, Central America, South America, Southeast Asia, and Oceania (Dunay et al. 2016). The number of severe human cases worldwide is estimated above 500,000. Incidences range from 0.1 to 1/100,000 per year in temperate climates, 10 to 100/100,000 per year in the humid tropics to more than 100/100,000 per year during outbreaks and in high-exposure risk groups (Musso and La Scola 2013). Epidemic outbreaks are common after rainfall or flooding. Endemic transmission occurs because of factors such as tropical humid environments and poor sanitation. Urban epidemics are reported in cities throughout the developing world and will likely intensify as the world's slum population doubles to two billion by 2030 (Costa et al. 2015). *Leptospira* persist in entire continent except Antarctica. Specifically, leptospiral ERU in North America is commonly associated with the species *L. interrogans* serovar Pomona type kennewicki, whereas in Europe, ERU case studies show the implication of species *L. kirschneri* serovar Grippotyphosa (Malalana et al. 2015). European investigators reported that vitreous samples from 78% of ERU clinical cases were positive for *Leptospira* spp. Interestingly, 81% of the horses positive with *Leptospira* showed no further recurrences after vitrectomy, whereas in 83% of horses negative with *Leptospira* further recurrences occurred (Witkowski et al. 2016). In the United Kingdom, where both the *L. interrogans* serovar Pomona and *L. kirschneri* serovar Grippotyphosa are found to be rare, but the species *L. interrogans* serovar Sejroe are common. *L. interrogans* serovars Australis, Canicola, Hardjo, and Icterohaemorrhagiae are less commonly associated with ERU (Malalana et al. 2015).

8.6 Infectious cycle

Autoimmunity plays an important role in the ocular pathogenesis. Leptospirosis occurs biphasic in which an acute or septicemic phase lasts about a week and is characterized by high fevers (39°C–41°C) for 7–9 days after the initial exposure; a second immune phase occurs during the second week of illness, in which the disappearance of the organism from the bloodstream coincides with the appearance of antibodies. After the initial bacteremia the leptospires are eliminated by the immune system from all host tissues except from immunologically privileged places like the brain or eyes, resulting in immunological pathology in the eyes like uveitis (2 days to 4 weeks). Leptospiral antibodies are first detectable in serum 4–8 days after exposure and may be maintained for at least 7 years. Conjunctival suffusion is seen in most patients in some series. Uveitis may present weeks, months, or occasionally years after the acute stage. Chronic visual disturbance can persist 20 years or more after the acute illness. Chronic uveitis develops after a few days of unrelenting severe inflammation or following multiple recurrent episodes of uveitis (Gilger 2016).

8.6.1 Pathogenic mechanism

All pathogenic serovars associated with animal leptospirosis can also be pathogenic to humans. Transmission to humans occurs through penetration of the organism into the bloodstream via cuts, skin abrasions, or mucus membranes. Mechanisms behind the leptospiral infection that trigger the ocular infections are not clearly understood. Leptospires penetrate mucus membranes and abraded skin and rapidly gain access to the vascular space, such as in the ocular structure that causes inflammation and damage to the small blood vessels and results in vasculitis with leakage and extravasation of cells, hemorrhages which directs to cytotoxicity and cell death, and exposing the immune system to an immune privileged site (Figure 8.1). Most of the complications of leptospirosis are associated with localization of leptospires within these immune privileged sites such as the placenta, renal tubules, and anterior and posterior chambers of the eyes during the immune phase (Frellstedt 2009). In early ERU, congestion of uveal vessels and inflammatory cellular infiltrates are observed. Neutrophils are the first cells infiltrating the uvea and can result in hypopyon when accumulated in the anterior chamber. They are soon replaced by lymphocytes, plasma cells, and macrophages. With time and further recurrence, organization of the lymphocyte infiltrate is evident. Nodules in the ciliary body and iris are composed of B lymphocytes in the center and T lymphocytes in the periphery (Deeg 2002). Humoral and cellular immune reaction occur when interphotoreceptor-retinoid binding protein (IRBP) and predominance of $CD4^+$ T-cell infiltrates the affected area. An immune response to cellular retinaldehyde-binding protein (CRALBP) was detected in a large percentage of ERU cases (Gilger 2016). Generally, leptospiral LPS stimulated adherence of neutrophils to endothelial cells and platelets, causing aggregation and suggesting a role in the development of thrombocytopenia. The other reason may be that some of the proteins LruA, LruB, and LruC have been reported to cross

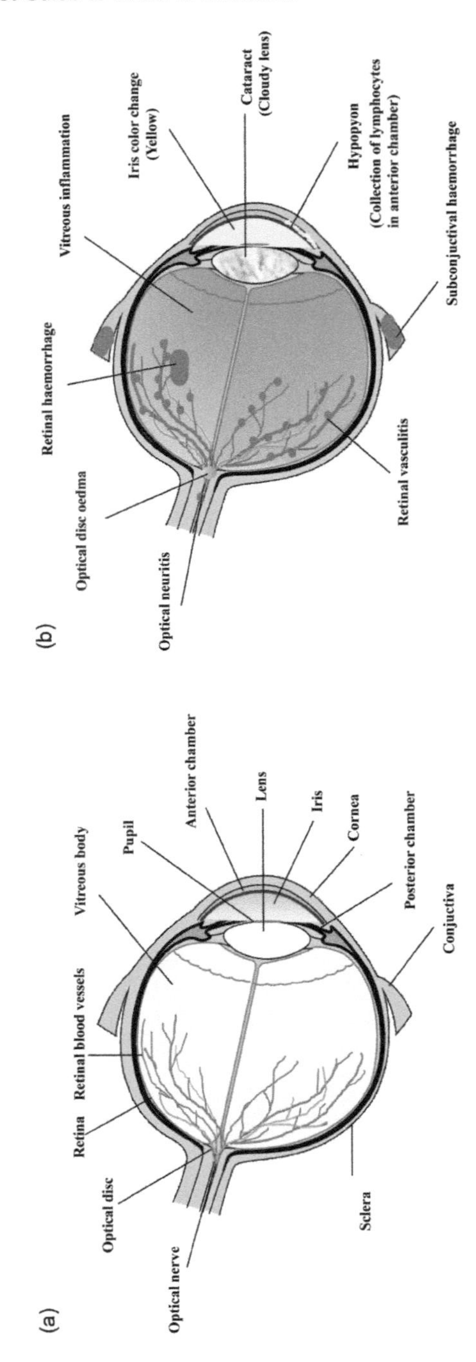

Figure 8.1 Schematic representation of (a) a normal healthy eye and (b) clinical complications due to infection with *Leptospira* spp.

react between the antibodies present in equine cornea, lens, and retina, which serve as an initiative factor for pathogenesis of leptospiral-induced uveitis. Due to the delayed development of uveitis, it is difficult to confirm a diagnosis of leptospiral uveitis (Malalana et al. 2015).

8.6.2 Risk factors

Though leptospirosis is a zoonotic disease, the carriage of *Leptospira* has been found in virtually all mammalian species human often infected accidentally. Transmission occurs in both industrialized and developing countries. Leptospirosis in humans is always acquired from an animal source; human-to-human transmission is for practical purposes nonexistent. Pathogenic *Leptospira* have been found in the proximal renal tubules of the kidneys of carriers, although other tissues and organs may also serve as a source of infection. From the kidneys, leptospires are excreted in urine and may then contaminate soil, surface water, streams, and rivers. The carriers may be wild or domestic animals, especially rodents and small marsupials, cattle, pigs, and dogs (Adler and De La Pena Moctezuma 2010). Leptospirosis is an occupational disease for veterinarians, farmers, abattoir workers, butchers, hunters, rodent control workers, and other occupations requiring contact with animals. Indirect contact with contaminated wet soil or water is responsible for the great majority of cases in the tropics, either through occupational exposure as in rice or taro farming, flooding after heavy rains, or exposure to damp soil and water during avocational activities (Musso and La Scola 2013). The expansion of urban slums worldwide has increased the chance of rat-borne transmission (Dunay et al. 2016). Genetic factors are strongly implicated in ERU. Uveitis occurs in all horse breeds, but the strong predilection of some breeds for ERU suggests a genetic link. ERU is most commonly seen in Appaloosas, European Warm-bloods, draught breeds, and Standard bred trotters (Witkowski et al. 2016).

8.7 Ocular manifestations

The prolonged symptom-free period between the systemic and ocular manifestations makes it difficult for the ophthalmologist to link uveitis to leptospirosis. The uveal tract is highly vascular, usually pigmented, and provides most of the blood supply to the eye. Because of the direct proximity to the peripheral vasculature, any disease of the systemic circulation may also affect the uveal tract (Malalana et al. 2015). Table 8.3 gives the broad spectrum of clinical manifestations involved during leptospiral ocular infection.

Table 8.3 Ocular Manifestations during Leptospiral Infection

Sl. No	Infection	Period of Infection	Signs and Symptoms
1.	Leptospiral uveitis	Anterior uveitis	Ocular pain, blepharospasm, lacrimation, chemosis, corneal changes (i.e., edema, vascularization, cellular infiltrate, keratic precipitates), aqueous flare, hypopyon, hyphema, miosis, iris color changes, and low intraocular pressure
		Posterior uveitis	Vitritis with liquefaction of the vitreous, the presence of vitreal floaters and retinal changes
2.	Leptospiral ERU	Classic ERU	Active intraocular inflammation, cataract, intraocular adhesion, and phthisis
		Insidious ERU	Low-grade inflammation with very few external signs
		Posterior ERU	Vitreal opacities, retinal inflammation and degeneration with no anterior sign of uveitis
3.	Common symptom of leptospiral ocular infection	Immunological phase (After 4 to 7 days of infection)	Conjunctival congestion without discharge, conjunctival chemosis, aqueous flare, fibrin, hyphema, miosis, cloudy-yellow-green vitreous, subconjunctival hemorrhages, panuveitis often accompanied with retinal periphlebitis and hypopyon
		Posterior segment manifestations	Vitritis, pars planitis, periphlebitis, choroiditis, papillitis, macular edema, retinal hemorrhages, retinal exudates, arteritis, retinal detachment resulting in blindness, corneal stromal opacities, inflammatory cells in the anterior chamber, hyperemic disc, optic neuritis, neuroretinitis, nongranulomatous uveitis hypopyon, cataract, vitreous inflammatory reaction, retinal vasculitis, retinal hemorrhages, phthisis bulbi or glaucoma and papillitis

ERU, equine recurrent uveitis.

8.8 Morbidity and mortality

Leptospirosis is a leading zoonosis cause of morbidity and mortality. Overall, leptospirosis was estimated to cause 1.03 million cases and 58,900 deaths each year. Although leptospirosis is a life-threatening disease and recognized as an important cause of pulmonary hemorrhage syndrome, the lack of global estimates for morbidity and mortality has contributed to its neglected disease status. Leptospiral ERU usually leads in to moon blindness among horses (Verma et al. 2013).

8.9 Complications

Cataracts tend to be the most common complication of ocular eye infection and occurs during steroid treatment of leptospiral uveitis. Steroids are the mainstay of treatment for leptospiral uveitis. Ocular involvement is seen both in the systemic bacteremic phase as well as in the immunological phase. The incidence of ocular signs during acute systemic phase varies from 2% to 90%. However, in some instances, the ocular manifestations may be subclinical or of such low order as to be overlooked; they are usually found by those who search for it (Rathinam 2005). One of the studies shows that 14% of sero-positive cases of leptospiral uveitis develop cataracts and of them, 76% develop significant cataract before the steroid treatment. Leptospiral uveitis usually responds promptly to treatment and cataract removal and intraocular lens implantation results in complete recovery of vision among human cases.

8.10 Diagnosis

Slit lamp bio-microscopic examination of the anterior segment of the eye revealed the presence of inflammatory cell collection or nongranulomatous keratic precipitates (KPs) at the back of the cornea. In case of severe inflammation, the cells gravitate down to form hypopyon, which has been noted to occur in 12% of patients with leptospiral uveitis. Leptospiral DNA can be detected in ocular fluids of affected horses even when they are seronegative. This means that lack of detectable serum antibodies does not rule out leptospirosis as a potential contributing factor (Witkowski et al. 2016). Other various laboratory procedures involved in the diagnosis has been represented in Table 8.4.

Misdiagnosis of cases leads in to unilateral uveitis with hypopyon and arthralgia as Behcet's disease, but the severe vitreous reaction, vasculitis, and hyperemic disc can differentiate leptospiral uveitis from other uveitis. The onset of leptospiral uveitis is acute and of shorter duration, whereas Behcet's uveitis is chronic, recurrent, and insidious.

Table 8.4 Various Diagnosis and Identification Methods for Leptospirosis

Sl. No.	Diagnostic Methods	Technique Used	References
1.	Bacteriological methods	Isolation, animal inoculation	Inada et al. (1915); Noguchi and Kligler (1920)
2.	Microscopic methods	Silver staining Direct dark-field microscopy,	Warthin and Starry (1992); Koshina et al. (1925)
3.	Serological diagnosis	CF Immunofluorescence Counter immunoelectrophoresis Microscopic agglutination test RIA ELISA Patoc slide agglutination. IgM dot-ELISA dipstick test	Sturdza et al. (1960); Torten et al. (1966); Terpstra et al. (1979); Terpstra et al. (1980); Tu et al. (1982)
4.	Molecular diagnosis	In-situ hybridization PCR Dot-blotting Real-Time PCR LAMP	Terpstra et al. (1980); Tu et al. (1982); Terpstra et al. (1987); Lilenbaum et al. (2009)
5.	Molecular typing	PFGE Ribotyping REA RFLP Detection of VNTR MLST AFLP Arbitrarily primed multiple locus sequence typing, insertion sequences-based typing	Ciceroni et al. (2002); Perolat et al. (1994); Brown and Levett (1997); Kawabata et al. (2001); Majed et al. (2005); Ahmed et al. (2006); Moreno et al. (2016)

(Continued)

Table 8.4 (*Continued*) Various Diagnosis and Identification Methods for Leptospirosis

Sl. No.	Diagnostic Methods	Technique Used	References
6.	Kit method	Leptospira-MC test (microcapsule agglutination assay) LeptoTek LFA (lateral flow assay) IHA Test-it (lateral flow assay) Leptocheck-WB (lateral flow assay) SD *Leptospira* LF (lateral flow assay) IgM LFA (lateral flow assay) LeptoTek DriDot (latex-card agglutination test) Leptorapide (latex-card agglutination test)	Arimitsu et al (1982); Smits et al. (2001); Levett et al. (2006)

AFLP, amplified fragment length polymorphism; CF, complement fixation; ELISA, enzyme-linked immunosorbent assay; IHA, indirect hemagglutination; MLST, multiple locus sequence typing; PCR, polymerase chain reaction; PFGE, pulsed field gel electrophoresis; REA, restriction endonucleases analysis; RIA, radioimmunoassay; RFLP, restriction fragment length polymorphism; VNTR, variable number of tandem repeats.

8.11 Treatment

It depends on the severity and duration of symptoms at the time of infection. Antibiotic treatment with doxycycline, vibramycin, oracea, adoxa, penicillin, oxytetracycline, streptomycin, cefotaxime, erythromycin, and fluoroquinolones is more effective when initiated early in the course of illness. Topical 1% atropine alone or combination with 10% phenylephrine is used to relieve ciliary muscle spasm and achieve mydriasis. Topical or systemic nonsteroidal anti-inflammatory drugs (NSAIDs) such as corticosteroids are used to decrease the inflammation, and dexamethasone and prednisolone are the two most commonly used topical corticosteroids (Jabs et al. 2005; Taylor et al. 2010). Topical steroids do not penetrate the posterior segment of the eye; it may be adequate for disease affecting the anterior uvea, but other strategies, such a systemic corticosteroid administration may be necessary in case of posterior uveitis. However, systemic administration of corticosteroids for ERU is less commonly performed because it can sometimes be associated with significant adverse effects in horses such as laminitis and flunixin meglumine has better ocular penetration than other NSAIDs. Subconjunctival and intraocular injections of triamcinolone acetonide have been reported in horses with no overt adverse effects observed (Yi et al. 2008 and Gilger 2016). Studies in horses have shown that intravitreal injection of rapamycin appears to be safe, although its efficacy in ERU has not yet been established. Injection of the anterior chamber with tissue plasminogen activator (TPA) can be performed to accelerate fibrinolysis in cases with severe fibrin accumulation in the anterior chamber. Surgical treatment includes the implantation of cyclosporine-releasing devices and vitrectomy to remove fibrin, inflammatory cells, and debris trapped in vitreous fluids to improve vision and delay the progression of the clinical signs (Witkowski et al. 2016). Another surgical option in the treatment of ERU is pars plana vitrectomy (PPV). The main goals of this procedure are the clearance of the ocular media and removal of cells and inflammatory mediators from the posterior segment. Recurrence of ERU was prevented in 73%–92% of horses in which PPV were performed. Long-term antibiotic therapy is not well established to work in a high transmission endemic setting. Phacoemulsification for the treatment of cataract secondary to ERU combined with cyclosporine implantation or PPV can be performed in an attempt to restore vision. When vision is lost, and the eye is still painful, enucleation might be necessary for the animal's comfort. Enucleation can be performed using a trans-palpebral or trans-conjunctival approach; a prosthesis can be placed within the orbit to improve the cosmetic outcome. Alternatively, 15 intrascleral prosthesis placement can be performed in cases in which the fibrous tunic of the eye is healthy enough to support the prosthesis (Malok et al. 1999; Deeg et al. 2002; Gilger 2016; Yi et al. 2008).

8.12 Prevention

Human vaccines have not been applied widely in Western countries. Immunization with polyvalent vaccines has been practiced in China and Japan, where large numbers of cases occur in rice field workers. In France, a monovalent vaccine containing serovar Icterohaemorrhagiae is licensed for human use. A vaccine containing serovars Canicola, Icterohaemorrhagiae, and Pomona has been developed recently in Cuba. Vaccines to prevent human leptospirosis are available in some countries, and large-scale clinical trials have been reported from Russia, China, Japan, and Vietnam, but they are serovar specific and require annual boosters. A vaccine against multiple serovars has been developed in Cuba and shown to be 78.1% effective, but it is currently in the early stages of clinic trials and safety testing (Bharti et al. 2003; Dunay et al. 2016). General measures such as rodent and infection control, self-sanitation approach, and avoiding of contaminated water reservoirs can prevent from leptospirosis. Prophylactic antibiotics beneficial for short-term, well-defined exposures such as those involved in military training or recreational sports like swimming, although long-term measure is difficult to practice in tropical countries.

Acknowledgments

Authors like to thank Dr. John F. Timoney, Keeneland Chair of Infectious Diseases, Gluck Equine Research Center, University of Kentucky, Lexington, USA, for the critical scientific suggestions for this work.

Financial support

This study was supported by the Indian Council of Medical Research (Sanction No:Leptos/34/2013-ECD-I), Ministry of Health (Sanction No: Leptos/33/2013-ECD-I), Department of Biotechnology (DBT) (No. BT/PR12133/ AAQ/3/707/2014), DBT-NER (No. BT/PR16685/NE/95/249/2015), Ministry of Science and Technology, Government of India, New Delhi.

References

Adler, B. and A. De La Pena Moctezuma. 2010. Leptospira and leptospirosis. *Veterinary Microbiology* 140:287–296.

Ahmed, N., S. M. Devi, M. De los A Valverde, P. Vijayachari, R. S. Machang'u, W. A. Ellis, and R. A. Hartskeerl. 2006. Multilocus sequence typing method for identification and genotypic classification of pathogenic Leptospira species. *Annals of Clinical Microbiology and Antimicrobials* 5 (1):28.

Arimitsu, Y. O., S. H. Kobayashi, K. I. Akama, and T. Y. Matuhasi. 1982. Development of a simple serological method for diagnosing leptospirosis: A microcapsule agglutination test. *Journal of Clinical Microbiology* 15 (5):835–841.

Bharti A. R., J. E. Nally, and J. N. Ricaldi. 2003. Leptospirosis: A zoonotic disease of global importance. *The Lancet Infectious Diseases* 12:757–771.

Brand, N. and H. Benmoshe. 1963. Human Leptospirosis associated with eye complications. *Israel Medical Journal* 22:182.

Brown, P. D., and P. N. Levett. 1997. Differentiation of Leptospira species and serovars by PCR-restriction endonuclease analysis, arbitrarily primed PCR and low-stringency PCR. *Journal of Medical Microbiology* 46 (2):173–181.

Ciceroni, L., S. Ciarrocchi, A. Ciervo, A. Petrucca, A. Pinto, A. Calderaro, and C. Chezzi. 2002. Differentiation of leptospires of the serogroup Pomona by monoclonal antibodies, pulsed-field gel electrophoresis and arbitrarily primed polymerase chain reaction. *Research in Microbiology* 153 (1):37–44.

Costa, F., J. E. Hagan, J. Calcagno et al. 2015. Global morbidity and mortality of Leptospirosis: A systematic review. *PLoS Neglected Tropical Diseases* 9 (9):e0003898.

Deeg, C. A., M. Ehrenhofer, S. R. Thurau, S. Reese, G. Wildner, and B. Kaspers. 2002. Immunopathology of recurrent uveitis in spontaneously diseased horses. *Experimental Eye Research* 75 (2):127–133.

Dikken, H. and E. Kmety 1978. Serological typing methods of leptospirosis. *Methods in Microbiology* 11:259–293.

Duke-Elder, S. and E. S. Perkins. 1966. Diseases of the uveal tract. In *System of Ophthalmology*, Vol 9, pp. 919. Edited by S. Duke-Elder, Maryland Heights, MO: The C. V. Mosby Company.

Dunay, S., J. Bass, and J. Stremick. 2016. Leptospirosis: A global health burden in review. *Emergency Medicine* 6:336.

Faine, S., B. Adler, C. Bolin, and P. Perolat. 1999. *Leptospira and Leptospirosis*, 2nd edn, Melbourne, Australia: MediSci.

Fletcher, W. 1928. Recent work on leptospirosis, tsutsugamushi disease and tropical typhus in the Federated Malay States. *Transaction of the Royal Society Tropical Medicine Hygiene* 21:265–287.

Fouts, D. E., M. A. Matthias, H. Adhikarla et al. 2016. What makes a bacterial species pathogenic? Comparative genomic analysis of the genus Leptospira. *PLoS Neglected Tropical Diseases* 10 (2):e0004403.

Frellstedt, L. 2009. Equine recurrent uveitis: A clinical manifestation of leptospirosis. *Equine Veterinary Education* 21 (10):546–552.

Gilger, B. C. 2016. *Equine Ophthalmology*, 3rd edn, pp. 318–349. Ames, IA: Wiley Blackwell.

Gilger, B. C., E. Malok, K. V. Cutter, T. Stewart, D. W. Horohov, and J. B. Allen. 1999. Characterization of T-lymphocytes in the anterior uvea of eyes with chronic equine recurrent uveitis. *Veterinary Immunological Immunopathology* 71:17–28.

Heath Jr, C. W., A. D. Alexander, and M. M. Galton. 1965. Leptospirosis in the United States: Analysis of 483 cases in man, 1949–1961. *New England Journal of Medicine* 273 (16):857–864.

Inada, R., Y. Ido, R. Hoki, R. Kaneko, and H. Ito. 1915. The etiology mode of infection and specific therapy of Weil's disease. *The Journal of Experimental Medicine* 23:377–402.

Jabs, D. A., R. B. Nussenblatt, and J. T. Rosenbaum. 2005. Standardization of Uveitis Nomenclature (SUN) Working Group. Standardization of uveitis nomenclature for reporting clinical data: Results of the First International Workshop. *American Journal of Ophthalmology* 140:509–516.

Kawabata, H., L. A. Dancel, S. Y. Villanueva, Y. Yanagihara, N. Koizumi, and H. Watanabe. 2001. flaB-Polymerase Chain Reaction (flaB-PCR) and Its Restriction Fragment Length Polymorphism (RFLP) Analysis Are an Efficient Tool for Detection and Identification of Leptospira spp. *Microbiology and Immunology* 45 (6):491–496.

Korthof, G. 1932. Experimentelles Schlammfieber beim Menschen. *Zentralbl Bakteriol Parasitenkd Infektionskr Hyg Abt I Orig* 125:429.

Koshina, M., S. Shiozawa, and K. Kitayama. 1925. Studies of Leptospira-hebdomadis. *Journal of Experimental Medicine* 42 (6):873–895.

Levett, P. N. 2001. Leptospirosis. *Clinical Microbiology Reviews* 14:296–326.

Levett, P. N., S. L. Branch, C. U. Whittington, C. N. Edwards, and H. Paxton. 2006. Two methods for rapid serological diagnosis of acute leptospirosis. *Clinical and Diagnostic Laboratory Immunology* 8 (2): 349–351.

Lilenbaum, W., R. Varges, P. Ristow, A. Cortez, S. O. Souza, L. J. Richtzenhain, and Vasconcellos, S. A. 2009. Identification of Leptospira spp. carriers among seroreactive goats and sheep by polymerase chain reaction. *Research in Veterinary Science* 87 (1):16–19.

Majed, Z., E. Bellenger, D. Postic, C. Pourcel, G. Baranton, and M. Picardeau. 2005. Identification of variable-number tandemrepeat loci in Leptospirainterroganssensustricto. *Journal of Clinical Microbiology* 43 (2): 539–545.

Malalana, F., A. Stylianides, and C. McGowan. 2015. Equine recurrent uveitis: Human and equine perspectives. *The Veterinary Journal* 206:22–29.

Mohammed, H., C. Nozha, K. Hakim, F. Abdelaziz, and B. Rekia. 2011. Leptospira: Morphology, classification and pathogenesis. *Journal of Bacteriology and Parasitology* 2:120.

Moreno, L. Z., F. Miraglia, W. Lilenbaum, J. S. Neto, J. C. Freitas, Z. M. Morais, R. A. Hartskeerl, B. L. Da Costa, S. A. Vasconcellos, and A. M. Moreno. 2016. Profiling of Leptospirainterrogans, L. santarosai, L. meyeri and L. borgpetersenii by SE-AFLP, PFGE and susceptibility testing—A continuous attempt at species and serovar differentiation. *Emerging Microbes & Infections* 5 (3):e17.

Musso, D. and B. La Scola. 2013. Laboratory diagnosis of leptospirosis: A challenge. *Journal of Microbiology, Immunology and Infection* 46:245e252.

Natarajaseenivasan, K., K. Vedhagiri, V. Sivabalan, S. G. Prabagaran, S. Sukumar, S. C. Artiushin, and J. F. Timoney. 2011. Seroprevalence of *Leptospira borgpetersenii* serovar javanica infection among dairy cattle, rats and humans in the Cauvery river valley of southern India. *South East Asian Journal of Tropical Medicine and Public Health* 42:679.

Natarajaseenivasan, K., S. Ratnam, P. Ramadass, and P. S. HelenManual. 1996. Persistence of dinger's rings by *Leptospirainterrogans* serovar *australis* in semi solid EMJH medium. *Indian Veterinary Journal* 73:571–572.

Noguchi, H. 1918. Morphological characteristics and nomenclature of *Leptospira (spirochaeta) icterohaemorrhagiae* (inada and ido). *Journal of Experimental Medicine* 27 (5):575–92.

Noguchi, H. 1919. Etiology of yellow fever: V. Properties of blood serum of yellow fever patients in relation to Leptospiraicteroides. *Journal of Experimental Medicine* 30 (1):9–12.

Noguchi, H. and I. J. Kligler. 1920. Immunological studies with a strain of leptospira isolated from a case of yellow fever in Merida, Yucatan. *Journal of Experimental Medicine* 32 (5):627–637.

Perolat, P., F. Merien, A. W. Ellis, and G. Baranton. 1994. Characterization of Leptospira isolates from serovarhardjo by ribotyping, arbitrarily primed PCR, and mapped restriction site polymorphisms. *Journal of Clinical Microbiology* 32 (8):1949–1957.

Planka, R. and D. Dean. 2000. Overview of the epidemiology, microbiology, and pathogenesis of *Leptospira* spp. in humans. *Microbes and Infection* 2 (10):1265–1276.

Prabhakaran, S. G., S. Shanmughapriya, S. Dhanapaul, A. James, and K. Natarajaseenivasan. 2014. Risk factors associated with rural and urban epidemics of Leptospirosis in Tiruchirappalli District of Tamilnadu, India. *Journal of Public Health* 22 (4):323–333.

Ramadass, P., B. D. Jarvis, R. J. Corner, D. Penny, and R. B. Marshall. 1992. Genetic characterization of pathogenic Leptospira species by DNA hybridization. *International Journal of Systematic and Evolutionary Microbiology* 42 (2):215–219.

Rathinam, S. R. 2005. Ocular manifestaions of Leptospirosis. *Journal of Postgraduate Medicine* 51:189–194.

Rathinam, S. R., L. Sureshbabu, K. Natarajaseenivasan. 1996. Leptospiral antibodies in patients with recurrent ophthalmic involvement. *The Indian Journal of Medical Research* 103:66–68.

Ren, S. X., G. Fu, X. G. Jiang et al. 2003. Unique physiological and pathogenic features of *Leptospira interrogans* revealed by wholegenome sequencing. *Nature* 422 (6934):888–893.

Smits, H. L., C. K. Eapen, S. Sugathan, M. Kuriakose, M. H. Gasem, C. Yersin, and G. C. Gussenhoven. 2001. Lateral-flow assay for rapid serodiagnosis of human leptospirosis. *Clinical and Diagnostic Laboratory Immunology* 8 (1):166–169.

Smythe, L. D., I. L. Smith, G. A. Smith, M. F. Dohnt, M. L. Symonds, L. J. Barnett, and D. B. McKay. 2002. A quantitative PCR (TaqMan) assay for pathogenic Leptospira spp. *BMC Infectious Diseases* 2 (1):13.

Sonthayanon, P., W. Chierakul, V. Wuthiekanun, J. Thaipadungpanit, T. Kalambaheti, S. Boonsilp, P. Amornchai, L. D. Smythe, D. Limmathurotsakul, N. P. Day, and S. J. Peacock. 2011. Accuracy of loop-mediated isothermal amplification for diagnosis of human leptospirosis in Thailand. *The American Journal of Tropical Medicine and Hygiene* 84 (4):614–620.

Stimson, A. M. 1907. Note on an organism found in yellow fever tissue. *Public Health Reports* 22:541.

Stuart, R. D. 1946. The preparation and use of a simple culture medium for leptospirae. *The Journal of pathology* 58 (3):343–349.

Sturdza, N., M. Elian, and G. Tulpan. 1960. Diagnosis of Human Leptospirosis by the Complement Fixation Test with a Single Antigen. Note I. *Archives Roumaines De Pathologie Experimentale et de Microbiologie* 19 (4):571–582.

Suwimonteerabutr, J., W. Chaicumpa, P. Saengjaruk, P. Tapchaisri, M. Chongsa-nguan, T. Kalambaheti, P. Ramasoota, Y. Sakolvaree, and P. Virakul. 2005. Evaluation of a monoclonal antibody–based dot-blot ELISA for detection of Leptospira spp. in bovine urine samples. *American Journal of Veterinary Research* 66 (5):762–766.

Taylor, S. R., H. Isa, L. Joshi, and S. Lightman. 2010. New developments in corticosteroid therapy for uveitis. *Ophthalmologica* 224:46–53.

Terpstra, W. J., G. J. Schoone, and G. S. Ligthart. 1979. Counterimmunoelectrophoresis in the diagnosis of human leptospirosis. *ZentralblBakteriolOrig A* 244 (2–03):285–290.

Terpstra, W. J., G. J. Schoone, G. S. Ligthart, and J. Ter Schegget. 1987. Detection of Leptospirainterrogans in clinical specimens by in situ hybridization using biotin-labelled DNA probes. *Microbiology* 133 (4):911–914.

Terpstra, W. J., G. S. Ligthartand, and G. J. Schoone. 1980. Serodiagnosis of human leptospirosis by enzyme-linked-immunosorbent-assay (ELISA). *ZentralblBakteriolOrig A* 247 (3):400–405.

Torten, M., E. Shenberg, and J. Van der Hoeden. 1966. The use of immunofluorescence in the diagnosis of human leptospirosis by a genus-specific antigen. *The Journal of Infectious Disease* spp. 116 (5):537–543.

Tu, V., B. Adler, and S. Faine. 1982. The role of macrophages in the protection of mice against Leptospirosis: In vitro and in vivo studies. *Pathology* 14 (4):463–468.

Uhlenhuth, P. and W. Formme. 1916. Quoted in Topley and Wilson's Principles of Bacteriology. *Virology and Immunity* 8 (2):617.

Verma, A. and B. Stevenson. 2012. Leptospiral uveitis-there is more to it than meets the eye! *Zoonoses Public Health* 59 (2):132–141.

Verma, A., B. Stevenson, and B. Adler. 2013. Leptospirosis in horses. Veterinary Microbiology 167 (1–2):61–66.

Voronina, O. L., M. S. Kunda, E. I. Aksenova et al. 2014. The characteristics of ubiquitous and unique *Leptospira* strains from the collection of Russian centre for leptospirosis. *BioMed Research International* 5:649034.

Warthin, A. S. and A. C. Starry. 1922. The staining of spirochetes in cover-glass smears by the silver-agar method. *The Journal of Infectious Diseases* 30 (6):592–600.

Weil, A. 1886. About a peculiar acute infectious disease associated with splenic tumor, icterus, and nephritis. *Deutsches Archiv für Klinische Medizin* 39:209–232.

Witkowski, L., A. Cywinska, K. Paschalis-Trela, M. Crisman, and J. Kita. 2016. Multiple etiologies of equine recurrent uveitis—A natural model for human autoimmune uveitis: A brief review. *Comparative Immunology, Microbiology and Infectious Diseases* 44:14–20.

Yi, N. Y., J. L. Davis, J. H. Salmon, and B. C. Gilger. 2008. Ocular distribution and toxicity of intravitreal injection of triamcinolone acetonide in normal equine eyes. *Veterinary Ophthalmology* 1:15–19.

9

Beneficial Lactic Acid Bacteria

Use of Lactic Acid Bacteria in Production of Probiotics

Galina Novik and Victoria Savich

Contents

9.1 Introduction

Lactic acid bacteria (LAB) are one of commercially valuable groups of microorganisms. LAB have a long application record starting from the ancient times. The bacteria have been employed in dairying, baking, fish, and meat processing. However, the first pure culture of LAB (*Bacterium lactis*, now known as *Lactococcus lactis*) was obtained by Lister in 1873 (Santer 2010). The development of microbiological methods allowing investigations of morphological, physiological, biochemical, and genetic properties of LAB led to deeper insight into their biology and originated new applications for this bacterial group. LAB have been shown as sources of bioactive compounds. These substances as well as LAB themselves provide for health benefits. Mechanisms of favorable action are diverse and play important roles in the modulation of immunological and gastrointestinal functions and control of pathogens. Bacteria do not necessarily possess similar features, so that the development of selection criteria for probiotics is vital for their further use. Besides medicine and food processing, LAB are applied in agricultural and industrial sectors.

9.2 Morphology, cytology and physiology of LAB

Metabolic pathways of lactic acid bacteria. Biosynthesis of bioactive secondary metabolites. Gene clusters for secondary metabolic pathways. LAB representatives are gram-positive, nonsporulating, catalase-negative, anaerobic or microaerophilic, acid-tolerant, organotrophic, and strictly fermentative rods or cocci producing lactic acid as a major end product (König and Fröhlich 2009). Cell wall of LAB has the typical gram-positive structure formed by a thick, multilayered peptidoglycan envelope decorated with proteins, teichoic acids, and polysaccharides and surrounded in genus *Lactobacillus* by an outer shell of proteins packed in a paracrystalline layer (S-layer) (Delcour et al. 1999). *Lactobacillus* S-layer proteins differ from those of other bacteria in their smaller size and highly predicted *pI*. The positive charge in S-layer proteins is concentrated in the more conserved cell wall binding domain, which can be either N- or C-terminal depending on the species. The more variable domain is responsible for the self-assembly of the monomers into a periodic structure (Hynönen and Palva 2013). Peptidoglycan is the main constituent of the gram-positive cell wall consisting of glycan chains composed of alternating N-acetylglucosamine and N-acetylmuramic acid and linked by β-1,4 bonds. In most bacterial species, peptidoglycan basic structure is partially altered; either glycan chains undergo N-deacetylation or O-acetylation or free carboxyl groups of the amino acids in the peptide chains are amidated. Teichoic acids, anionic polymers made up of alditol-phosphate repeating units, also can be modified by replacing free hydroxyl groups of the alditol-phosphate chains with various sugars or D-alanin. Bacterial polysaccharides of cell wall exhibit great diversity in sugar composition, linkage, branching, and substitution. The structural divergence shown by these cell wall components may underlie differences in processes such as autolysis and characteristics such as stress resistance, probiotic properties, or phage sensitivity (Chapot-Chartier and Kulakauskas 2014). The extracellular and surface-associated proteins also can be involved in cell wall metabolism, degradation, and uptake of nutrients, communication, and binding to substrates or hosts (Zhou et al. 2010).

The genomes of LAB and bifidobacteria have low and high guanine-cytosine (GC) contents and vary in size from 1.3 to 3.3 and 1.9 to 2.9 Mb, respectively (Klaenhammer et al. 2005; Wu et al. 2017). In bifidobacterial genomes adapted for colonization of the gastrointestinal tract, 1,604–2,588 genes have been identified (Pokusaeva 2011). The number of predicted protein-encoding genes in LAB ranges from about 1,700 to about 2,800. The distinctive feature of LAB genomes is gene loss due to adaptation to nutritionally rich environments (Makarova et al. 2006). *Lactobacillales* is presumed to have lost 600–1200 genes inherited from its *Bacilli* ancestor, including genes encoding biosynthetic enzymes and sporulation (Makarova and Koonin 2007). Some key genes for LAB survival in a new environment could be acquired via horizontal gene transfer (Wu et al. 2017). The presence of multiple pseudogenes when compared with other groups of bacteria is also evidence of genome reduction. Loss of carbohydrate transport and metabolism genes and amino acid biosynthesis genes accompanies adaptation to the environment rich in lactose and protein (Schroeter and Klaenhammer 2009). Plasmids found in many LAB vary in size and gene content. Some plasmids carry genes that potentially contribute to adaptation of the host cell and code for bacteriocins, amino acid or sugar transporters, and restriction-modification systems (Wu et al. 2017). With progress of genomic research, key features of LAB genomes continue to be discovered leading to better understanding of the physiology and metabolism of this microbial group.

LAB do not possess a functional respiratory system, so they derive the energy required for their metabolism from the oxidation of chemical compounds, mainly sugars. Sugars are fermented by LAB via homofermentative or heterofermentative pathways. Homofermentative bacteria produce lactic acid as the only product of glucose fermentation through glycolysis or Embden–Meyerhof–Parnas pathway. Heterofermentative bacteria use the pentose phosphate pathway generating carbon dioxide (CO_2) and ethanol or acetate, besides lactic acid. Other hexoses are also fermented by LAB after preliminary isomerization or phosphorylation. In addition, LAB with heterofermentative type of fermentation successfully metabolize pentoses. Disaccharides are split enzymatically into monosaccharides entering the appropriate pathways (Von Wright and Axelsson 2011). Genus *Bifidobacterium* degrades hexose sugars through a particular metabolic pathway or *bifid shunt* allowing to produce more energy in the form of ATP. Bifidobacterial pathway yields 2.5 mol of ATP, 1.5 mol of acetate, and 1 mol of lactate from 1 mol of fermented glucose, while the homofermentative LAB produce 2 mol of ATP and 2 mol of lactic acid and heterofermentative LAB produce 1 mol each of lactic acid, ethanol, and ATP per 1 mol of fermented glucose. Fructose-6-phosphoketolase enzyme plays a key role in this pathway and is considered to be a taxonomic marker for the family of *Bifidobacteriaceae* (Pokusaeva 2011). Recently a novel metabolic pathway (galacto-N-biose (GNB)/lacto-N-biose (LNB) I pathway) that utilizes both human milk oligosaccharides and host glycoconjugates and is essential for colonization of the infant gastrointestinal tract was found in genus *Bifidobacterium* (Fushinobu 2010). This route was suggested to be specific for *Bifidobacterium*; however, further studies showed ability of some LAB to metabolize LNB and GNB. Nevertheless, metabolic pathways responsible for catabolism of these compounds in LAB are completely different from those described for *Bifidobacterium* species (Bidart et al. 2014).

The proteolytic system of LAB provides amino acids essential for bacterial growth by protein conversion. It is also engaged in generation of flavor compounds, accounting for the development of organoleptic properties of fermented food (Liu et al. 2010). Two major pathways convert amino acids to flavor compounds: elimination reactions catalyzed by lyases and pathways initiated by aminotransferases. Lyases take part in the production of methanethiol from methionine, while aminotransferases convert amino acids to corresponding α-keto acids. The α-keto acids are key intermediates in aroma generation and can be further transformed into other compounds: α-hydroxyacids, acetyl-CoA derivatives, and aldehydes while the latter turn into alcohols and carboxylic acids (Steele et al. 2013). The proteolytic system of LAB contains cell wall–bound proteinase degrading milk proteins into oligopeptides, peptide transporters transferring peptides into the cell, and various intracellular peptidases breaking down the peptides into shorter peptides and amino acids (Liu et al. 2010).

Lipid metabolism is the enzymatic break down of lipids into fatty acids and glycerol by lipases with either intracellular or extracellular localization in LAB strains. The latter are able to perform unique fatty acid transformation reactions, including isomerization, hydration, dehydration, and saturation (Hayek and Ibrahim 2013). Such products of lipid metabolism as conjugated fatty acids have beneficial effects on health, making them a target of intensive study. LAB were found to successfully produce conjugated linoleic acid (CLA) through two consecutive reactions: hydration of linoleic acid to 10-hydroxy-12-octadecenoic acid and dehydrating isomerization of the hydroxy fatty acid to CLA. Ricinoleic acid also can be transformed into CLA. On the other hand, linoleic acid can be used in production of conjugated trienoic acid through alkali-isomerization (Ogawa et al. 2005). *Bifidobacterium* species show ability to conduct isomerization of linoleic acid to CLA (Raimondi et al. 2016).

Besides the aforementioned compounds, LAB are sources of many other substances: bacteriocins, vitamins, enzymes, exopolysaccharides, and sweeteners. A wide range of LAB are able to produce capsular or extracellular polysaccharides with various chemical composition and properties. The term *exopolysaccharide* (EPS) often denotes all forms of polysaccharides located outside of the microbial cell wall. Polysaccharides are divided into two groups: homopolysaccharides composed of only one type of monosaccharide and heteropolysaccharides containing two or more types of sugars. Two pathways for biosynthesis of exocellular polysaccharides have been described for LAB: the Wzy-dependent pathway and the extracellular glycosyltransferase pathway for synthesis of glucans and fructans. Genes encoding Wzy-dependent proteins in LAB are typically organized in a cluster with an operon structure and can be chromosomal as well as plasmid-borne. These gene clusters are extremely diverse, and their nucleotide sequences are among the most variable sequences in LAB genomes. Genes in the operon can be categorized into several groups: modulatory genes, polysaccharide assembly machinery genes, genes encoding glycosyltransferase involved in the assembly of the repeating units, and genes required for the synthesis of activated sugar precursors and modification of the sugar residues. Clusters are usually 15–20 Kb in size and comprise less than 30 genes. Genes of LAB have the same orientation and are transcribed as a single mRNA. Extracellular glycosyltransferase pathway is a simple biochemical route employing a specific glucansucrase or fructansucrase and an extracellular sugar donor for the synthesis of glucans or fructans (Zeidan et al. 2017). All EPS-producing bifidobacteria appear to synthesize EPS through internal heteropolysaccharide pathways. Their EPS biosynthesis is concentrated in the 25.6 Kb region composed of 20 genes, 18 of which are positioned in oppositely directed but adjacent transcriptional loci. Regulatory genes have not been identified in *Bifidobacterium* spp. to date (Ryan et al. 2015).

LAB are known as producers of vitamins, mainly K, and some vitamins of the B group (Patel et al. 2013). Riboflavin (vitamin B_2) plays a significant role in cellular metabolism, acting as the precursor of electron carriers in oxidation–reduction reactions. Biosynthesis of riboflavin dependent on the precursors guanosine triphosphate and D-ribulose 5-phosphate occurs through seven enzymatic steps. B_2 synthesis genes in LAB form a single operon, including riboflavin-specific deaminase and reductase (*ribG*), riboflavin synthase alpha subunit (*ribB*), a bifunctional enzyme catalyzing the formation of 3,4-dihydroxy-2-butanone 4-phosphate from ribulose 5-phosphate (*ribA*), and riboflavin synthase beta subunit (*ribH*). Riboflavin biosynthesis and transport are controlled by conserved regulatory region located upstream of the operon (LeBlanc et al. 2011; Capozzi et al. 2012). Folate (vitamin B_{11}) is involved in essential functions, like DNA replication, repair, methylation, and synthesis of nucleotides, vitamins, and some amino acids. The biosynthetic pathway in LAB includes several consecutive steps, wherein the precursor guanosine triphosphate is converted into tetrahydrofolate (Capozzi et al. 2012). The folate biosynthetic genes of *L. lactis* MG1363 are organized in a folate gene cluster, consisting of six genes (*folA*, *folB*, *folKE*, *folP*, *ylgG*, and *folC*) (Sybesma et al. 2003). The similar structure of the cluster was revealed for *Lactobacillus* species (Santos et al. 2008). Cobalamin (vitamin B_{12}) is one of the vital vitamins whose deficiency affects hematopoietic, neurological, and cardiovascular systems. Cobalamin is synthesized only by prokaryotes via aerobic and anaerobic pathways. The biosynthetic route of the vitamin from 5-aminolaevulinic acid is divided into three sections: biosynthesis of uroporphyrinogen III from 5-aminolaevulinic acid; conversion of uroporphyrinogen III into the ring-contracted, deacylated intermediate precorrin 6 or cobalt-precorrin 6; transformation of the intermediate to form adenosyl cobalamin (Scott and Roessner 2002). Production of B_{12} was observed mainly in other groups of bacteria, although the analyses revealed the presence of 32 open reading frames related to coenzyme B_{12} production (*cbi*, *cob*, *hem*, and *cbl* gene cluster) in *Lactobacillus coryniformis* (Torres et al. 2016). *Lactobacillus reuteri* JCM 1112T has a unique cluster of 58 genes encoding biosynthesis of reuterin and cobalamin. At least two independent insertion events can be engaged in formation of the cluster (Morita et al. 2008). Vitamin K plays a significant role in blood clotting, bone, and kidney function, tissue calcification, and atherosclerotic plaque prevention. The vitamin is usually available in two forms: phylloquinone (K_1) and menaquinone (K_2). The latter is produced by some intestinal bacteria, like LAB (Patel et al. 2013). Biosynthesis of menaquinone was studied in other groups of bacteria, starting from chorismite converted from shikimate (Meganathan and Kwon 2009). However, operon *men* encoding menaquinone synthesis was also found in some LAB (Bolotin et al. 2001; Wegmann et al. 2007).

The other substances produced by LAB are mannitol, sorbitol, tagatose, and xylitol used as sweeteners in food industry. Mannitol is a six-carbon sugar alcohol synthesized by bacteria from fructose using mannitol dehydrogenase (Patra et al. 2009; Papagianni 2012). The research revealed that one-third of fructose could be replaced with glucose, maltose, galactose, mannose, raffinose, or starch with glucoamylase, and two-thirds of fructose could be replaced with sucrose for mannitol production (Saha and Nakamura 2003). Tagatose is an isomer of fructose showing prebiotic effect and antioxidant activity, and it can be used for control of diabetes and obesity. D-tagatose can be produced from D-galactose by L-arabinose isomerase (*araA*) (Chouayekh et al. 2007; Patra et al. 2009). Sorbitol is another six-carbon sugar alcohol produced by catalytic hydrogenation of glucose, with applications in the food and pharmaceutical industries. Only a few organisms are able to synthesize sorbitol. LAB strains are often subjected to metabolic engineering to achieve sorbitol hyperexpression (Patra et al. 2009; Papagianni 2012). Xylitol is a five-carbon sugar alcohol produced by reduction of xylose. LAB have not been reported to produce xylitol naturally, but recombinant strains with xylose reductase were able to generate this compound (Papagianni 2012). Bacteriocins are considered in the separate chapter.

9.3 Growth. Optimization of medium composition and physicochemical conditions

Substrate pretreatment. Selection of supplements, autoregulators and autoinducers. LAB are industrially attractive microorganisms that are able to produce a number of valuable compounds. Bacteria require appropriate medium and physicochemical conditions to grow and express normal metabolic activities. LAB are known as fastidious microorganisms that demand specific cultural conditions. It urges optimization of medium composition and cultivation factors. Optimal physicochemical parameters (e.g., temperature, pH, water activity, and redox potential) are diverse among LAB strains. Bacteria of this group can grow in pH range 3.5–9.6 and temperatures 5°C–45°C (Abdel-Rahman et al. 2013). Nevertheless, temperature rise led to the decrease of growth yield based on ATP production as well as specific growth rate of bacteria, whereas specific lactate production rate remained constant or even increased. pH fall resulted in inhibition of both growth rate and lactate production (Adamberg et al. 2003). Autolysis of *Lactococcus lactis* strains is necessary for cheese ripening and is highly strain dependent. However, autolysis in most of the strains is favored by low NaCl concentrations (0.17 M) and acidic pH (5.4) (Ramírez-Nuñez et al. 2011). Some *Lactobacillus* strains successfully ferment glucose and produce lactic acid even at 8% salt level (Rao et al. 2004).

LAB cannot grow on simple mineral media supplemented only with a carbon source. These microorganisms often demand various free amino acids, peptides, nucleic acid derivatives, fatty acid esters, minerals, vitamins, and buffering agents in the medium. As a carbon source glucose is commonly preferred by the majority of LAB cultures, but some strains have opted for alternative sugars (Hayek and Ibrahim 2013).

To cheapen production of various compounds generated by LAB, especially lactic acid, new substrates have been tested because of high cost of the raw materials, such as starch and refined sugars. Lignocellulosics due to their abundance, low price, high polysaccharide content, and renewability have been chosen as potential carbohydrate feedstock; however, LAB were not able to use these substrates without pretreatment. Various physical, chemical and biological methods were used to remove lignin, separate cellulose and hemicellulose, increase the accessible surface area, partially depolymerize cellulose, and enhance porosity of the materials to promote the subsequent action of the hydrolytic enzymes. Enzymatic hydrolysis converts the polysaccharides remaining after pretreatment in the water-insoluble solid fraction into soluble sugars further utilized by LAB (Abdel-Rahman et al. 2011). Compounds toxic to fermentative organisms such as furfural, phenolic derivatives, and inorganic acids are also released during pretreatment, urging to seek resistant bacteria or carry out detoxification (Guo et al. 2010). Other wastes can be used as substrates for LAB; however, they are often subjected to either pretreatment or supplementation of missing compounds as carbon sources or minerals (Dumbrepatil et al. 2008; Pacheco et al. 2009; Panesar et al. 2010; Özyurt et al. 2017).

Besides external factors, bacteria of the same or different species are able to conduct regulation via quorum sensing. Quorum-sensing bacteria produce and release chemical signal molecules termed autoinducers altering gene expression and behavior in detecting bacterium (Waters and Bassler 2005). Three-component regulatory operon involving quorum-sensing mechanism was found in *Lactobacillus plantarum* NC8. The presence of specific bacteria could act as an environmental signal able to switch on bacteriocin production in *L. plantarum* NC8 mediated by the induction factor PLNC8IF. Supernatants of other species (*Lactococcus lactis* MG1363) expressing this gene also promoted bacteriocin production in NC8 (Maldonado et al. 2004). *Lactobacillus plantarum* DC400 synthesized pheromone PlnA both under mono- or co-culture conditions. PlnA represents an induction factor for gene regulation (pheromone behavior), and it acts as an antimicrobial peptide. Its biosynthesis was induced to different extent depending on microbial partnership with *Lactobacillus sanfranciscensis* DPPMA174 influencing the highest yield of PlnA. The mixed culture carried out biosynthesis and the concentrations of specific volatile organic compounds (e.g., furanone B and decanoic acid) were suggested to act as signal molecules (Di Cagno et al. 2010). Autoinducer-2 (AI-2) was used as signaling molecule in one of the primary bacterial interspecies communication mechanisms known as the luxS-mediated universal signaling system. The *Lactobacillus sakei* NR28 produced a significant reducing effect on the expression of virulence factors in enterohaemorrhagic *Escherichia coli* by AI-2 signaling inhibition (Park et al. 2014). On the other hand, luxS-mediated system of bifidobacteria takes part in gut colonization and protection from pathogens (Christiaen et al. 2014). Production of some compounds can be regulated by product itself. External nisin has been shown to induce its own synthesis, functioning as signaling molecule (Kuipers et al. 1995; Qiao et al. 1996). PepR1 of *L. bulgaricus* was involved in regulation of the prolidase PepQ biosynthesis. In the absence of glucose PepR1 does not stimulate pepQ transcription and is constantly synthesized. In the presence of glucose PepR1 blocks transcription of its own gene and induces *pepQ* transcription (Morel et al. 2001).

Co-cultivation may exert positive effect on growth of the both bacteria. *Streptococcus thermophilus* in combination with *L. bulgaricus* (basonym *Lactobacillus delbrueckii* ssp. *bulgaricus*) showed up-regulation of peptides and amino acid transporters and of specific amino acid biosynthetic pathways, notably for sulfur amino acids as well as genes and proteins involved in the metabolism of various sugars (Herve-Jimenez et al. 2008). Further research of interactions between these bacteria showed that formic acid, folic acid, and fatty acids were provided by *S. thermophilus*. The cleavage of casein into peptides by membrane-resident protease of *L. bulgaricus* and the enhanced expression of peptidases in *S. thermophilus* supported increased growth rates of both species in mixed culture. Genes involved in iron uptake by *S. thermophilus* were affected and genes coding for exopolysaccharide production in both organisms were up-regulated in mixed culture as compared to monocultures (Sieuwerts et al. 2010).

9.4 Production of bacteriocins. Genetic regulation of bacteriocin production

Bacteriocins are ribosomally synthesized peptides possessing antimicrobial activity. They often demonstrate activity toward a specific group of bacteria over a wide pH range in contrast to antibiotics. Bacteriocins are also readily degraded by proteolytic enzymes due to their proteinaceous nature, which makes them harmless to the human body and the surrounding environment and useful for food and clinical applications (Perez et al. 2014). Bacteriocins are generally divided into several classes.

1. Class I (lantibiotics) unites small thermostable peptides (<5 kDa) possessing unusual post-translationally modified residues such as lanthionine or 3-methyllanthionine. Such atypical residues form covalent bonds between amino acids, resulting in internal "rings" and giving lantibiotics their characteristic structural features.
2. Class II is represented by unmodified heat stable bacteriocins (<10 kDa) lacking unusual modifications.

3. Class III includes unmodified heat unstable bacteriocins sized over 10 kDa with bacteriolytic or non-lytic mechanism of action (Perez et al. 2014; Alvarez-Sieiro et al. 2016).
4. Class IV embraces complex bacteriocins containing lipid or carbohydrate moieties (Ahmad et al. 2017). Recently the members were re-classified as bacteriolysins (i.e., hydrolytic polypeptides), leaving only three classes of bacteriocins (Mokoena 2017).

In turn, the classes are divided into subclasses based on the biosynthesis mechanism and biological activity (Alvarez-Sieiro et al. 2016).

Mode of action of these compounds depends on group of bacteriocins. Nevertheless, they often affect cell membrane. Different charges of membrane and bacteriocins lead to electrostatic interaction between them, facilitating attraction of the molecules to the membranes. Bacteriocins often influence gram-positive bacteria, whereas gram-negative cells contain extra lipopolysaccharide outer membrane, so that additional agents are required to compromise its integrity. Bacteriocins are able to form pores leading to the dissipation of the membrane potential, take part in the efflux of cell metabolites, or induce membrane permeabilization (Perez et al. 2015).

Majority of the genes encoding bacteriocins are clustered in operons lying in the bacterial chromosome, plasmids, or transposons. The expression needs at least two genes: gene directly encoding bacteriocin and gene of immunity protein providing protection from the compound (some operons of class II bacteriocins). In most cases, the production is also dependent on specific export machinery and regulation factors. The operons of lantibiotics are more complex because they require additional enzymes for posttranslational modifications. Bacteriocins are mainly synthesized as propeptides with leader sequences (Dimov et al. 2005). Some bacteriocins possess more complex structure consisting of several peptides (Stephens et al. 1998).

Nisin is the best-studied compound among lantibiotics as well as bacteriocins. The nisin gene cluster contains 11 genes (*nisABTCIPRKFEG*). *nisA* encodes the precursor peptide of 57 amino acid residues; *nisB* and *nisC* encode putative enzymes involved in the posttranslational modification reactions; *nisT* encodes a putative transport protein of the ABC translocator family, which is probably engaged in the extrusion of modified nisin precursor; *nisP* encodes extracellular protease responsible for precursor processing; *nisI* encodes lipoprotein involved in the producer self-protection against nisin; *nisFEG* encodes putative transporter proteins which are also implied in immunity mechanism. *nisR* and *nisK* take part in the regulation of nisin biosynthesis. NisR is a response regulator, and NisK is a sensor histidine kinase which belongs to the class of two-component regulatory systems (Kuipers et al. 1995). Biosynthesis of bacteriocin goes the following way. *nisA* is translated to pre-nisin A followed by transformation to precursor nisin A with formation of several disulfide bridges and modification of some amino-acids. Then the precursor nisin A is exported out of the cell, while the leader peptide is cleaved and the final product nisin A is obtained (Dimov et al. 2005). Added to regulatory genes of the operon, transcription of nisin can be stimulated by nisin itself, nisin mutants, or analogs, but not by the unmodified precursor peptide or by other antimicrobial peptides (Kuipers et al. 1995).

Among class II bacteriocins pediocin PA-1 was most thoroughly studied biochemically and genetically and, unlike nisin, pediocin PA-1 is heterologously expressed in other genera (Moon et al. 2006). The bacteriocin operon is represented by only four genes *pedA*, *pedB*, *pedC*, and *pedD*. *pedA* encodes the precursor of pediocin PA-1; *pedB* encodes the immunity protein; products of *pedC* and *pedD* take part in bacteriocin transport. PedC is suggested to be involved in the channel formation. PedD shows homology with other bacteriocin ABC-transporters and also is capable of processing pediocin by cleavage of the leader sequence (Venema et al. 1995).

Besides direct genetic regulation, various factors influence bacteriocin production. The optimal conditions for producing bacterial strains are also favorable for generation of antimicrobial compounds. Temperature, pH, the presence or lack of certain substances were able to affect bacteriocin production (Diep et al. 2000; Li et al. 2002; Leroy and De Vuyst 2005; Van den Berghe et al. 2006). Class I and II bacteriocin regulation relies on signal transduction systems mostly differentiated with regard to peptide inducer. The bacteriocin of class I has a two-component regulatory system activating auto expression. Class II regulation is almost identical to class I regulation pathways, but it is generally associated with peptide pheromone induction as opposed to autoregulation (Snyder and Worobo 2014).

9.5 Biosafety assessment of lactic acid bacteria

LAB have a long application history. They find use in production and preservation of fermented food and probiotics and have "qualified presumption of safety" (QPS) status and are "generally recognized as safe" (GRAS) microorganisms by the European Food Safety Authority (EFSA) and Food and Agriculture Organization of the United Nations (FAO), respectively. Nevertheless, there are several criteria to ensure safety of LAB used as probiotics for the consumers.

1. Correct assessment of taxonomic identity should be defined in the screening process for new probiotic strain. It allows to affiliate the studied strain with the identified variants and obtain scientific and technological information, including data on growth conditions, metabolic and genomic characteristics.
2. The strain should adhere to colonization site and multiply with beneficial impact on the host organism.
3. The strain should not carry transmissible antibiotic resistance genes and virulence factors and should not induce adverse effects in the host body.
4. The strain should be viable and genetically stable to ensure that specific health-promoting characteristics and functionalities are not affected during long-term preservation and production.
5. Administration of D(–)-lactate-producing probiotics should be carefully traced in patients with risk of developing D-lactic acidosis, in cases of bowel surgery and subsequent short gut syndrome, in the newborn category.

In addition to probiotic strain characteristics, evaluation of biosafety should take into account purity of the culture from contaminating microbes or other substances, including allergenic materials, physiological status of the consumers, supplied dose, and method of administration.

Assessment of LAB safety is carried out by various methods, but general health status of the animal models and the specific parameters are mainly studied (Sanders et al. 2010).

9.6 Biochemistry and genetics of antibiotic resistance

Antibiotics are antimicrobial agents killing or inhibiting growth of bacteria. Intensive pursuit of substances targeting only disease-causing microbes began in the twentieth century, although antibacterial properties of molds have been known since ancient times. Penicillin discovery by Fleming and the following description of its purification process led to mass production of antibiotics and search of new ones. In the course of time resistant bacteria have been revealed. Abuse of antibiotics increased antibiotic resistance, especially since the sixties. Moreover, bacteria not susceptible to several antimicrobials (multidrug-resistant bacteria) have emerged (Aminov 2010; Davies and Davies 2010; Ventola 2015).

Antibiotics demonstrate their antibacterial activity via different ways. They may inhibit synthesis of proteins, nucleic acids, cell wall components, disorganize membrane, and so on. Biological mechanisms of antibiotic resistance are diverse, but they all can be summarized:

1. Antibiotic degradation or transformation: Bacteria can produce one or more enzymes decomposing or chemically modifying antimicrobial agents.
2. Active efflux that pumps out the antibiotic molecules penetrating into the cell until they reach concentration subminimal for antibacterial activity.
3. Receptor modification interfering with binding function and leading to loss of antibacterial effect.
4. Changes in permeability of cell wall restricting antimicrobial access to target sites.
5. Acquisition of metabolic pathways alternative to those inhibited by the drug.
6. Hyperproduction of the target enzyme (Alanis 2005; Van Hoek et al. 2011).

Antibiotic resistance can occur in two ways: via mutation of regulatory or structural genes or acquisition of a resistance gene from an exogenous source (horizontal gene transfer). LAB can be carriers of antibiotic resistance genes to pathogenic species, although LAB are not generally targeted by antibiotic treatments since they are considered to be non-pathogenic. It makes them objects of studies to secure safety of probiotics.

Antibiotic resistance of LAB can be related to the absence of the target. Most species are resistant to metronidazole because they do not possess hydrogenase activity. Insensitivity to sulfonamides and trimethoprim is determined by LAB limited biosynthetic capabilities and lack of the folic acid synthesis. Antibiotic profiles of different representatives of this group are quite various. *Lactobacillus* species are generally resistant to glycopeptides, such as vancomycin, whereas *Lactococcus* is usually susceptible to this antibiotic (Ammor et al. 2007). To detect transmissible antibiotic resistance genes, screening by PCR technique and nucleotide sequencing were carried out. However, antibiotic resistance can be encoded by several genes, as in tetracycline case. At least 40 different tetracycline resistance genes (*tet*) have been characterized (Roberts 2005). Different bacterial groups carried diverse sets of *tet* determinants (Roberts 1996).

9.7 Taxonomy of LAB. Significance of LAB systematics

The first pure culture of LAB was isolated late in the nineteenth century (Santer 2010). After publication of the monograph by Orla-Jensen in 1919, the principles of modern LAB classification were formulated. Taxonomic affiliation of the bacteria was based on cellular morphology, mode of glucose fermentation, growth temperatures, and range of sugar utilization

(Orla-Jensen 1919). Taxonomic classification has long been focused solely on phenotypic characteristics, with genetic data analysis introduced only in the 1960s. Since the 1980s, the development of PCR technique and sequencing of the 16S rRNA gene led to major changes in prokaryotic systematics (Sentausa and Fournier 2013).

At present most of LAB belong to phylum *Firmicutes*, order *Lactobacillales*, including genera *Aerococcus*, *Alloiococcus*, *Carnobacterium*, *Enterococcus*, *Lactobacillus*, *Lactococcus*, *Leuconostoc*, *Oenococcus*, *Pediococcus*, *Streptococcus*, *Symbiobacterium*, *Tetragenococcus*, *Vagococcus*, and *Weissella*. Species of *Bifidobacterium* genus from phylum *Actinobacteria* are referred to LAB in some cases because of their ability to produce lactic acid, but these bacterial groups are phylogenetically distinct (Biavati 2001; Liu et al. 2014).

Due to industrial and medical implications, correct identification and taxonomic affiliation are significant for further LAB application. Systematic approach allows to classify bacteria into groups or taxa on the basis of their mutual similarity or evolutionary relatedness, distinguish the known strains and recognize novel ones. A proper identification of the probiotic strain defines safety risks associated with the specific microorganism and rules out the inclusion of potential pathogenic species in commercial product formulas (Gueimonde et al. 2006).

Identification of lactic acid bacteria. Instrumental Technique. Identification on generic and species level. Gene sequencing: 16S rRNA and constitutive genes. Identification to species, subspecies, strain level. Molecular typing: REP-PCR, ERIC-, BOX-, (GTG)5-PCR, RAPD-PCR.

Due to their extensive use, especially in food fermentations, LAB have been thoroughly characterized. This bacterial group is one of the well-studied microbial objects. The correct identification of LAB is essential to improve technological aspects and provide safety and quality. Currently various identification techniques are available. Phenotypic methods are traditionally used in bacterial description and classification. These methods include morphological and physiological characterization, carbohydrate fermentation patterns, and protein profiling. They usually show a relatively poor reproducibility and low taxonomic resolution, often allowing differentiation only at the genus level. Phenotypic methods do not afford unique descriptions for each bacterium and strains of various genera may possess similar features that not necessarily represent evolutionary relationships between species. However, these methods do not require special equipment, and the experiments can be carried out at any laboratory. Moreover, multiple phenotypic techniques can be combined to correct identification (Temmerman et al. 2004; Moore et al. 2010).

Molecular-genetic techniques exhibit various levels of discriminatory power, from species level to differentiation of individual strains (typing). Many methods are based on PCR, conducting the selective amplification of targeted DNA fragments using specific oligonucleotide primers. DNA-DNA hybridization (DDH) is one of the first genomic methods used for the comparison of bacteria. A DDH similarity of approximately 70% serves as the recommended demarcation value for bacterial species. Nevertheless, the method is tedious and complicated for wide application (Moore et al. 2010). DDH should be performed in cases where the new taxon contains more than a single strain or when strains share more than 97% of 16S rRNA gene sequence similarity (Mattarelli et al. 2014).

Nowadays 16S rRNA gene sequence analysis is the most practiced molecular-genetic technique. Popularity of the method is supported by ubiquitous distribution of the gene, its functional stability and large size. Besides, 16S rRNA exhibits both evolutionarily conserved regions and highly variable structural elements (Ludwig and Klenk 2005). The gene structure allows to design universal primers to identify different species as well as primers to distinguish separate species and strains, including LAB (Chagnaud et al. 2001; Caro et al. 2015). However, because of the conserved nature of the gene, in some cases 16S rRNA analysis is not sufficient for differentiation between LAB species, like *Enterococcus* spp. (Moraes et al. 2013). The 16S rRNA gene can be present in multiple copies, which might cause identification problems. *Lactobacillus* strains have been shown to possess usually from four to seven copies of the gene, while copy number of rRNA per bifidobacterial genome can vary from 1 to 5 (Candela et al. 2004; Lee et al. 2008).

Internal transcribed spacer (ITS) separating 16S and 23S rRNA genes is also used in bacterial identification. 16S-23S ITS displays considerable variation in both the length and the nucleotide sequence suitable for application in molecular-genetic techniques (Gürtler and Stanisich 1996). 16S-23S region exhibits larger variation than rRNA genes, which is appropriate to discriminate taxonomically proximal species. 16S rRNA analysis cannot always cope with distinguishing closely related *Lactobacillus* species as compared to DDH procedure or Southern type hybridization. PCR with the primers tuned to 16S-23S region allows the same accuracy of species identification as the latter techniques (Berthier and Ehrlich 1998). Moreover, spacer sequence identification showed the advantage of distinguishing between *Lactobacillus rhamnosus* and *Lactobacillus casei* strains, which could not be accomplished by comparison of 16S V2-V3 region sequences (Tannock et al. 1999).

Housekeeping genes encode products essential for cell survival, and therefore, they undergo changes rarely. However, these genes have been reported to evolve much faster than rRNAs; hence, they can be engaged in bacterial identification (Ochman and Wilson 1987). The phenylalanyl-tRNA synthase alpha subunit (*pheS*) and the RNA polymerase alpha subunit (*rpoA*) partial gene sequences can be used as genomic markers alternative to 16S rRNA gene sequences, and they have a higher discriminatory power for LAB. Strains of the same enterococcal species have at least 99% *rpoA* and 97% *pheS* gene sequence similarity, whereas different enterococcal species have at maximum 97% *rpoA* and 86% *pheS* gene sequence similarity (Naser et al. 2005). The *pheS* gene sequence analysis provided the highest discrimination for the identification of different species of lactobacilli. *pheS* provided an interspecies gap, which normally exceeded 10% divergence and an intraspecies variation up to 3%, while *rpoA* revealed a lower resolution with an interspecies gap normally exceeding 5% and an intraspecies variation up to 2% (Naser et al. 2007). Additionally, the *pheS* and *rpoA* genes were successfully used in identification of novel species (Yi et al. 2013; Chang et al. 2015; Kadri et al. 2015).

To discriminate different bacterial strains of the same species, methods of molecular typing are preferred. Pulsed field gel electrophoresis (PFGE) is the technique used for separating larger pieces of DNA by applying electrical current that periodically changes direction (three directions) in a gel matrix, unlike the conventional gel electrophoresis where the current flows only in one direction (Adzitey et al. 2013). PFGE allows to study genotypic diversity of species, select and optimize cultures with desired properties for industry (Psoni et al. 2007; Kahala et al. 2008; Bouchard et al. 2015).

Restriction fragment length polymorphism (RFLP) is characterized by the use of restriction enzymes to digest DNA and following separation of the restriction fragments according to their length by agarose gel electrophoresis (Adzitey et al. 2013). Single digestion with *AciI* of *rpoB* gene (coding for RNA polymerase β-subunit) in *Leuconostoc*, *Oenococcus*, *Pediococcus*, and two or three digestions (*AciI*, *HinfI* and *MseI*) in *Lactobacillus* spp. allowed to identify LAB species commonly isolated from wine (Claisse et al. 2007). Restriction patterns of the *tuf* gene, encoding the elongation factor Tu and universally distributed in gram-positive bacteria, derived by enzymes *AluI* and *HaeIII* could effectively differentiate closely related *Lactobacillus* species (Park et al. 2012). However, sometimes strain variations could not be demonstrated by the RFLP analysis. Morphologic differences (colony shape and size) were evident between *Lactobacillus kefir* strains ATCC 35411 and ATCC 8007, but genotypic results failed to differentiate them (Mainville et al. 2006).

Random amplified polymorphism DNA (RAPD) is the method in which arbitrary primers (typically 10-mer primers) are used to randomly amplify segments of target DNA under low-stringency PCR conditions with generation of the set of finger printing patterns of different sizes specific to each strain (Adzitey et al. 2013). The technique was successfully tested in identification of *Lactobacillus*, *Lactococcus*, *Enterococcus*, and *Streptococcus* strains (Cocconcelli et al. 1995; Samarzija et al. 2002; Rossetti and Giraffa 2005). Additionally, RAPD can follow and study the progression of starter cultures in food fermentations (Plengvidhya et al. 2004; Siragusa et al. 2009).

Amplified fragment length polymorphism (AFLP) involves the use of restriction enzymes to digest total genome DNA followed by amplification of a subset of selected restriction fragments (Adzitey et al. 2013). The AFLP method enables to delineate closely related strains, like *Lactococcus lactis*. Advanced analysis of the AFLP profiles in combination with genome analysis can facilitate the recognition of genetic markers responsible for specific phenotypic traits, contributing further to development of rapid and predictive screening procedure for culture collections (Kütahya et al. 2011).

Some methods of molecular typing are based on use of oligonucleotide primers complementary to repetitive sequences. Diverse regions of DNA flanked by the rep sequences are amplified, leading to amplicon patterns. The conserved repetitive sequences are divided into four types: the repetitive extragenic palindromic (REP), the enterobacterial repetitive intergenic consensus (ERIC), the BOX, and the polytrinucleotide $(GTG)_5$ sequences (Mohapatra et al. 2007). Repetitive extragenic palindromic (REP) primers, the enterobacterial repetitive intergenic consensus (ERIC) primers, and the $(GTG)_5$ primer can be used in the typing of *Lactobacillus* strains. DNA concentration and quality did not affect the ERIC-PCR profiles, indicating that this method, unlike other high-resolution methods, can be adapted to high-throughput analysis of isolates. Also, ERIC-PCR simultaneously types isolates to the strain and species levels, compared to PFGE that can type only to the strain level (Stephenson et al. 2009). PCR using the BOXAIR primers provides differentiation at species, subspecies and strain level, acting as the tool confirming phenotypic identification (Mohammed et al. 2009).

Thus, there are a lot of techniques available for identification of LAB. Methods differ in taxonomic resolution, labor expense, and cost. Phenotypic techniques are simple and affordable, but they do not possess high discriminatory power. Molecular-genetic methods provide more accurate identification, allowing to distinguish even separate strains. However, in most situations identification to the species level is required, where 16S rRNA sequencing or DDH suffice. Typing methods delineate strains and permit to study and select LAB with set characteristics for industrial processes.

9.8 Beneficial lactic acid bacteria

Selection criteria for probiotic strains. Industrial strain development. Technological properties. Beneficial properties of LAB are diverse. They are able to improve nutritional value of foodstuffs, stimulate lactose digestion, control infections, allergic reactions and some types of cancer, and enhance immune system functions (Gilliland 1990; Wedajo 2015). Valuable properties of products largely depend on selected LAB strains. There are several important selection criteria allowing to choose strains, which can be further used in manufacturing of probiotic products with desired characteristics.

1. Origin of strain: Probiotic bacterium should originate in microflora of host organism. It will guarantee safety and better attachment to intestinal wall as compared with LAB from other groups of organisms.
2. Accurate identification to genus, species, and strain level allows to screen variants possessing favorable properties and to avoid pathogenic species.
3. Probiotic should be safe for consumers.
4. The strain should be resistant to bile acids and the gut environment to survive in the gastrointestinal tract and exercise its beneficial properties on the organism.
5. Probiotic culture should be able to adhere to and colonize intestinal epithelium. It prevents cells from wash-out and ensures immune modulation, competitive expulsion of pathogens, production of enzymes, lactic acid, and vitamins by LAB.
6. The probiotic strain should be capable of producing antimicrobial substances such as organic acids, hydrogen peroxide, bacteriocins, and so on. Synthesis of these compounds is one of the mechanisms of LAB beneficial action.
7. LAB should stimulate immune response and provide protection from various types of diseases (Shewale et al. 2014).

Compliance with these criteria will allow variants with beneficial properties to be picked up. Nevertheless, coupled to health-promoting effects, the LAB strains should possess certain technological properties ensuring their use in industrial processes.

1. The cultures should be stable and viable. LAB are chosen for the ability to survive in fermentation process, during manipulations and storage, to achieve maximum biomass concentration in a simple, and cheap nutrient medium. Poor endurance of bacteria restrains scope of LAB application, even strains with good selected characteristics.

2. Scale-up of probiotic technology is indispensable for mass fabrication of the product.
3. The strain should have good organoleptic properties contributing to attractive product taste, flavor, appearance, and texture.
4. The strain should be resistant to phages. Sensitivity to bacteriophages is one of the most grave challenges in food industry. LAB viruses cause great economic losses due to fermentation failure; 0.1% to 10% of all milk fermentations are negatively affected by virulent phages. Their presence results in deterioration of product quality parameters, like taste, flavor, texture, and development of contaminating microbiota (Szczepankowska et al. 2013; Shewale et al. 2014; Wedajo 2015).

Applications of probiotics face many problems. The proper selection criteria and technological properties allow to choose strains with beneficial properties sustained during fermentation process. The probiotic product favorably influences host health. Mechanisms of LAB-positive action are diverse and some compounds produced by LAB are especially valuable to the organism.

Lactic acid bacteria as sources of bioactive compounds. Biologically active polar lipids and polysaccharides. Chemical composition and structure. Health benefits of bioactive compounds.

LAB employed in the food industry usually produce biologically active compounds, giving the final product an additional nutritional and health-promoting value. LAB are sources of vitamins, bacteriocins, exopolysaccharides, enzymes, conjugated linoleic acid, and so on. These compounds differ in chemical structure and beneficial properties. Biosynthesis and composition of some substances (e.g., bacteriocins, vitamins, and sweeteners) were discussed in previous chapters.

γ-aminobutyric acid (GABA) is the amino acid acting as the major inhibitory neurotransmitter in the mammalian central nervous system. GABA displays hypotensive, tranquilizing, diuretic, and antidiabetic effects and additionally upgrades plasma concentration, growth hormones and protein synthesis in the brain. The biosynthesis of GABA is catalyzed by glutamate decarboxylase transforming glutamate to the bioactive compound (Dhakal et al. 2012). The ability to produce GABA was found in some *Lactobacillus*, *Lactococcus*, *Leuconostoc*, and *Weissella* strains (Kim and Kim 2012; Lacroix et al. 2013; Kook and Cho 2013). In LAB, GABA takes part in acid resistance mechanism (Sanders et al. 1998).

Conjugated fatty acids, products of lipid metabolism, are important bioactive compounds, especially CLA. CLA belongs to the family of isomers of octadecadienoic acid (18:2) carrying a pair of conjugated double bonds along the alkyl chain. As mentioned previously, it is synthesized through two consecutive reactions from linoleic acid. Many LAB demonstrate the ability to produce CLA isomers; hence meat and milk from ruminants and the derived products are the natural sources of these bioactive compounds. Properties of CLA administration depend on isomer, doses administered, and the period of study. Biological activities of CLA are expressed as inhibition of various types of cancer, immunoregulatory, antioxidant, anti-osteoporotic, and anti-atherosclerotic effects, and decrease in body fat mass. However, not all isomers are absorbed to the same extent, and there are some reports of possible adverse impact. Pro-carcinogenic effects and increased production of prostaglandins attributed to CLA 10-trans and 12-cis isomers have been reported. A negative alteration in the serum lipid profile and probability of developing insulin resistance have been demonstrated (Van Nieuwenhove et al. 2012; Lehnen et al. 2015; Kuhl and De Dea Lindner 2016).

EPS form a diverse group of compounds divided into homo- and heteropolysaccharides, depending on the number of monosaccharide types in their structure. Homopolysaccharides are made from sucrose using glucansucrase or levansucrase, while the synthesis of heteropolysaccharides involves sugar transportation, sugar nucleotide synthesis, repeating unit synthesis, and polymerization of the repeating units. In bacterial cells, EPS take part in protection from various adverse factors such as phage attack, toxic metal ions, and desiccation (Harutoshi 2013). Extensive EPS applications are determined by the specific compound. They can be used as adjuvants, emulsifiers, carriers, stabilizers, humectants, bio-thickeners, prebiotics, sweeteners, plasma substitutes, matrices of chromatography columns, anticoagulants, and so on. EPS may be applied not only in food processing and pharmaceutics, but in paper industry, metal-plating, and oil recovery. However, in some cases, EPS cause food spoilage. Mass production of EPS requires correct knowledge of EPS biosynthesis mechanisms and optimized bioprocess technology (Patel et al. 2012).

Bioactive peptides are produced from proteins during LAB fermentation and have positive influence on the organism. These peptides exert a wide range of effects. They control arterial blood pressure through the contraction of smooth muscles of blood vessels; express free radical-scavenging activity and promote synthesis of antioxidants, retarding lipid peroxidation; reduce or inhibit the formation of blood clots; repress the reabsorption of bile acid in the ileum, decreasing blood cholesterol level; show agonistic or antagonistic action toward opiate receptor; suppress the appetite, preventing weight build-up; display bacterial membrane-lytic activities; exert immunomodulatory and cytomodulatory effects; transfer different minerals by forming soluble organophosphate salts; and play a growth-promoting role. Milk and colostrum of dairy species are considered as the most important sources of bioactive peptides (Park and Nam 2015).

LAB are producers of many ferments that can be used in various fields. Enzymes degrade caseins yielding key flavor components, which contribute to the sensory perception of dairy products (Smit et al. 2005). These bacteria also possess a broad array of enzymatic activities influencing wine composition and quality of wine (Matthews et al. 2004). Some representatives of genera *Lactobacillus*, *Lactococcus*, and *Streptococcus* demonstrated amylolytic activity. It provides opportunity to produce lactic acid directly from starch as a carbon source (Petrova et al. 2013). β-galactosidase, or lactase, is extensively used in food and pharmaceutical industries due to its capability to hydrolyze lactose to monosaccharide and eliminate lactose intolerance problem (de Vrese et al. 2001).

The main distinctive feature of LAB is the ability to produce lactic acid, conferring the group name. Since its discovery in 1780 by Scheele, the compound has become the important chemical used in food, cosmetic, pharmaceutical, and chemical industries. Lactic acid is regarded as the raw material in manufacturing of a number of products such as lactate ester, propylene glycol, 2,3-pentanedione, propanoic acid, acrylic acid, acetaldehyde, dilactide, and even biodegradable polymer polylactic acid mainly applied in packaging. Lactic acid functions as a descaling agent, pH regulator, neutralizer, chiral intermediate, solvent, humectant, cleaning aid, skin-lightening and rejuvenating substance, moisturizer, slow acid-releaser, metal complexing and antimicrobial agent. It is also applied in tableting, prostheses, surgical sutures, controlled drug delivery systems, as electrolyte in solutions (Wee et al. 2006). The food industry is the main consumer of lactic acid accounting for approximately 85% of its total demand (John et al. 2007). It is widely used in almost every segment of the food industry for flavoring, pH regulation, mineral fortification, increasing shelf life, and better control of foodborne pathogens (Wee et al. 2006). The estimated global lactic acid demand of 714.2 kilo tons in 2013 is expected to reach 1,960.1 kilo tons by 2020 (SpecialChem 2014). The compound can be produced by two ways: via chemical synthesis or microbial fermentation. LAB produce lactic acid by homofermentative or heterofermentative pathways, whereas *Bifidobacterium* cultures use special metabolic route.

Mechanisms of beneficial action of lactic acid bacteria. Gastroenterological effects. Regulation of lipid metabolism. Immunity enhancement. Cancer prevention.

LAB play a fundamental role in health maintenance. LAB and the derived products can be used in treatment of various infections and cancer, immunity enhancement, and regulation of diverse functions of the organism. Positive action of the bacteria can be achieved by several mechanisms.

Pathogenic microorganisms can cause adverse effects on the body, leading to death ultimately. LAB helps to prevent proliferation and growth of pathogens. One of the mechanisms of antimicrobial activity is produced by secretion of bioactive compounds. Some chemicals, like bacteriocins demonstrate activity toward specific group, species, or strain of bacteria. Eliminating pathogenic bacteria, the broad-spectrum bacteriocins may change microbiota diversity (Rea et al. 2011). Short-chain fatty acids (SCFA), like lactic acid, affect bacterial fitness via acid stress, additionally modulating host immune functions and serving as metabolic substrates (Sun and O'Riordan 2013). Lactic acid, besides pH reduction, also functions as a permeabilizer of the gram-negative bacterial outer membrane and may act as a potentiator of the effects of other antimicrobial substances (Alakomi et al. 2000). Hydrogen peroxide shows microbicidal properties by damaging cell structure. Some studies demonstrated that hydrogen peroxide displayed enhanced killing activity in the presence of lactic acid, while in other cases pathogens could be suppressed with acid but not peroxide (Atassi and Servin 2010; O'Hanlon et al. 2011).

Other LAB strategies to prevent spread of pathogens are displacement, exclusion, and competition. The studies show that LAB were able to attach to mucosa of intestinal epithelial cells, blocking pathogen adhesion, reducing colonization and invasion and preventing infection. Bacteria act as a barrier to avoid direct contact between pathogens and epithelial cells, protecting thereby cells from the damage inflicted by pathogenic species (Jankowska et al. 2008; Abdel-Daim et al. 2013). LAB may not elicit bactericidal effect on pathogens, but decrease their toxin production. Shiga-toxin-producing *Escherichia coli* inhibits protein synthesis in eukaryotic cells and plays a role in hemorrhagic colitis and hemolytic uremic syndrome. *Bifidobacterium*, *Lactobacillus*, and *Pediococcus* strains have been shown to down-regulate shiga toxin expression, but the probiotic effect was strain-specific and might be related to pH effect as a result of organic acid production by LAB (Carey et al. 2008).

Nutritional requirements may be shared by pathogens and microflora, so that they may potentially compete for growth-limiting resources. In case pathogen competes with host microbiota for only one limiting nutrient, the pathogen will be eliminated if its R* (steady-state resource concentration balancing the pathogen's birth rate versus mortality loss) exceeds the corresponding values of the host competing microflora. Where key indigenous microbes show low abundance, pathogen control by indigenous microorganisms can be enhanced by the introduction of novel populations of non-pathogenic competitors (Smith and Holt 1996).

Positive effect of LAB can be expressed via direct action on the organism, like enhancement of barrier function. The intestinal epithelium acts as a selectively permeable barrier regulating the absorption of nutrients, electrolytes and water, and providing effective defense against toxins, antigens, and enteric flora (Groschwitz and Hogan 2009). Breakdown of intestinal barrier function plays a crucial role in development of such pathologies as infectious enteritis and inflammatory bowel diseases (Halpern and Denning 2015). Mucins, large complex glycoproteins, protect intestinal mucosal surfaces by limiting access of environmental matter to their epithelial cells. It was shown that *Lactobacillus* strains increased extracellular secretion of mucin, leading to reduced adherence of enteropathogen *E. coli* during coincubation experiments (Mack et al. 2003). The increase in intestinal permeability induced translocation of antigens across the epithelium, provoking inflammation. *Bifidobacterium* strains possess the capacity to prevent disruption of intestinal epithelial barrier and to promote its integrity. The up-regulation of the production of SCFA (acetate and formate) restores the barrier (Hsieh et al. 2015).

Additionally, LAB can exert an immunomodulatory effect on the organism. The intestine contains 70%–80% of all immunoglobulin A (IgA) producing cells. Macrophages, regulatory T cells, and effector B and T lymphocytes induce the protective IgA function-associated with the mucosal surfaces. The inductive sites are represented by the Peyer's patches (PP), the appendix and the small lymphoid nodules in the large intestine (Perdigón et al. 2001). It was shown that concentrations of IgA^{+} cells and IL-6-producing cells increased after 7 days of *Lactobacillus casei* administration. IL-6 released by epithelial cells or macrophages takes part in the enhancement of IgA secretion by inducing the terminal development of B cells in plasmatic cells, which express the corresponding immunoglobulin (Galdeano and Perdigón 2006). The use of high doses of antibiotic kanamycin resulted in increased

IgE levels and decrease in IgA with the number of PP cells, but LAB treatment led to the reverse situation (Kim and Jeung 2016). The administration of *L. bulgaricus* raised both interferon γ and interleukin 17 production by $CD4^+$ T cells from PP. The population of $CD4^+$ T cells is about 10% in PP, and they are in charge of responding to exogenous antigens by building up anti-inflammatory and anti-infectious functions (Kamiya et al. 2017). Dendritic cells are crucial immune cells linking innate immune response and acquired immunity by distinct capacity to recognize pathogenic and endogenous inflammatory signals. These cells vary in tissue distribution, pattern of cytokine/chemokine production, and interactions with other immune cells. Plasmacytoid dendritic cells (pDC) take part in various processes ranging from the enhancement of anti-viral immunity to augmentation of differentiation of $CD4^+$ induced regulatory T cells. *Lactococcus lactis* stimulated pDC capacity to induce $CD4^+CD25^+$ regulatory T cell generation (Jounai et al. 2012). LAB were even shown to promote the production of cytokines in macrophage cells (Hong et al. 2009).

LAB are able to influence lipid metabolism and synthesis. Visceral fat accumulation may spur up progress of several diseases, including diabetes, hyperlipidemia, hypertension, and arteriosclerosis. Sterol regulatory element-binding protein (SREBP) expression leads to the transcriptional activation of lipogenic genes in the liver and the development of beta-cell dysfunction in the pancreas caused by elevated levels of free fatty acids. Administration of *Lactobacillus gasseri* exerted anti-lipogenic effects manifested as decrease in expression of the mRNA SREBP and fatty acid synthase gene in the liver and reduction of free fatty acids in the blood (Yonejima et al. 2013). *Lactobacillus plantarum* displayed multiple effects on lipid metabolism. It significantly lowed intracellular triglyceride deposits and glycerol-3-phosphate dehydrogenase (GPDH) activity, mRNA expression of transcription factors, like peroxisome proliferator-activated receptor γ and CCAAT/enhancer-binding protein α involved in adipogenesis, the expression level of adipogenic markers, like adipocyte fatty acid binding protein, leptin, GPDH, and fatty acid translocase (CD36). Thus, the bacterium inhibited lipid accumulation in the differentiated adipocyte by down-regulating the expression of adipogenic transcription factors and other specific genes responsible for lipid metabolism (Park et al. 2013). On the other hand, 10-oxo-12(Z)-octadecenoic acid produced by LAB induced adipocyte differentiation via stimulation of peroxisome proliferator activated receptors γ predominantly expressed in white adipose tissue, and increased adiponectin production regulating glucose levels as well as fatty acid breakdown and insulin-stimulated glucose uptake. As a consequence, LAB fatty acids could be involved in the regulation of host energy metabolism (Goto et al. 2015). Application of mixed culture of LAB in the other experiment resulted in inhibition of fat absorption. It also inhibited 3-hydroxy-3-methyl glutaryl-CoA reductase activity, a major regulatory enzyme in cholesterol biosynthesis. It was also presumed that bacteria facilitated conversion of cholesterol to bile acids (Banjoko et al. 2012). *Lactobacillus fermentum* demonstrated ability to remove cholesterol from the cultural medium by assimilation (Pereira and Gibson 2002).

Another remarkable feature of LAB is cancer treatment potential. Cancer is one of the major causes of morbidity and mortality worldwide, with approximately 14 million new cases in 2012 and 8.8 million deaths in 2015. Nearly one in six deaths is due to cancer. Disease incidence is expected to rise by about 70% over the next two decades, making it an acute global challenge (WHO 2017). Some LAB strains have been shown to activate antitumor mechanisms regulating the host immune response. Probiotic *Lactobacillus acidophilus* was able to promote apoptosis, genetically programmed cell death, in murine colon adenocarcinoma cells (Chen et al. 2012). *Lactobacillus kefiri* selectively induced apoptosis in gastric cancer cells in a dose-dependent manner, showing no effects in breast cancer cells and human peripheral blood mononuclear cells (Ghoneum and Felo 2015). Administration of *Lactobacillus casei* and its extracts revealed anti-proliferative and pro-apoptotic effects in regard to colon carcinoma cells (Tiptiri-Kourpeti et al. 2016). Reactive oxygen species (ROS) can provoke carcinogenesis. LAB possess antioxidant properties and may prevent neoplasm development. However, studies indicate that impact of LAB on DNA damage is ambivalent. Majority of strains showed protective action against oxidative stress, while some of them induced DNA damage in untreated cells probably triggered by release of hydrogen peroxide from bacterial cells (Koller et al. 2008). Some compounds initiating cancer (carcinogens) may get into food consumed by humans. LAB can bind or degrade these substances preventing cancer development. The binding is a physical phenomenon, mostly expressed via cation-exchange mechanism. Intact cell wall and peptidoglycan show higher binding activity than the bacterial cells. However, binding process does not entail drastic changes in absorption and distribution of carcinogens. It is rapidly reversed in the gut by unfavorable conditions or factors inhibiting binding (Bolognani et al. 1997; Rajendran and Ohta 1998). *Lactobacillus delbrueckii* ssp. *bulgaricus* and *Streptococcus salivarius* ssp. *thermophiles* prevented DNA damage induced by N-methyl-N′-nitro-N-nitroso guanidine in isolated primary rat colon cells. Possible mechanism of this protection is associated with thiol-containing products of protein breakdown catalyzed by bacterial proteases (Wollowski et al. 1999).

The immune system takes part in control of cancer promotion and progression. Some LAB strains display immunomodulatory effect. and therefore they may contribute to development of anti-tumor effect (Takagi et al. 2001).

Mechanisms of LAB beneficial action are diverse. They are manifested as antimicrobial activity, enhancement of barrier function, immunomodulatory, and anti-cancer effects, influence on lipid metabolism of host organism. Health-promoting action of LAB facilitates its potential use in medicine and formulation of functional food.

9.9 Medical application of lactic acid bacteria

Beneficial features of LAB make them an attractive target in diverse spheres, including medicine. Many studies deal with treatment of various diseases by LAB application. LAB can be used to control a wide range of diseases: diarrhea of various etiology, allergy, inflammatory bowel diseases, cancer, and so on. Inflammatory bowel diseases represent a group of inflammatory dysfunctions of the colon and small intestine, with Crohn disease (CD) and ulcerative colitis (UC) as the principal types. Increase in the ratio of harmful bacteria and reduction in the levels of beneficial bacteria is commonly associated with inflammatory bowel diseases. Apart from it, abnormal host response to luminal antigens, including the resident microflora, and enhanced mucosal permeability characterize this condition. As mentioned previously, LAB demonstrate antimicrobial activity, .change intestinal permeability and modulate immune response, so that LAB and bifidobacterial probiotics can provide a remedy by improving clinical symptoms. The use of probiotics produced a favorable effect on treatment and maintenance of UC, while effectiveness for CD control was less significant (Bai and Ouyang 2006; Saez-Lara et al. 2015).

Diarrhea results from disequilibrium in the water flows in the gut mainly generated by Na-solutes cotransport systems (Na-glucose) or chloride secretion through the apical membrane of enterocytes. The first system is related to water absorption, while the latter determines water secretion in the intestinal lumen. These transporters or channels are highly regulated structures and various factors may affect their performance leading to diarrhea development. LAB are used in treatment of the disease caused by viruses, pathogenic bacteria, antibiotics, and radiation. The same LAB properties that improve condition of patients suffering from inflammatory bowel diseases are able to mitigate effects of diarrhea of various etiologies. Lactose intolerance also may lead to this disorder. In the latter case positive action of probiotics is related to the presence of β-galactosidase hydrolyzing lactose to monosaccharide and eliminating intolerance problem (Heyman 2000; Samaržija et al. 2009).

Allergies are widespread health problems in the world. It is estimated that as much as 30%–40% of global population is susceptible to allergenic agents (Żukiewicz-Sobczak et al. 2014). Allergic state emerges as the result of an inappropriate reaction to usually innocuous substances. Immune hypersensitivity reactions are mediated predominantly by IgE antibodies or T cells (Schnyder and Pichler 2009). As LAB are able to modulate immune system, they can diminish effects of allergic reactions. Type I allergy is characterized by shift in the T helper cell type 1 (Th1) and 2 (Th2) balance towards Th2-dominated response, with increase in the levels of Th2 cytokines IL-4 and IL-5. *Lactobacillus plantarum* strain inhibits allergic response through modulation of Th1/Th2 balance and promotion of regulatory T cells (Ai et al. 2016). The use of recombinant LAB producing the major birch pollen allergen Bet v 1 leads to reduced allergen-specific IgE level concomitantly with increased allergen-specific IgA concentration and offers a promising approach to prevent systemic and local allergic immune responses (Daniel et al. 2006). Even heat-killed strains show stimulating effect on IL-12p70 production, which in turn shifts the balance between the T helper type 1 and 2 cell response (Sashihara et al. 2006). LAB inducers of IL-12p70 and IL-10 in dendritic cells, supporting IFN-and IL-10 production in $CD4^+$ T cells reduce hyperresponsiveness, bronchial inflammation, and proliferation of specific T cells in cervical lymph nodes (Van Overtvelt et al. 2010).

Hepatic encephalopathy is a common and usually reversible neurocognitive syndrome occurring in patients with cirrhosis. It manifests itself as a spectrum of changes from the state of low-level cognitive dysfunction detectable in up to 70% of the patients leading to the plausible risk of cerebral edema and death. Treatment of hepatic encephalopathy with *Streptococcus thermophilus* and strains of *Lactobacillus* and *Bifidobacterium* exerted long-term positive effects (Shavakhi et al. 2014). Strain *Enterococcus faecium* SF68 demonstrated the same efficiency as lactulose in treatment of chronic hepatic encephalopathy, with no adverse symptoms and a 2-week remission (Loguercio et al. 1995).

As mentioned previously, LAB are able to regulate lipid metabolism in the body and prevent cancer development. Various mechanisms of action provide opportunity to apply these bacteria in treatment of obesity, hypercholesterolemia, colon cancer, and so on. Owing to production of a number of beneficial compounds and immunomodulatory action, LAB probiotics may be used as bioactive food supplements.

Production and application of probiotics and prebiotics. Management and research. Market demand of probiotics and prebiotics. Current market situation. Strategies.

Probiotics are health-promoting microbial agents. The global probiotics market totaled $31.8 billion and $34.0 billion in 2014 and 2015, respectively. The market capacity should reach $50.0 billion by 2020, growing at compound annual growth rate of 8.0% from 2015 to 2020 (Kumar 2016). Expansion of probiotics market is driven by rising application in the animal feed sector and consumer demand for natural products with health benefits represented by beverages, probiotics, and supplemented foods (Global Industry Analysts Inc. 2016). Functional foodstuffs and beverages segment accounted for more than 80% of the probiotics market share in 2015. Dairy products, cereals, baked food, fermented meat, and dry food were the major commodities. The experts expect significant growth in the functional food segment owing to rising demand instigated by a range of diseases (Global Market Insights Inc. 2016). Prebiotics are usually nondigestible food materials for supporting probiotic growth. Prebiotics consumption favors growth of probiotic cultures, helping to fight chronic pathologies. Prebiotics market size was above USD 3.5 billion in 2016 and may exceed USD 7 billion in 2024, with consumption rate exceeding 1.4 million tons (Global Market Insights Inc. 2017).

Production of probiotics is carried out in the following way. Using selection and technological criteria, a probiotic strain with defined features is chosen. Composition of the medium and cultural conditions for bacterial growth also pass through the optimization stage. A pretreatment phase is essential to adjust substrate characteristics and eliminate microorganisms able to interfere with fermentation process. After the pretreatment stage, bacterial cultures are inoculated into the medium and incubated for the definite period. After fermentation process, the product may be subjected to processing and upgrading to acquire additional characteristics, like flavor, taste, and so on. Finally the product is packed for further storage and market supply (Novik et al. 2017).

9.10 Functional food and nutritional supplements

Functional food includes products imparting some health benefits to the consumers. The presence of functional microorganisms such as LAB and *Bifidobacterium* provides valuable properties of food. Fermented products are used from ancient times; however, the first proposal of deliberate bacterial use in food as probiotic was formulated by Metchnikoff (Mackowiak 2013). Now functional foodstuffs are manufactured worldwide with the global market worth about USD 129.39 billion in 2015. This market is expected to reach USD 255.10 billion by 2024 (Grand View Research Inc. 2016).

Bacteria in food demonstrate a whole spectrum of properties. Microorganisms transform chemical constituents during food fermentation, enhancing accessibility of nutrients, enriching food flavor and taste, rendering bio-preservative quality, upgrading food safety, and degrading toxic components and anti-nutritive factors (Tamang et al. 2016). They may take part in prevention and treatment of diarrhea, inflammatory bowel disease, allergies, cholesterolemia and lactose intolerance, immunomodulation and cause anticancer effects via diverse mechanisms of action. Moreover, probiotic bacteria produce health-promoting bioactive compounds. Summing up, LAB and bifidobacteria used in food industry should meet certain selection criteria and possess technological properties.

Biotechnological products. Agricultural and agro-industrial applications of lactic acid bacteria.

LAB find wide use in food processing as sources of probiotics and bioactive compounds. As mentioned previously, the main part of probiotics is applied in functional food and beverages, but some of them can be used in maintenance of health and meat quality as well as treatment and prevention of various animal diseases. *Lactobacillus pentosus* strain LB-31 distinguished by the best characteristics concerning growth parameters, lactic acid production, acidic pH and bile salts tolerance, cell surface hydrophobicity, antimicrobial susceptibility, and antagonistic activity proved beneficial to broilers due to ability of modulating the immune response and upgrading morpho-physiological, productive, and health parameters of fowl (García-Hernández et al. 2016). Commercial product FM-B11 containing eleven LAB poultry gut isolates significantly reduced viability of *Salmonella* in day-of-hatch broilers. Possible mechanisms of action involve competitive exclusion or stimulation of a host innate immune response (Higgins et al. 2007). Wheat germ agglutinin (WGA) lectins are components that protect wheat from insects, yeast, and bacteria. WGA has toxic effects on intestinal epithelial cells obtained from 14-d-old broilers depending on time of exposure and lectin concentration. Adherent and nonadherent strains of LAB could avoid eukaryotic cells-WGA interactions by different mechanisms. Nonadherent bacteria could capture WGA in the intestinal lumen, reducing the amount of free lectin able to interact with epithelial cells. Adherent bacteria attached to intestinal cells interfered with the interaction between WGA and eukaryotic cells. Despite bacterial binding to epithelial cell surface, in some cases the latter remains vulnerable to damage (Babot et al. 2017). Supplementation of *Lactobacillus* strains into broiler rations could improve the body weight gain and feed conversion rate from 1 to 42 days of age and was effective in cutting abdominal fat deposition but only after 28 days of age. Such diet additionally reduced serum total cholesterol, low-density lipoprotein cholesterol and triglycerides in broilers from 21 to 42 days of age, but there was no significant difference in serum high-density lipoprotein cholesterol and in the weight of organs between control and *Lactobacillus*-fed broilers (Kalavathy et al. 2003). Other experiments showed that the addition of probiotic *Lactobacillus* spp. to the feed did not significantly improve weight gain, feed intake, and feed conversion rate of broiler chickens, but tended to increase the total number of anaerobic bacteria in the ileum and caeca, the number of LAB in the caeca and to significantly raise the weight of small intestine (jejunum and ileum), reducing the number of *Enterobacteria* in the ileum, when compared with the control (Olnood et al. 2015).

Probiotics fed to ruminant livestock have been shown to decrease scours in neonatal calves, to promote milk yields in dairy cows, decrease morbidity in newly weaned calves and new calves at the feedlot, and increase daily gains and carcass weight in feedlot cattle. Moreover, strains of *Lactobacillus acidophilus* were shown to reduce fecal shedding of *E. coli* by feedlot cattle at harvest time (Krehbiel et al. 2003). Feeding microbial inoculum of *Lactobacillus casei*, *Lactobacillus salivarius*, and *Pediococcus acidilactici* with milk replacer, when young calves consumed a large quantity of spray-dried whey powder generating intestinal imbalance, promoted earlier consumption of starter, and indirectly, may have stimulated earlier development of the rumen, omasum, and reticulum, favoring early weaning. Inoculated calves showed better growth performance, which could be related to improved digestion of lactose and spray-dried whey proteins (Frizzo et al. 2010). LAB are able to inhibit *E. coli*, preventing metritis development in dairy postpartum cows mainly by acid production (Otero et al. 2006). Steers fed with *Enterococcus faecium* EF212 had numerically lower concentrations of blood CO_2 than control steers, which is consistent with a reduced risk of metabolic acidosis (Ghorbani et al. 2002).

LAB display a lot of beneficial effects on newborn and weaned piglets, growing pigs, and sows. Studies demonstrated probiotic ability to increase average dairy and weight gains, feed conversion, to modulate immune system, to show antibacterial effects, to promote apparent ileal digestibility of crude protein, crude fiber, and organic matter to alleviate diarrhea, to raise the ratio of monounsaturated and polyunsaturated fatty acids to upgrade meat quality (Yang et al. 2015).

The growth of aquaculture has accelerated over the past decades. Aquaculture allows a selective increase in the production of species used for human consumption, industry, or sport fishing. Viral, bacterial, and fungal infections cause devastating economic losses worldwide. Use of chemical additives and veterinary medicines, especially antibiotics, generates significant risks to public health by promoting the selection, propagation, and persistence of bacterial-resistant strains. Therefore, probiotics are considered as the means to prevent and treat various diseases as well as to stimulate growth of aquatic organisms and feed conversion efficiency (Martínez Cruz et al. 2012). *Lactobacillus rhamnosus* reduced the mortality of rainbow trout (*Oncorhynchus mykiss*) significantly from 52.6% to 18.9% from furunculosis caused by *Aeromonas salmonicida* ssp. *salmonicida* (Nikoskelainen et al. 2001). Blue shrimps *Litopenaeus stylirostrisfed* treated with probiotic *Pediococcus acidilactici* displayed lower infection (20% instead of 45% in the control group) and mortality (25% instead of 41.7% in the control group) rates under exposure to *Vibrio nigripulchritudo*. Compared to the infected control group, probiotic-fed shrimp exhibited higher antioxidant status and lower oxidative stress level (Castex et al. 2010). The rations supplemented with 0.01% *Lactobacillus acidophilus* powder caused positive influence on growth, feed utilization, and survival of snakehead (*Channa striata*) fingerlings (Munir et al. 2016). Administration of *Lactobacillus plantarum* for 60 days exerted favorable effects on the specific growth rate and feed utilization efficiency of *Labeo rohita* juveniles, additionally increasing the serum lysozyme and alternative complement pathway activities, phagocytosis, and respiratory burst activity against *Aeromonas hydrophila* infection. The serum IgM levels were considerably higher in the experimental groups as compared to the control group after 30 days of feeding. The treated fish displayed enhanced survival rate (77.7%) (Giri et al. 2013). Specific and relative growth rate, protein efficiency, and feed-conversion ratio, survival, blood parameters, and total immunoglobulin concentrations were much better in African catfish (*Clarias gariepinus*) fed the ration supplemented with *L, acidophilus* when compared with the control (Al-Dohail et al. 2009). *Dicentrarchus labrax* (European sea bass) juveniles supplied with LAB showed 81% increment of body weight in a long-treated group (59 days) and 28% rise in short-term experiment (25 days) with respect to the control. Probiotics decreased cortisol levels in treated animals and affected the transcription of two antagonistic genes involved in the regulation of body growth—IGF-I and myostatin (MSTN). IGF-I transcription was increased, while MSTN was inhibited (Carnevali et al. 2006).

Some experiments concerned the use of LAB in treatment of plant diseases. Plant pathogen *Ralstonia solanacearum* causes bacterial wilt. *Lactobacillus* sp. strain KLF01 isolated from rhizosphere of tomato reduced disease severity of tomato and red pepper as compared to nontreated plants (Shrestha et al. 2009a). *Lactobacillus* KLF01 and *Lactococcus* KLC02 strains showed 55% and 60% bio-control efficacy, respectively, in regard to *Pectobacterium carotovorum* subsp. *carotovorum*, soft rot pathogen, on Chinese cabbage

(Shrestha et al. 2009b). These LAB significantly reduced bacterial spot caused by *Xanthomonas campestris* pv. *vesicatoria* on pepper plants in comparison with untreated plants in both greenhouse and field experiments. Additionally, LAB are able to colonize roots, produce indole-3-acetic acid, siderophores, and solubilize phosphates (Shrestha et al. 2014). LAB are effective in the removal of the root-knot nematodes. The decreased pH levels in agricultural soil due to lactic acid produced by bacteria are correlated with reduced population of nematodes (Takei et al. 2008). Microalgae are used as feed for live prey (rotifers, *Artemia*), larvae and adult fish, mollusks, and crustaceans. The growth of microalgae *Isochrysis galbana* was enhanced by LAB, both in the absence and in the presence of nutrients in the culture. The highest final biomass concentration was achieved by adding *Pediococcus acidilactici*, whereas *Leuconostoc mesenteroides* spp. *mesenteroides* and *Carnobacterium piscicola* provided for maximal growth rates. However, the latter species also showed inhibitory effect on *Moraxella* (Planas et al. 2015).

Agriculture, medicine, and food industry are not the only application fields for LAB and their products. Many cosmetic ingredients have been developed using LAB and bifidobacteria. Supernatants of these bacteria contain lactate and amino acids, which contribute to the hydration of the skin. Studies revealed that skim milk fermented by *Streptococcus thermophilus* had skin hydrating, antioxidative, cytoprotective, and pH control effects. *Aloe vera* fermented by *Lactobacillus plantarum* possessed four times higher skin moistening effect than nonfermented *A. vera* juice. Soybean milk fermented by *Bifidobacterium breve* demonstrated the potential to enhance hyaluronic acid production in human cell culture. *Streptococcus thermophilus* YIT 2084 proved capable to produce hyaluronic acid used as conventional cosmetic ingredient (Izawa and Sone 2014). Lactic acid itself is primarily used as moisturizer and pH regulator, additionally possessing multiple other properties such as antimicrobial activity, skin lightening, and hydration (Vijayakumar et al. 2008).

In the chemical industry, lactic acid undergoes a variety of chemical conversions into potentially useful chemicals and takes part in diverse technological processes. The compound is used in dyeing of silks and other textile fabrics, as a mordant in printing of woolens, in bating and plumping of leathers, in deliming of hides, in vegetable tanning, and as a flux for soft solders (Vijayakumar et al. 2008). Recently lactic acid has drawn attention as substrate for the production of polylactic acid (PLA) or biodegradable plastic. Lactic acid can be used to produce PLA of variable molecular weight, but usually the polymer with high molecular weight has the superior commercial value. There are three main methods to synthesize PLA:

direct condensation polymerization; direct polycondensation in an azeotropic solution; polymerization through lactide formation. The major PLA application sphere today is packaging (nearly 70%); however, the polymer is used in other fields. It is applied in implants and medical devices. Because PLA degrades with time, the removal of implanted detail is not required. Nevertheless, in some cases, such implants may cause rejection reaction from the human host. In medical devices, PLA is considered as an alternative to metal implants responsible for possible corrosion and distortion of magnetic resonance images. The polymer can be readily processed into fibers due to its ability to absorb organic compounds and its wicking properties. Some other potential applications of PLA include paints, cigarette filters, three-dimensional printing, parts for space exploration, and environmental remediation (Jamshidian et al. 2010; Castro-Aguirre et al. 2016).

9.11 Conclusions and future prospects

LAB are gram-positive rods or cocci mostly belonging to order *Lactobacillales*. These bacteria are fastidious microorganisms requiring definite cultivation conditions and various components in the medium. In turn, LAB produce a number of valuable compounds such as bacteriocins, vitamins, enzymes, exopolysaccharides, sweeteners, lactic acid, and others. Provided the defined cultural conditions and nutrient medium composition, diverse strains display the ability to ferment substrates and synthesize bioactive compounds. To demonstrate vital properties and gain maximum benefits from them, LAB cultures are selected according to the established criteria.

LAB have long history of application, and they are usually considered safe microorganisms for use in medicine and food industry. LAB can contribute as probiotics into maintenance of human and animal health, prevention, and treatment of diarrhea of various etiology, allergy, inflammatory bowel diseases, and cancer. They are widely used in production of fermented foodstuffs, especially dairy products, cereals, fermented meat, and baked and dry food. The market of probiotics grows steadily and is expected to reach USD 50.0 billion by 2020 due to rising demand in the animal feed sector and consumer interest in health-promoting beneficial food. Some studies revealed favorable effect of bacteria on plants. Moreover, LAB and the derived products demonstrate attractive commercial prospects in other fields, like chemistry and the manufacture of plastics. New LAB applications are likely to emerge in the near future.

References

Abdel-Daim, A., N. Hassouna, M. Hafez, M. S. A. Ashor, and M. M. Aboulwafa. 2013. Antagonistic activity of *Lactobacillus* isolates against *Salmonella* typhi in vitro. *BioMed Research International* 2013:680605. doi:10.1155/2013/680605.

Abdel-Rahman, M. A., Y. Tashiro, and K. Sonomoto. 2011. Lactic acid production from lignocellulose-derived sugars using lactic acid bacteria: Overview and limits. *Journal of Biotechnology* 156(4):286–301. doi:10.1016/j.jbiotec.2011.06.017.

Abdel-Rahman, M. A., Y. Tashiro, and K. Sonomoto. 2013. Recent advances in lactic acid production by microbial fermentation processes. *Biotechnology Advances* 31(6):877–902. doi:10.1016/j.biotechadv.2013.04.002.

Adamberg, K., S. Kask, T. M. Laht, and T. Paalme. 2003. The effect of temperature and pH on the growth of lactic acid bacteria: A pH-auxostat study. *International Journal of Food Microbiology* 85(1–2):171–183. doi:10.1016/S0168-1605(02)00537-8.

Adzitey, F., N. Huda, and G. R. R. Ali. 2013. Molecular techniques for detecting and typing of bacteria, advantages and application to foodborne pathogens isolated from ducks. *3 Biotech* 3(2):97–107. doi:10.1007/s13205-012-0074-4.

Ahmad, V., M. S. Khan, Q. M. Jamal, M. A. Alzohairy, M. A. Al Karaawi, and M. U. Siddiqui. 2017. Antimicrobial potential of bacteriocins: In therapy, agriculture and food preservation. *International Journal of Antimicrobial Agents* 49(1):1–11. doi:10.1016/j.ijantimicag.2016.08.016.

Ai, C., N. Ma, Q. Zhang, et al. 2016. Immunomodulatory effects of different lactic acid bacteria on allergic response and its relationship with in vitro properties. *PLoS One* 11(10):e0164697. doi:10.1371/journal.pone.0164697.

Alakomi, H.-L., E. Skyttä, M. Saarela, T. Mattila-Sandholm, K. Latva-Kala, and I. M. Helander. 2000. Lactic acid permeabilizes gram-negative bacteria by disrupting the outer membrane. *Applied and Environmental Microbiology* 66(5):2001–2005. doi:10.1128/AEM.66.5.2001-2005.2000.

Alanis, A. J. 2005. Resistance to antibiotics: Are we in the post-antibiotic era? *Archives of Medical Research* 36(6):697–705. doi:10.1016/j.arcmed.2005.06.009.

Al-Dohail, M. A., R. Hashim, and M. Aliyu-Paiko. 2009. Effects of the probiotic, *Lactobacillus acidophilus*, on the growth performance, haematology parameters and immunoglobulin concentration in African Catfish (*Clarias gariepinus*, Burchell 1822) fingerling. *Aquaculture Research* 40(14):1642–1652. doi:10.1111/j.1365-2109.2009.02265.x.

Alvarez-Sieiro, P., M. Montalbán-López, D. Mu, and O. P. Kuipers. 2016. Bacteriocins of lactic acid bacteria: Extending the family. *Applied Microbiology and Biotechnology* 100(7):2939–2951. doi:10.1007/s00253-016-7343-9.

Aminov, R. I. 2010. A brief history of the antibiotic era: Lessons learned and challenges for the future. *Frontiers in Microbiology* 1:134. doi:10.3389/fmicb.2010.00134.

Ammor, M. S., A. B. Flórez, and B. Mayo. 2007. Antibiotic resistance in non-enterococcal lactic acid bacteria and bifidobacteria. *Food Microbiology* 24(6):559–570. doi:10.1016/j.fm.2006.11.001.

Atassi, F., and A. L. Servin. 2010. Individual and co-operative roles of lactic acid and hydrogen peroxide in the killing activity of enteric strain *Lactobacillus johnsonii* NCC933 and vaginal strain *Lactobacillus gasseri* KS120.1 against enteric, uropathogenic and vaginosis-associated pathogens. *FEMS Microbiology Letters* 304(1):29–38. doi:10.1111/j.1574-6968.2009.01887.x.

Babot, J. D., E. Argañaraz Martínez, M. J. Lorenzo-Pisarello, M. C. Apella, and A. Perez Chaia. 2017. Lactic acid bacteria isolated from poultry protect the intestinal epithelial cells of chickens from in vitro wheat germ agglutinin-induced cytotoxicity. *British Poultry Science* 58(1):76–82. doi:10.1080/00071668.2016.1251574.

Bai, A., and Q. Ouyang. 2006. Probiotics and inflammatory bowel diseases. *Postgraduate Medical Journal* 82(968):376–382. doi:10.1136/pgmj.2005.040899.

Banjoko, I. O., M. M. Adeyanju, O. Ademuyiwa, et al. 2012. Hypolipidemic effects of lactic acid bacteria fermented cereal in rats. *Lipids in Health and Disease* 11:170. doi:10.1186/1476-511X-11-170.

Berthier, F., and S. D. Ehrlich. 1998. Rapid species identification within two groups of closely related lactobacilli using PCR primers that target the 16S/23S rRNA spacer region. *FEMS Microbiology Letters* 161(1):97–106. doi:10.1111/j.1574-6968.1998.tb12934.x.

Biavati, B. 2001. Bifidobacteria. In *Microorganisms as health supporters, vol. 3. Probiotics and bifidobacteria*, ed. B. Biavati, V. Bottazzi, L. Morelli, and C. Schiavi, 10–33. Novara: Mofin-Alce.

Bidart, G. N., J. Rodríguez-Díaz, V. Monedero, and M. J. Yebra. 2014. A unique gene cluster for the utilization of the mucosal and human milk-associated glycans galacto-N-biose and lacto-N-biose in *Lactobacillus casei*. *Molecular Microbiology* 93:521–538. doi:10.1111/mmi.12678.

Bolognani, F., C. J. Rumney, and I. R. Rowland. 1997. Influence of carcinogen binding by lactic acid-producing bacteria on tissue distribution and in vivo mutagenicity of dietary carcinogens. *Food and Chemical Toxicology* 35(6):535–545. doi:10.1016/S0278-6915(97)00029-X.

Bolotin, A., P. Wincker, S. Mauger, et al. 2001. The complete genome sequence of the lactic acid bacterium *Lactococcus lactis* ssp. *lactis* IL1403. *Genome Research* 11(5):731–753. doi:10.1101/gr.169701.

Bouchard, D. S., B. Seridan, T. Saraoui, et al. 2015. Lactic acid bacteria isolated from bovine mammary microbiota: Potential allies against bovine mastitis. *PLoS One* 10(12):e0144831. doi:10.1371/journal.pone.0144831.

Candela, M., B. Vitali, D. Matteuzzi, and P. Brigidi. 2004. Evaluation of the *rrn* operon copy number in *Bifidobacterium* using real-time PCR. *Letters in Applied Microbiology* 38(3):229–232. doi:10.1111/j.1472-765X.2003.01475.x.

Capozzi, V., P. Russo, M. T. Dueñas, P. López, and G. Spano. 2012. Lactic acid bacteria producing B-group vitamins: A great potential for functional cereals products. *Applied Microbiology and Biotechnology* 96(6):1383–1394. doi:10.1007/s00253-012-4440-2.

Carey, C. M., M. Kostrzynska, S. Ojha, and S. Thompson. 2008. The effect of probiotics and organic acids on Shiga-toxin 2 gene expression in enterohemorrhagic *Escherichia coli* O157:H7. *Journal of Microbiological Methods* 73(2):125–132. doi:10.1016/j.mimet.2008.01.014.

Carnevali, O., L. de Vivo, R. Sulpizio, et al. 2006. Growth improvement by probiotic in European sea bass juveniles (*Dicentrarchus labrax*, L.), with particular attention to IGF-1, myostatin and cortisol gene expression. *Aquaculture* 258(1):430–438. doi:10.1016/j.aquaculture.2006.04.025.

Caro, I., G. Bécares, L. Fuentes, et al. 2015. Evaluation of three PCR primers based on the 16S rRNA gene for the identification of lactic acid bacteria from dairy origin. *CyTA Journal of Food* 13(2):181–187. doi:10.1080/19476337.2014.934297.

Castex, M., P. Lemaire, N. Wabete, and L. Chim. 2010. Effect of probiotic *Pediococcus acidilactici* on antioxidant defences and oxidative stress of *Litopenaeus stylirostris* under *Vibrio nigripulchritudo* challenge. *Fish and Shellfish Immunology* 28(4):622–631. doi:10.1016/j.fsi.2009.12.024.

Castro-Aguirre, E., F. Iñiguez-Franco, H. Samsudin, X. Fang, and R. Auras. 2016. Poly(lactic acid)-Mass production, processing, industrial applications, and end of life. *Advanced Drug Delivery Reviews* 107:333–366. doi:10.1016/j.addr.2016.03.010.

Chagnaud, P., K. Machinis, L. A. Coutte, A. Marecat, and A. Mercenier. 2001. Rapid PCR-based procedure to identify lactic acid bacteria: Application to six common *Lactobacillus* species. *Journal of Microbiological Methods* 44(2):139–148. doi:10.1016/S0167-7012(00)00244-X.

Chang, C. H., Y. S. Chen, T. T. Lee, Y. C. Chang, and B. Yu. 2015. *Lactobacillus formosensis* sp. nov., a lactic acid bacterium isolated from fermented soybean meal. *International Journal of Systematic and Evolutionary Microbiology* 65(Pt 1):101–106. doi:10.1099/ijs.0.070938-0.

Chapot-Chartier, M.-P., and S. Kulakauskas. 2014. Cell wall structure and function in lactic acid bacteria. *Microbial Cell Factories* 13(Suppl 1):S9. doi:10.1186/1475-2859-13-S1-S9.

Chen, C. C., W. C. Lin, M. S. Kong, et al. 2012. Oral inoculation of probiotics *Lactobacillus acidophilus* NCFM suppresses tumour growth both in segmental orthotopic colon cancer and extra-intestinal tissue. *British Journal of Nutrition* 107(11):1623–1634. doi:10.1017/S0007114511004934.

Chouayekh, H., W. Bejar, M. Rhimi, K. Jelleli, M. Mseddi, and S. Bejar. 2007. Characterization of an L-arabinose isomerase from the *Lactobacillus plantarum* NC8 strain showing pronounced stability at acidic pH. *FEMS Microbiology Letters* 277(2):260–267. doi:10.1111/j.1574-6968.2007.00961.x.

Christiaen, S. E., M. O'Connell Motherway, F. Bottacini, et al. 2014. Autoinducer-2 plays a crucial role in gut colonization and probiotic functionality of *Bifidobacterium breve* UCC2003. *PLoS One* 9(5):e98111. doi:10.1371/journal.pone.0098111.

Claisse, O., V. Renouf, and A. Lonvaud-Funel. 2007. Differentiation of wine lactic acid bacteria species based on RFLP analysis of a partial sequence of *rpoB* gene. *Journal of Microbiological Methods* 69(2):387–390. doi:10.1016/j.mimet.2007.01.004.

Cocconcelli, P. S., D. Porro, S. Galandini, and L. Senini. 1995. Development of RAPD protocol for typing of strains of lactic acid bacteria and enterococci. *Letters in Applied Microbiology* 21(6):376–379. doi:10.1111/j.1472-765X.1995.tb01085.x.

Daniel, C., A. Repa, C. Wild, et al. 2006. Modulation of allergic immune responses by mucosal application of recombinant lactic acid bacteria producing the major birch pollen allergen Bet v 1. *Allergy* 61:812–819. doi:10.1111/j.1398-9995.2006.01071.x.

Davies, J., and D. Davies. 2010. Origins and evolution of antibiotic resistance. *Microbiology and Molecular Biology Reviews* 74(3):417–433. doi:10.1128/MMBR.00016-10.

de Vrese, M., A. Stegelmann, B. Richter, S. Fenselau, C. Laue, and J. Schrezenmeir. 2001. Probiotics - compensation for lactase insufficiency. *The American Journal of Clinical Nutrition* 73(2 Suppl):421S–429S.

Delcour, J., T. Ferain, M. Deghorain, E. Palumbo, and P. Hols. 1999. The biosynthesis and functionality of the cell-wall of lactic acid bacteria. *Antonie Van Leeuwenhoek* 76(1–4):159–184. doi:10.1007/978-94-017-2027-4_7.

Dhakal, R., V. K. Bajpai, and K.-H. Baek. 2012. Production of gaba (γ–aminobutyric acid) by microorganisms: A review. *Brazilian Journal of Microbiology* 43(4):1230–1241. doi:10.1590/S1517-83822012000400001.

Di Cagno, R., M. De Angelis, M. Calasso, et al. 2010. Quorum sensing in sourdough *Lactobacillus plantarum* DC400: Induction of plantaricin A (PlnA) under co-cultivation with other lactic acid bacteria and effect of PlnA on bacterial and Caco-2 cells. *Proteomics* 10(11):2175–2190. doi:10.1002/pmic.200900565.

Diep, D. B., L. Axelsson, C. Grefsli, and I. F. Nes. 2000. The synthesis of the bacteriocin sakacin A is a temperature-sensitive process regulated by a pheromone peptide through a three-component regulatory system. *Microbiology* 146(Pt 9):2155–2160. doi:10.1099/00221287-146-9-2155.

Dimov, S., P. Ivanova, and N. Harizanova. 2005. Genetics of bacteriocins biosynthesis by lactic acid bacteria. *Biotechnology and Biotechnological Equipment* 19(Sup2):4–10. doi:10.1080/13102818.2005.10817270.

Dumbrepatil, A., M. Adsul, S. Chaudhari, J. Khire, and D. Gokhale. 2008. Utilization of molasses sugar for lactic acid production by *Lactobacillus delbrueckii* subsp. *delbrueckii* mutant Uc-3 in batch fermentation. *Applied and Environmental Microbiology* 74(1):333–335. doi:10.1128/AEM.01595-07.

Frizzo, L. S., L. P. Soto, M. V. Zbrun, et al. 2010. Lactic acid bacteria to improve growth performance in young calves fed milk replacer and spray-dried whey powder. *Animal Feed Science and Technology* 157(3):159–167. doi:10.1016/j.anifeedsci.2010.03.005.

Fushinobu, S. Unique sugar metabolic pathways of bifidobacteria. 2010. *Bioscience, Biotechnology, and Biochemistry* 74(12):2374–2384. doi:10.1271/bbb.100494.

Galdeano, C. M., and G. Perdigón. 2006. The probiotic bacterium *Lactobacillus casei* induces activation of the gut mucosal immune system through innate immunity. *Clinical and Vaccine Immunology* 13(2):219–226. doi:10.1128/CVI.13.2.219-226.2006.

García-Hernández, Y., T. Pérez-Sánchez, R. Boucourt, et al. 2016. Isolation, characterization and evaluation of probiotic lactic acid bacteria for potential use in animal production. *Research in Veterinary Science* 108:125–132. doi:10.1016/j.rvsc.2016.08.009.

Ghoneum, M., and N. Felo. 2015. Selective induction of apoptosis in human gastric cancer cells by *Lactobacillus kefiri* (PFT), a novel kefir product. *Oncology Reports* 34(4):1659–1666. doi:10.3892/or.2015.4180.

Ghorbani, G. R., D. P. Morgavi, K. A. Beauchemin, and J. A. Leedle. 2002. Effects of bacterial direct-fed microbials on ruminal fermentation, blood variables, and the microbial populations of feedlot cattle. *Journal of Animal Science* 80(7):1977–1985. doi:10.2527/2002.8071977x.

Gilliland, S. E. 1990. Health and nutritional benefits from lactic acid bacteria. *FEMS Microbiology Reviews* 7(1–2):175–188. doi:10.1016/0378-1097(90)90705-U.

Giri, S. S., V. Sukumaran, and M. Oviya. 2013. Potential probiotic *Lactobacillus plantarum* VSG3 improves the growth, immunity, and disease resistance of tropical freshwater fish, *Labeo rohita. Fish and Shellfish Immunology* 34(2):660–666. doi:10.1016/j.fsi.2012.12.008.

Global Industry Analysts Inc. 2016. MCP-1084: Probiotics – a global strategic business report. http://www.strategyr.com/MCP-1084.asp (Accessed September 20, 2017).

Global Market Insights Inc. 2016. Probiotics Market Size By End Use (Human, Animal), By Application (Functional Foods & Beverages [Dairy, Non-dairy, Cereals, Baked Goods, Fermented Meat Products, Dry Foods], Dietary Supplements [Food, Nutritional, Specialty, Infant Formula], Animal Feed Probiotics), Industry Analysis Report, Regional Outlook (U.S., Germany, UK, China, Japan, India, Brazil), Application Potential, Price Trends, Competitive Market Share & Forecast, 2016–2023. https://www.gminsights.com/industry-analysis/probiotics-market (accessed September 20, 2017).

Global Market Insights Inc. 2017. Prebiotics Market Size By Ingredient (Inulin, GOS, FOS, MOS), By Application (Animal Feed, Food & Beverages [Dairy, Cereals, Baked Goods, Fermented Meat, Dry Foods], Dietary Supplements [Food, Nutrition, Infant Formulations]), Industry Analysis Report, Regional Outlook, Application Potential, Price Trends, Competitive Market Share & Forecast, 2017–2024. https://www.gminsights.com/industry-analysis/prebiotics-market (accessed September 20, 2017).

Goto, T., Y. I. Kim, T. Furuzono, et al. 2015. 10-oxo-12(Z)-octadecenoic acid, a linoleic acid metabolite produced by gut lactic acid bacteria, potently activates PPARγ and stimulates adipogenesis. *Biochemical and Biophysical Research Communications* 459(4):597–603. doi:10.1016/j.bbrc.2015.02.154.

Grand View Research Inc. 2016. Functional Foods Market Analysis By Product (Carotenoids, Dietary Fibers, Fatty Acids, Minerals, Prebiotics & Probiotics, Vitamins), By Application, By End-Use (Sports Nutrition, Weight Management, Immunity, Digestive Health) And Segment Forecasts, 2014–2024. http://www.grandviewresearch.com/industry-analysis/functional-food-market (accessed September 20, 2017).

Groschwitz, K. R., and S. P. Hogan. 2009. Intestinal barrier function: Molecular regulation and disease pathogenesis. *The Journal of Allergy and Clinical Immunology* 124(1):3–22. doi:10.1016/j.jaci.2009.05.038.

Gueimonde, M., R. Frias, and A. Ouwehand. 2006. Assuring the continued safety of lactic acid bacteria used as probiotics. *Biologia* 61(6):755–760. doi:10.2478/s11756-006-0153-2.

Guo, W, W. Jia, Y. Li, and S. Chen. 2010. Performances of *Lactobacillus brevis* for producing lactic acid from hydrolysate of lignocellulosics. *Applied Biochemistry and Biotechnology* 161(1–8):124–136. doi:10.1007/s12010-009-8857-8.

Gürtler, V., and V. A. Stanisich. 1996. New approaches to typing and identification of bacteria using the 16S-23S rDNA spacer region. *Microbiology* 142(Pt 1):3–16. doi:10.1099/13500872-142-1-3.

Halpern, M. D., and P. W. Denning. 2015. The role of intestinal epithelial barrier function in the development of NEC. *Tissue Barriers* 3(1–2):e1000707. doi:10.1080/21688370.2014.1000707.

Harutoshi, T. 2013. Exopolysaccharides of lactic acid bacteria for food and colon health applications. In *Lactic acid bacteria - R & D for food, health and livestock purposes*, ed. J. Marcelino Kongo. InTech. doi:10.5772/50839. https://www.intechopen.com/books/lactic-acid-bacteria-r-d-for-food-health-and-livestock-purposes/exopolysaccharides-of-lactic-acid-bacteria-for-food-and-colon-health-applications (accessed September 20, 2017).

Hayek, S. A., and S. A. Ibrahim. 2013. Current limitations and challenges with lactic acid bacteria: A review. *Food and Nutrition Sciences* 4(11):73–87. doi:10.4236/fns.2013.411A010.

Herve-Jimenez, L., I. Guillouard, E. Guedon, et al. 2008. Physiology of *Streptococcus thermophilus* during the late stage of milk fermentation with special regard to sulfur amino-acid metabolism. *Proteomics* 8(20):4273–4286. doi:10.1002/pmic.200700489.

Heyman, M. 2000. Effect of lactic acid bacteria on diarrheal diseases. *Journal of the American College of Nutrition* 19(2 Suppl):137S–146S. doi:10.1080/07315724.2000.10718084.

Higgins, J. P., S. E. Higgins, J. L. Vicente, A. D. Wolfenden, G. Tellez, and B. M. Hargis. 2007. Temporal effects of lactic acid bacteria probiotic culture on *Salmonella* in neonatal broilers. Poultry Science 86(8):1662–1666. doi:10.1093/ps/86.8.1662.

Hong, W. S., H. C. Chen, Y. P. Chen, and M. J. Chen. 2009. Effects of kefir supernatant and lactic acid bacteria isolated from kefir grain on cytokine production by macrophage. *International Dairy Journal* 19(4):244–251. doi:10.1016/j.idairyj.2008.10.010.

Hsieh, C.-Y., T. Osaka, E. Moriyama, Y. Date, J. Kikuchi, and S. Tsuneda. 2015. Strengthening of the intestinal epithelial tight junction by *Bifidobacterium bifidum*. *Physiological Reports* 3(3):e12327. doi:10.14814/phy2.12327.

Hynönen, U., and A. Palva. 2013. *Lactobacillus* surface layer proteins: Structure, function and applications. *Applied Microbiology and Biotechnology* 97(12):5225–5243. doi:10.1007/s00253-013-4962-2.

Izawa, N., and T. Sone. 2014. Cosmetic ingredients fermented by lactic acid bacteria. In *Microbial production*, ed. H. Anazawa, S. Shimizu, 233–42. Springer: Tokyo, Japan. doi:10.1007/978-4-431-54607-8_20.

Jamshidian, M., E. A. Tehrany, M. Imran, M. Jacquot, and S. Desobry. 2010. Poly-lactic acid: Production, applications, nanocomposites, and release studies. *Comprehensive Reviews in Food Science and Food Safety* 9(5):552–571. doi:10.1111/j.1541-4337.2010.00126.x.

Jankowska, A., D. Laubitz, H. Antushevich, R. Zabielski, and E. Grzesiuk. 2008. Competition of *Lactobacillus paracasei* with *Salmonella enterica* for adhesion to Caco-2 Cells. *Journal of Biomedicine and Biotechnology* 2008:357964. doi:10.1155/2008/357964.

John, R. P., K. M. Nampoothiri, and A. Pandey. 2007. Fermentative production of lactic acid from biomass: An overview on process developments and future perspectives. *Applied Microbiology and Biotechnology* 74(3):524–534. doi:10.1007/s00253-006-0779-6.

Jounai, K., K. Ikado, T. Sugimura, Y. Ano, J. Braun, and D. Fujiwara. 2012. Spherical lactic acid bacteria activate plasmacytoid dendritic cells immunomodulatory function via TLR9-dependent crosstalk with myeloid dendritic cells. *PLoS One* 7(4):e32588. doi:10.1371/journal.pone.0032588.

Kadri, Z., F. Spitaels, M. Cnockaert, et al. 2015. *Enterococcus bulliens* sp. nov., a novel lactic acid bacterium isolated from camel milk. *Antonie Van Leeuwenhoek* (5):1257–1265. doi:10.1007/s10482-015-0579-z.

Kahala, M., M. Mäki, A. Lehtovaara, et al. 2008. Characterization of starter lactic acid bacteria from the Finnish fermented milk product viili. *Journal of Applied Microbiology* 105(6):1929–1938. doi:10.1111/j.1365-2672.2008.03952.x.

Kalavathy, R., N. Abdullah, S. Jalaludin, and Y. W. Ho. 2003. Effects of *Lactobacillus* cultures on growth performance, abdominal fat deposition, serum lipids and weight of organs of broiler chickens. *British Poultry Science* 44(1):139–144. doi:10.1080/0007166031000085445.

Kamiya, T., Y. Watanabe, S. Makino, H. Kano, and N. M. Tsuji. 2017. Improvement of intestinal immune cell function by lactic acid bacteria for dairy products. *Microorganisms* 5(1):1. doi:10.3390/microorganisms5010001.

Kim, S. H., W. Jeung, I. D. 2016. Choi, et al. Lactic acid bacteria improves peyer's patch cell-mediated immunoglobulin A and tight-junction expression in a destructed gut microbial environment. *Journal of Microbiology and Biotechnology* 26(6):1035–1045. doi:10.4014/jmb.1512.12002.

Kim, M. J., and K. S. Kim. 2012. Isolation and identification of γ-aminobutyric acid (GABA)-producing lactic acid bacteria from kimchi. *Journal of the Korean Society for Applied Biological Chemistry* 55(6):777–785. doi:10.1007/s13765-012-2174-6.

Klaenhammer, T. R., R. Barrangou, B. L. Buck, M. A. Azcarate-Peril, and E. 2005. Altermann. Genomic features of lactic acid bacteria effecting bioprocessing and health. *FEMS Microbiology Reviews* 29(3):393–409. doi:10.1016/j.femsre.2005.04.007.

Koller, V. J., B. Marian, R. Stidl, et al. 2008. Impact of lactic acid bacteria on oxidative DNA damage in human derived colon cells. *Food and Chemical Toxicology* 46(4):1221–1229. doi:10.1016/j.fct.2007.09.005.

Kook, M. C., and S. C. Cho. 2013. Production of GABA (gamma amino butyric acid) by lactic acid bacteria. *Korean Journal for Food Science of Animal Resources* 33(3):377–389. doi:10.5851/kosfa.2013.33.3.377.

König, H., and J. Fröhlich. 2009. Lactic Acid Bacteria. In *Biology of microorganisms on grapes, in must and in wine*, ed. H. König, G. Unden, J. Fröhlich, 3–29. Springer-Verlag Berlin Heidelberg. doi:10.1007/978-3-540-85463-0.

Krehbiel, C. R., S. R. Rust, G. Zhang, and S. E. Gilliland. 2003. Bacterial direct-fed microbials in ruminant diets: Performance response and mode of action. *Journal of Animal Science* 81(14_suppl_2):E120–E132. doi:10.2527/2003.8114_suppl_2E120x.

Kuhl, G. C., and J. De Dea Lindner. 2016. Biohydrogenation of linoleic acid by lactic acid bacteria for the production of functional cultured dairy products: A review. *Foods* 5(1):13. doi:10.3390/foods5010013.

Kuipers, O. P., M. M. Beerthuyzen, P. G. de Ruyter, E. J. Luesink, and W. M. de Vos. 1995. Autoregulation of nisin biosynthesis in *Lactococcus lactis* by signal transduction. *Journal of Biological Chemistry* 270(45):27299–27304. doi:10.1074/jbc.270.45.27299.

Kumar, A. 2016. The probiotics market: Ingredients, supplements, foods. https://www.bccresearch.com/market-research/food-and-beverage/probiotics-market-ingredients-supplements-foods-report-fod035e.html (accessed September 20, 2017).

Kütahya, O. E., M. J. C. Starrenburg, J. L. W. Rademaker, et al. 2011. High-resolution amplified fragment length polymorphism typing of *Lactococcus lactis* strains enables identification of genetic markers for subspecies-related phenotypes. *Applied and Environmental Microbiology* 77(15):5192–5198. doi:10.1128/AEM.00518-11.

Lacroix, N., D. St-Gelais, C. P. Champagne, and J. C. Vuillemard. 2013. Gamma-aminobutyric acid-producing abilities of lactococcal strains isolated from old-style cheese starters. *Dairy Science and Technology* 93(3):315–327. doi:10.1007/s13594-013-0127-4.

LeBlanc, J. G., J. E. Laiño, M. J. del Valle, et al. 2011. B-group vitamin production by lactic acid bacteria—current knowledge and potential applications. *Journal of Applied Microbiology* 111(6):1297–1309. doi:10.1111/j.1365-2672.2011.05157.x.

Lee, C. M., C. C. Sieo, N. Abdullah, and Y. W. Ho. 2008. Estimation of 16S rRNA gene copy number in several probiotic *Lactobacillus* strains isolated from the gastrointestinal tract of chicken. *FEMS Microbiology Letters* 287(1):136–141. doi:10.1111/j.1574-6968.2008.01305.x.

Lehnen, T. E., M. R. da Silva, A. Camacho, A. Marcadenti, and A. M. Lehnen. 2015. A review on effects of conjugated linoleic fatty acid (CLA) upon body composition and energetic metabolism. *Journal of the International Society of Sports Nutrition* 12:36. doi:10.1186/s12970-015-0097-4.

Leroy, F., and L. De Vuyst. 2005. Simulation of the effect of sausage ingredients and technology on the functionality of the bacteriocin-producing *Lactobacillus sakei* CTC 494 strain. *International Journal of Food Microbiology* 100(1–3):141–152. doi:10.1016/j.ijfoodmicro.2004.10.011.

Li, C., J. Bai, Z. Cai, and F. Ouyang. 2002. Optimization of a cultural medium for bacteriocin production by *Lactococcus lactis* using response surface methodology. *Journal of Biotechnology* 93(1):27–34. doi:10.1016/S0168-1656(01)00377-7.

Liu, M., J. R. Bayjanov, B. Renckens, A. Nauta, and R. J. Siezen. 2010. The proteolytic system of lactic acid bacteria revisited: A genomic comparison. *BMC Genomics* 11:36. doi:10.1186/1471-2164-11-36.

Liu, W., H. Pang, H. Zhang, and Y. Cai. 2014. Biodiversity of lactic acid bacteria. In *Lactic Acid Bacteria*, ed. H. Zhang, Y. Cai, 103–203. Springer: the Netherlands. doi:10.1007/978-94-017-8841-0_2.

Loguercio, C., R. Abbiati, M. Rinaldi, A. Romano, C. Del Vecchio Blanco, and M. Coltorti. 1995. Long-term effects of *Enterococcus faecium* SF68 versus lactulose in the treatment of patients with cirrhosis and grade 1-2 hepatic encephalopathy. *Journal of Hepatology* 23(1):39–46. doi:10.1016/0168-8278(95)80309-2.

Ludwig, W., and H. P. Klenk. 2005. Overview: A phylogenetic backbone and taxonomic framework for prokaryotic systematics. In *Bergey's Manual of Systematic Bacteriology, Vol. 2: The Proteobacteria, Part A, introductory essays*, eds. G. M. Garrity, D. J. Bean, N.R. Krieg, and J.T. Staley, 49–65. Springer US.

Mack, D. R., S. Ahrne, L. Hyde, S. Wei, and M. A. Hollingsworth. 2003. Extracellular MUC3 mucin secretion follows adherence of *Lactobacillus* strains to intestinal epithelial cells in vitro. *Gut* 52(6):827–833. doi:10.1136/gut.52.6.827.

Mackowiak, P. A. 2013. Recycling Metchnikoff: Probiotics, the intestinal microbiome and the quest for long life. *Frontiers in Public Health* 1:52. doi:10.3389/fpubh.2013.00052.

Mainville, I., N. Robert, B. Lee, and E. R. Farnworth. 2006. Polyphasic characterization of the lactic acid bacteria in kefir. *Systematic and Applied Microbiology* 29(1):59–68. doi:10.1016/j.syapm.2005.07.001.

Makarova, K., A. Slesarev, Y. Wolf, et al. 2006. Comparative genomics of the lactic acid bacteria. *Proceedings of the National Academy of Sciences of the United States of America* 103(42):15611–15616. doi:10.1073/pnas.0607117103.

Makarova, K. S., and E. V. Koonin. 2007. Evolutionary genomics of lactic acid bacteria. *Journal of Bacteriology* 189(4):1199–1208. doi:10.1128/JB.01351-06.

Maldonado, A., R. Jiménez-Díaz, and J. L. Ruiz-Barba. 2004. Induction of plantaricin production in *Lactobacillus plantarum* NC8 after coculture with specific gram-positive bacteria is mediated by an autoinduction mechanism. *Journal of Bacteriology* 186(5):1556–1564. doi:10.1128/JB.186.5.1556-1564.2004.

Martínez Cruz, P., A. L. Ibáñez, O. A. Monroy Hermosillo, and H. C. Ramírez Saad. 2012. Use of probiotics in aquaculture. *ISRN Microbiology* 2012:916845. doi:10.5402/2012/916845.

Mattarelli, P., W. Holzapfel, C. M. Franz, et al. 2014. Recommended minimal standards for description of new taxa of the genera *Bifidobacterium*, *Lactobacillus* and related genera. *International Journal of Systematic and Evolutionary Microbiology* 64(Pt 4):1434–1451. doi:10.1099/ijs.0.060046-0.

Matthews, A., A. Grimaldi, M. Walker, E. Bartowsky, P. Grbin, and V. Jiranek. 2004. Lactic acid bacteria as a potential source of enzymes for use in vinification. *Applied and Environmental Microbiology* 70(10):5715–5731. doi:10.1128/AEM.70.10.5715-5731.2004.

Meganathan, R., and O. Kwon. 2009. Biosynthesis of menaquinone (vitamin K2) and ubiquinone (coenzyme Q). *EcoSal Plus* 3(2). doi:10.1128/ecosalplus.3.6.3.3.

Munir, M. B., R. Hashim, M. S. Abdul Manaf, and S. A. M. Nor. 2016. Dietary prebiotics and probiotics influence the growth performance, feed utilisation, and body indices of snakehead (*Channa striata*) fingerlings. *Tropical Life Sciences Research* 27(2):111–125. doi:10.21315/tlsr2016.27.2.9.

Mohammed, M., H. Abd El-Aziz, N. Omran, S. Anwar, S. Awad, and M. El-Soda. 2009. Rep-PCR characterization and biochemical selection of lactic acid bacteria isolated from the Delta area of Egypt. *International Journal of Food Microbiology* 128(3):417–423. doi:10.1016/j.ijfoodmicro.2008.09.022.

Mohapatra, B. R., K. Broersma, and A. Mazumder. 2007. Comparison of five rep-PCR genomic fingerprinting methods for differentiation of fecal *Escherichia coli* from humans, poultry and wild birds. *FEMS Microbiology Letters* 277(1):98–106. doi:10.1111/j.1574-6968.2007.00948.x.

Mokoena, M. P. 2017. Lactic acid bacteria and their bacteriocins: Classification, biosynthesis and applications against uropathogens: A mini-review. *Molecules* 22(8):1255. doi:10.3390/molecules22081255.

Moon, G. S., Y. R. Pyun, and W. J. Kim. 2006. Expression and purification of a fusion-typed pediocin PA-1 in *Escherichia coli* and recovery of biologically active pediocin PA-1. *International Journal of Food Microbiology* 108(1):136–140. doi:10.1016/j.ijfoodmicro.2005.10.019.

Moore, E. R., S. A. Mihaylova, P. Vandamme, M. I. Krichevsky, and L. Dijkshoorn. 2010. Microbial systematics and taxonomy: Relevance for a microbial commons. *Research in Microbiology* 161(6):430–438. doi:10.1016/j.resmic.2010.05.007.

Moraes, P. M., L. M. Perin, A. S. Júnior, and L. A. Nero. 2013. Comparison of phenotypic and molecular tests to identify lactic acid bacteria. *Brazilian Journal of Microbiology* 44(1):109–112. doi:10.1590/S1517-83822013000100015.

Morel, F., M. Lamarque, I. Bissardon, D. Atlan, and A. Galinier. 2001. Autoregulation of the biosynthesis of the CcpA-like protein, PepR1, in *Lactobacillus delbrueckii* subsp *bulgaricus*. *Journal of Molecular Microbiology and Biotechnology* 3(1):63–66.

Morita, H., H. Toh, S. Fukuda, et al. 2008. Comparative genome analysis of *Lactobacillus reuteri* and *Lactobacillus fermentum* reveal a genomic island for reuterin and cobalamin production. *DNA research* 15(3):151–161. doi:10.1093/dnares/dsn009.

Naser, S. M., F. L. Thompson, B. Hoste, et al. 2005. Application of multilocus sequence analysis (MLSA) for rapid identification of *Enterococcus* species based on *rpoA* and *pheS* genes. *Microbiology* 151(Pt 7):2141–2150. doi:10.1099/mic.0.27840-0.

Naser, S. M., P. Dawyndt, B. Hoste, et al. 2007. Identification of lactobacilli by *pheS* and *rpoA* gene sequence analyses. *International Journal of Systematic and Evolutionary Microbiology* 57(Pt 12):2777–2789. doi:10.1099/ijs.0.64711-0.

Nikoskelainen, S., A. Ouwehand, S. Salminen, and G. Bylund. 2001. Protection of rainbow trout (*Oncorhynchus mykiss*) from furunculosis by *Lactobacillus rhamnosus*. *Aquaculture* 198(3):229–236. doi:10.1016/S0044-8486(01)00593-2.

Novik, G., O. Meerovskaya, and V. Savich. 2017. Waste degradation and utilization by lactic acid bacteria: Use of lactic acid bacteria in production of food additives, bioenergy and biogas. In *Food Additives*, ed.

D. N. Karunaratne. InTech. doi:10.5772/intechopen.69284. https://www.intechopen.com/books/food-additives/waste-degradation-and-utilization-by-lactic-acid-bacteria-use-of-lactic-acid-bacteria-in-production- (accessed September 20, 2017)

Ochman, H., and A. C. Wilson. 1987. Evolution in bacteria: Evidence for a universal substitution rate in cellular genomes. *Journal of Molecular Evolution* 26(1–2):74–86. doi:10.1007/BF02111283.

Ogawa, J., S. Kishino, A. Ando, S. Sugimoto, K. Mihara, and S. Shimizu. 2005. Production of conjugated fatty acids by lactic acid bacteria. *Journal of Bioscience and Bioengineering* 100(4):355–364. doi:10.1263/jbb.100.355.

O'Hanlon, D. E., T. R. Moench, and R. A. Cone. 2011. In vaginal fluid, bacteria associated with bacterial vaginosis can be suppressed with lactic acid but not hydrogen peroxide. *BMC Infectious Diseases* 11:200. doi:10.1186/1471-2334-11-200.

Olnood, C. G., S. S. Beski, M. Choct, and P. A. Iji. 2015. Novel probiotics: Their effects on growth performance, gut development, microbial community and activity of broiler chickens. *Animal Nutrition* 1(3):184–191. doi: 10.1016/j.aninu.2015.07.003.

Orla-Jensen, S. The lactic acid bacteria. 1919. Host and Son: Copenhagen.

Otero, M.C., L. Morelli, and M.E. Nader-Macías. 2006. Probiotic properties of vaginal lactic acid bacteria to prevent metritis in cattle. *Letters in Applied Microbiology* 43(1):91–97. doi:10.1111/j.1472-765X.2006.01914.x.

Özyurt, G., A. S. Özkütük, M. Boğa, M. Durmuş, and E. K. Boğa. 2017. Biotransformation of seafood processing wastes fermented with natural lactic acid bacteria; the quality of fermented products and their use in animal feeding. *Turkish Journal of Fisheries and Aquatic Sciences* 17(3):543–555. doi:10.4194/1303-2712-v17_3_11.

Pacheco, N., M. Garnica-González, and J. Y. Ramírez-Hernández, et al. 2009. Effect of temperature on chitin and astaxanthin recoveries from shrimp waste using lactic acid bacteria. *Bioresource Technology* 100(11):2849–2854. doi:10.1016/j.biortech.2009.01.019.

Panesar, P. S., J. F. Kennedy, C. J. Knill, and M. Kosseva. 2010. Production of L (+) lactic acid using *Lactobacillus casei* from whey. *Brazilian archives of Biology and Technology* 53(1):219–226. doi:10.1590/S1516-89132010000100027.

Papagianni, M. 2012. Metabolic engineering of lactic acid bacteria for the production of industrially important compounds. *Computational and Structural Biotechnology Journal* 3:e201210003. doi:10.5936/csbj.201210003.

Park, H., S. Yeo, Y. Ji, et al. 2014. Autoinducer-2 associated inhibition by *Lactobacillus sakei* NR28 reduces virulence of enterohaemorrhagic *Escherichia coli* O157:H7. *Food Control* 45:62–69. doi:10.1016/j.foodcont.2014.04.024.

Park, J.-E., S.-H. Oh, and Y.-S. Cha. 2013. *Lactobacillus plantarum* LG42 isolated from gajami sik-hae inhibits adipogenesis in 3T3-L1 adipocyte. *BioMed Research International* 2013:460927. doi:10.1155/2013/460927.

Park, S. H., J. H. Jung, D. H. Seo, et al. 2012. Differentiation of lactic acid bacteria based on RFLP analysis of the *tuf* gene. *Food Science and Biotechnology* 21(3):911–915. doi:10.1007/s10068-012-0119-9.

Park, Y. W., and M. S. Nam. 2015. Bioactive peptides in milk and dairy products: A review. *Korean Journal for Food Science of Animal Resources* 35(6):831–840. doi:10.5851/kosfa.2015.35.6.831.

Patel, A., N. Shah, and J. B. Prajapati. 2013. Biosynthesis of vitamins and enzymes in fermented foods by lactic acid bacteria and related genera-A promising approach. *Croatian Journal of Food Science and Technology* 5(2):85–91.

Patel, S., A. Majumder, and A. Goyal. 2012. Potentials of exopolysaccharides from lactic acid bacteria. *Indian Journal of Microbiology* 52(1):3–12. doi:10.1007/s12088-011-0148-8.

Patra, F., S. K. Tomar, and S. Arora. 2009. Technological and functional applications of low-calorie sweeteners from lactic acid bacteria. *Journal of Food Science* 74(1):R16–R23. doi:10.1111/j.1750-3841.2008.01005.x.

Perdigón, G., R. Fuller, and R. Raya. 2001. Lactic acid bacteria and their effect on the immune system. *Current Issues in Intestinal Microbiology* 2(1):27–42.

Pereira, D. I. A., and G. R. Gibson. 2002. Cholesterol assimilation by lactic acid bacteria and *Bifidobacteria* isolated from the human gut. *Applied and Environmental Microbiology* 68(9):4689–4693. doi:10.1128/AEM.68.9.4689-4693.2002.

Perez, R. H., T. Zendo, and K. Sonomoto. 2014. Novel bacteriocins from lactic acid bacteria (LAB): Various structures and applications. *Microbial Cell Factories* 13(Suppl 1):S3. doi:10.1186/1475-2859-13-S1-S3.

Perez, R. H., M. T. M. Perez, and F. B. Elegado. 2015. Bacteriocins from lactic acid bacteria: A review of biosynthesis, mode of action, fermentative production, uses, and prospects. *International Journal of Philippine Science and Technology* 8(2):61–67.

Petrova, P., K. Petrov, and G. Stoyancheva. 2013. Starch-modifying enzymes of lactic acid bacteria – structures, properties, and applications. *Starch/Stärke* 65:34–47. doi:10.1002/star.201200192.

Planas, M., J. A. Vázquez, and B. Novoa. 2015. Stimulative effect of lactic acid bacteria in the growth of the microalgae *Isochrysis galbana*. *Journal of Coastal Life Medicine* 3(12):925–930. doi:10.12980/jclm.3.2015j5-174.

Plengvidhya, V., F. Jr. Breidt, and H. P. Fleming. 2004. Use of RAPD-PCR as a method to follow the progress of starter cultures in sauerkraut fermentation. *International Journal of Food Microbiology* 93(3):287–296. doi:10.1016/j.ijfoodmicro.2003.11.010.

Pokusaeva, K., G. F. Fitzgerald, and D. van Sinderen. 2011. Carbohydrate metabolism in *Bifidobacteria*. *Genes and Nutrition* 6(3):285–306. doi:10.1007/s12263-010-0206-6.

Psoni, L., C. Kotzamanidis, M. Yiangou, N. Tzanetakis, and E. Litopoulou-Tzanetaki. 2007. Genotypic and phenotypic diversity of *Lactococcus lactis* isolates from Batzos, a Greek PDO raw goat milk cheese. *International Journal of Food Microbiology* 114(2):211–220. doi:10.1016/j.ijfoodmicro.2006.09.020.

Qiao, M., S. Ye, O. Koponen, et al. 1996. Regulation of the nisin operons in *Lactococcus lactis* N8. *Journal of Applied Microbiology* 80(6):626–634. doi:10.1111/j.1365-2672.1996.tb03267.x.

Raimondi, S., A. Amaretti, A. Leonardi, A. Quartieri, C. Gozzoli, and M. Rossi. 2016. Conjugated linoleic acid production by *Bifidobacteria*: Screening, kinetic, and composition. *BioMed Research International* 2016:8654317. doi:10.1155/2016/8654317.

Rajendran, R., and Y. Ohta. 1998. Binding of heterocyclic amines by lactic acid bacteria from miso, a fermented Japanese food. *Canadian Journal of Microbiology* 44(2):109–115. doi:10.1139/cjm-44-2-109.

Ramírez-Nuñez, J., R. Romero-Medrano, G. V. Nevárez-Moorillón, and N. Gutiérrez-Méndez. 2011. Effect of pH and salt gradient on the autolysis of *Lactococcus lactis* strains. *Brazilian Journal of Microbiology* 42(4):1495–1499. doi: 10.1590/S1517-83822011000400036.

Rao, M. S., J. Pintado, W. F. Stevens, and J. P. Guyot. 2004. Kinetic growth parameters of different amylolytic and non-amylolytic *Lactobacillus* strains under various salt and pH conditions. *Bioresource Technology* 94(3):331–337. doi:10.1016/j.biortech.2003.11.028.

Rea, M. C., A. Dobson, O. O'Sullivan, et al. 2011. Effect of broad- and narrow-spectrum antimicrobials on *Clostridium difficile* and microbial diversity in a model of the distal colon. *Proceedings of the National Academy of Sciences of the United States of America* 108(Suppl 1):4639–4644. doi:10.1073/pnas.1001224107.

Roberts, M. C. 1996. Tetracycline resistance determinants: Mechanisms of action, regulation of expression, genetic mobility, and distribution. *FEMS Microbiology Reviews* 19(1):1–24. doi:10.1016/0168-6445(96)00021-6.

Roberts, M. C. 2005. Update on acquired tetracycline resistance genes. *FEMS Microbiology Letters* 245(2):195–203. doi:10.1016/j.femsle.2005.02.034.

Rossetti, L., and G. Giraffa. 2005. Rapid identification of dairy lactic acid bacteria by M13-generated, RAPD-PCR fingerprint databases. *Journal of Microbiological Methods* 63(2):135–144. doi:10.1016/j.mimet.2005.03.001.

Ryan, P. M., R. P. Ross, G. F. Fitzgerald, N. M. Caplice, and C. Stanton. 2015. Sugar-coated: Exopolysaccharide producing lactic acid bacteria for food and human health applications. *Food and Function* 6(3):679–693. doi:10.1039/c4fo00529e.

Saez-Lara, M. J., C. Gomez-Llorente, J. Plaza-Diaz, and A. Gil. 2015. The role of probiotic lactic acid bacteria and bifidobacteria in the prevention and treatment of inflammatory bowel disease and other related diseases: A systematic review of randomized human clinical trials. *BioMed Research International* 2015:505878. doi:10.1155/2015/505878.

Saha, B. C., and L. K. Nakamura. 2003. Production of mannitol and lactic acid by fermentation with *Lactobacillus intermedius* NRRL B-3693. *Biotechnology and Bioengineering* 82(7):864–871. doi:10.1002/bit.10638.

Samaržija, D., M. Tudor, T. Prtilo, I. Dolenčić Špehar, Š. Zamberlin, and J. Havranek. 2009. Probiotic bacteria in prevention and treatment of diarrhea. *Mljekarstvo* 59(1): 28–32.

Samarzija, D., S. Sikora, S. Redzepović, N. Antunac, and J. Havranek. 2002. Application of RAPD analysis for identification of *Lactococcus lactis* subsp. *cremoris* strains isolated from artisanal cultures. *Microbiological Research* 157(1):13–17. doi:10.1078/0944-5013-00126.

Sanders, J. W., K. Leenhouts, J. Burghoorn, J. R. Brands, G. Venema, and J. Kok. 1998. A chloride-inducible acid resistance mechanism in *Lactococcus lactis* and its regulation. *Molecular Microbiology* 27(2):299–310. doi:10.1046/j.1365-2958.1998.00676.x.

Sanders, M. E., L. M. Akkermans, D. Haller, et al. 2010. Safety assessment of probiotics for human use. *Gut Microbes* 1(3):164–185. doi:10.4161/gmic.1.3.12127.

Santer, M. 2010. Joseph Lister: First use of a bacterium as a 'model organism' to illustrate the cause of infectious disease of humans. *Notes and Records of the Royal Society of London* 64(1):59–65. doi:10.1098/rsnr.2009.0029.

Santos, F., A. Wegkamp, W. M. de Vos, E. J. Smid, and J. Hugenholtz. 2008. High-level folate production in fermented foods by the B12 producer *Lactobacillus reuteri* JCM1112. *Applied and Environmental Microbiology* 74(10):3291–3294. doi:10.1128/AEM.02719-07.

Sashihara, T., N. Sueki, S. Ikegami. 2006. An analysis of the effectiveness of heat-killed lactic acid bacteria in alleviating allergic diseases. *Journal of Dairy Science* 89(8):2846–2855. doi:10.3168/jds.S0022-0302(06)72557-7.

Schnyder, B., and W. J. Pichler. 2009. Mechanisms of drug-induced allergy. *Mayo Clinic Proceedings* 84(3):268–272. doi:10.1016/S0025-6196(11)61145-2.

Schroeter, J., and T. Klaenhammer. 2009. Genomics of lactic acid bacteria. *FEMS Microbiology Letters* 292(1):1–6. doi:10.1111/j.1574-6968.2008.01442.x.

Scott, A. I., and C. A. Roessner. 2002. Biosynthesis of cobalamin (vitamin B(12)). *Biochemical Society Transactions* 30(4):613–620. doi:10.1042/bst0300613.

Sentausa, E., and P. E. Fournier. 2013. Advantages and limitations of genomics in prokaryotic taxonomy. *Clinical Microbiology and Infection* 19(9):790–795. doi: 10.1111/1469-0691.12181.

Shavakhi, A., H. Hashemi, E. Tabesh, et al. 2014. Multistrain probiotic and lactulose in the treatment of minimal hepatic encephalopathy. *Journal of Research in Medical Sciences* 19(8): 703–708.

Shewale, R. N., P. D. Sawale, C. D. Khedkar, and A. Singh. 2014. Selection criteria for probiotics: A review. *International Journal of Probiotics and Prebiotics* 9(1/2):17–22.

Shrestha, A., B. S. Kim, and D. H. Park. 2014. Biological control of bacterial spot disease and plant growth-promoting effects of lactic acid bacteria on pepper. *Biocontrol Science and Technology* 24(7):763–779. doi:10.1080/09583157.2014.894495.

Shrestha, A., E. C. Kim, K. C. Lim, S. Cho, H. J. Hur, and D. H. Park. 2009b. Biological control of soft rot on Chinese cabbage using beneficial bacterial agents in greenhouse and field. *The Korean Journal of Pesticide Science* 13(4):325–331.

Shrestha, A., K. U. Choi, C. K. Lim, J. H. Hur, and S. Y. Cho. 2009a. Antagonistic effect of *Lactobacillus* sp. strain KLF01 against plant pathogenic bacteria *Ralstonia solanacearum*. *The Korean Journal of Pesticide Science* 13(1):45–53.

Sieuwerts, S., D. Molenaar, S. A. van Hijum, et al. 2010. Mixed-culture transcriptome analysis reveals the molecular basis of mixed-culture growth in *Streptococcus thermophilus* and *Lactobacillus bulgaricus*. *Applied and Environmental Microbiology* 76(23):7775–7784. doi:10.1128/AEM.01122-10.

Siragusa, S., R. Di Cagno, D. Ercolini, F. Minervini, M. Gobbetti, and M. De Angelis. 2009. Taxonomic structure and monitoring of the dominant population of lactic acid bacteria during wheat flour sourdough type

I propagation using *Lactobacillus sanfranciscensis* starters. *Applied and Environmental Microbiology* 75(4):1099–1109. doi:10.1128/AEM.01524-08.

Smit, G., B. A. Smit, and W. J. Engels. 2005. Flavour formation by lactic acid bacteria and biochemical flavour profiling of cheese products. *FEMS Microbiology Reviews* 29(3):591–610. doi:10.1016/j.femsre.2005.04.002.

Smith, V. H., and R. D. Holt. 1996. Resource competition and within-host disease dynamics. *Trends in Ecology and Evolution* 11(9):386–389. doi:10.1016/0169-5347(96)20067-9.

Snyder, A. B., and R. W. Worobo. 2014. Chemical and genetic characterization of bacteriocins: Antimicrobial peptides for food safety. *Journal of the Science of Food and Agriculture* 94(1):28–44. doi:10.1002/jsfa.6293.

SpecialChem. 2014. Global lactic acid market to grow at a CAGR of 15.5% from 2014–20: Grand View Research. http://www.specialchem4bio.com/news/2014/05/23/global-lactic-acid-market-to-grow-at-a-cagr-of-15-5-from-2014-20-grand-view-research#sthash. McKc8RdH.dpuf (Accessed September 20, 2017).

Steele, J., J. Broadbent, and J. Kok. 2013. Perspectives on the contribution of lactic acid bacteria to cheese flavor development. *Current Opinion in Biotechnology* 24(2):135–141. doi:10.1016/j.copbio.2012.12.001.

Stephens, S. K., B. Floriano, D. P. Cathcart, et al. 1998. Molecular analysis of the locus responsible for production of plantaricin S, a two-peptide bacteriocin produced by *Lactobacillus plantarum* LPCO10. *Applied and Environmental Microbiology* 64(5):1871–1877.

Stephenson, D. P., R. J. Moore, and G. E. Allison. 2009. Comparison and utilization of repetitive-element PCR techniques for typing Lactobacillus isolates from the chicken gastrointestinal tract. *Applied and Environmental Microbiology* 75(21):6764–6776. doi:10.1128/AEM.01150-09.

Sun, Y., and M. X. D. O'Riordan. 2013. Regulation of bacterial pathogenesis by intestinal short-chain fatty acids. *Advances in Applied Microbiology* 85:93–118. doi:10.1016/B978-0-12-407672-3.00003-4

Sybesma, W., M. Starrenburg, M. Kleerebezem, I. Mierau, W. M. de Vos, and J. Hugenholtz. 2003. Increased production of folate by metabolic engineering of *Lactococcus lactis*. *Applied and Environmental Microbiology* 69(6):3069–3076. doi:10.1128/AEM.69.6.3069-3076.2003.

Szczepankowska, A.K., R.K. Górecki, P. Kołakowski, and J.K. Bardowski. 2013. Lactic acid bacteria resistance to bacteriophage and prevention techniques to lower phage contamination in dairy fermentation. In *Lactic acid bacteria - R & D for food, health and livestock purposes*, ed. J. Marcelino Kongo. InTech. doi:10.5772/51541. https://www.

intechopen.com/books/lactic-acid-bacteria-r-d-for-food-health-and-livestock-purposes/lactic-acid-bacteria-resistance-to-bacteriophage-and-prevention-techniques-to-lower-phage-contaminat (Accessed September 20, 2017).

Takagi, A., T. Matsuzaki, M. Sato, K. Nomoto, M. Morotomi, and T. Yokokura. 2001. Enhancement of natural killer cytotoxicity delayed murine carcinogenesis by a probiotic microorganism. *Carcinogenesis* 22(4):599–605. doi:10.1093/carcin/22.4.599.

Takei, T., M. Yoshida, Y. Hatate, K. Shiomori, and S. Kiyoyama. 2008. Lactic acid bacteria-enclosing poly(epsilon-caprolactone) microcapsules as soil bioamendment. *Journal of Bioscience and Bioengineering* 106(3):268–272. doi:10.1263/jbb.106.268.

Tamang, J. P., D.-H. Shin, S.-J. Jung, and S.-W. Chae. 2016. Functional properties of microorganisms in fermented foods. *Frontiers in Microbiology* 7:578. doi:10.3389/fmicb.2016.00578.

Tannock, G.W., A. Tilsala-Timisjarvi, S. Rodtong, J. Ng, K. Munro, T. Alatossava. 1999. Identification of *Lactobacillus* isolates from the gastrointestinal tract, silage, and yoghurt by 16S-23S rRNA gene intergenic spacer region sequence comparisons. *Applied and Environmental Microbiology* 65(9):4264–4267.

Temmerman, R., G. Huys, and J. Swings. 2004. Identification of lactic acid bacteria: Culture-dependent and culture-independent methods. *Trends in Food Science and Technology* 15(7):348–359. doi:10.1016/j.tifs.2003.12.007.

Tiptiri-Kourpeti, A., K. Spyridopoulou, V. Santarmaki, et al. 2016. *Lactobacillus casei* exerts anti-proliferative effects accompanied by apoptotic cell death and up-regulation of TRAIL in colon carcinoma cells. *PLoS One* 11(2):e0147960. doi:10.1371/journal.pone.0147960.

Torres, A. C., V. Vannini, J. Bonacina, G. Font, L. Saavedra, and M. P. Taranto. 2016. Cobalamin production by *Lactobacillus coryniformis*: Biochemical identification of the synthetized corrinoid and genomic analysis of the biosynthetic cluster. *BMC Microbiology* 16:240. doi:10.1186/s12866-016-0854-9.

Van den Berghe, E., G. Skourtas, E. Tsakalidou, and L. De Vuyst. 2006. *Streptococcus macedonicus* ACA-DC 198 produces the lantibiotic, macedocin, at temperature and pH conditions that prevail during cheese manufacture. *International Journal of Food Microbiology* 107(2):138–147. doi:10.1016/j.ijfoodmicro.2005.08.023.

Van Hoek, A. H. A. M., D. Mevius, B. Guerra, P. Mullany, A. P. Roberts, and H. J. M. Aarts. 2011. Acquired antibiotic resistance genes: An overview. *Frontiers in Microbiology* 2:203. doi:10.3389/fmicb.2011.00203.

Van Nieuwenhove, C. P., V. Terán and S. N. González. 2012. Conjugated linoleic and linolenic acid production by bacteria: Development of functional foods. In *Probiotics*, ed. E. C. Rigobelo. InTech. doi:10.5772/50321. https://www.intechopen.com/books/probiotics/conjugated-linoleic-and-linolenic-acid-production-by-bacteria-development-of-functional-foods (accessed September 20, 2017).

Van Overtvelt, L., H. Moussu, S. Horiot, et al. 2010. Lactic acid bacteria as adjuvants for sublingual allergy vaccines. *Vaccine* 28(17):2986–2992. doi:10.1016/j.vaccine.2010.02.009.

Venema, K., J. Kok, J. D. Marugg, et al. 1995. Functional analysis of the pediocin operon of *Pediococcus acidilactici* PAC1.0: PedB is the immunity protein and PedD is the precursor processing enzyme. *Molecular Microbiology* 17(3):515–522. doi:10.1111/j.1365-2958.1995.mmi_17030515.x.

Ventola, C. L. 2015. The Antibiotic resistance crisis: Part 1: Causes and threats. *Pharmacy and Therapeutics* 40(4):277–283.

Vijayakumar, J., R. Aravindan, and T. Viruthagiri. 2008. Recent trends in the production, purification and application of lactic acid. *Chemical and Biochemical Engineering Quarterly* 22(2):245–264.

Von Wright, A., and L. Axelsson. 2011. Lactic acid bacteria: An introduction. In *Lactic acid bacteria: Microbiological and functional aspects,* 4th ed., ed. S. Lahtinen, A. C. Ouwehand, S. Salminen, A. von Wright, 1–16. CRC Press.

Waters, C. M., and B. L. Bassler. 2005. Quorum sensing: Cell-to-cell communication in bacteria. *Annual Review of Cell and Developmental Biology* 21:319–346. doi:10.1146/annurev.cellbio.21.012704.131001.

Wedajo, B. 2015. Lactic acid bacteria: Benefits, selection criteria and probiotic potential in fermented food. *Journal of Probiotics and Health* 3:129. doi:10.4172/2329-8901.1000129.

Wee, Y. J., J. N. Kim, and H. W. Ryu. 2006. Biotechnological production of lactic acid and its recent applications. *Food Technology and Biotechnology* 44(2):163–172.

Wegmann, U., M. O'Connell-Motherway, A. Zomer, et al. 2007. Complete genome sequence of the prototype lactic acid bacterium *Lactococcus lactis* subsp. *cremoris* MG1363. *Journal of Bacteriology* 189(8):3256–3270. doi:10.1128/JB.01768-06.

WHO. 2017. Cancer. http://www.who.int/mediacentre/factsheets/fs297/en/ (accessed September 20, 2017).

Wollowski, I., S. T. Ji, A. T. Bakalinsky, C. Neudecker, and B. L. Pool-Zobel. 1999. Bacteria used for the production of yogurt inactivate carcinogens and prevent DNA damage in the colon of rats. *Journal of Nutrition* 129(1):77–82.

Wu, C., J. Huang, and R. Zhou. 2017. Genomics of lactic acid bacteria: Current status and potential applications. *Critical Reviews in Microbiology* 43(4):393–404. doi:10.1080/1040841X.2016.1179623.

Yang, F., C. Hou, X. Zeng, and S. Qiao. 2015. The use of lactic acid bacteria as a probiotic in swine diets. *Pathogens* 4(1):34–45. doi:10.3390/pathogens4010034.

Yi, E. J., J. E. Yang, J. M. Lee, et al. 2013. *Lactobacillus yonginensis* sp. nov., a lactic acid bacterium with ginsenoside converting activity isolated from Kimchi. *International Journal of Systematic and Evolutionary Microbiology* 63(Pt 9):3274–3279. doi:10.1099/ijs.0.045799-0.

Yonejima, Y., K. Ushida, and Y. Mori. 2013. Effect of lactic acid bacteria on lipid metabolism and fat synthesis in mice fed a high-fat diet. *Bioscience of Microbiota, Food and Health* 32(2):51–58. doi:10.12938/bmfh.32.51.

Zeidan, A. A., V. K. Poulsen, T. Janzen, et al. 2017. Polysaccharide production by lactic acid bacteria: From genes to industrial applications. *FEMS Microbiology Reviews* 41(Supp_1):S168–S200. doi:10.1093/femsre/fux017.

Zhou, M., D. Theunissen, M. Wels, and R. J. Siezen. 2010. LAB-Secretome: A genome-scale comparative analysis of the predicted extracellular and surface-associated proteins of lactic acid bacteria. *BMC Genomics* 11:651. doi:10.1186/1471-2164-11-651.

Żukiewicz-Sobczak, W., P. Wróblewska, P. Adamczuk, and W. Silny. 2014. Probiotic lactic acid bacteria and their potential in the prevention and treatment of allergic diseases. *Central-European Journal of Immunology* 39(1):104–108. doi:10.5114/ceji.2014.42134.

10

Role of Bacteria in Dermatological Infections

Thirukannamangai Krishnan Swetha and
Shunmugiah Karutha Pandian

Contents

10.1 Introduction

Skin serves as an effective barrier that comes into immediate action as the first line of defense by providing shelter to the inside of an organism from environmental assaults such as ultraviolet (UV) irradiations, chemical toxins, oxidative stress, and so on (Proksch et al., 2008). It also harbors a diversified microbiota including bacteria, fungi, and viruses that render the host with a protective shield against colonization of pathogenic microorganisms (Chiller et al., 2001). Any breach in the dermal layer disgruntles the barrier action of the skin and aids easy entry of pathogens, which in turn, lead to the development of skin and skin structure infections (SSSIs; also known as skin and soft tissue infections [SSTIs]) (Proksch et al., 2008; Ibrahim et al., 2015). Dermatological dysfunctions, with special emphasis on those influenced by bacterial attack, cover a wide spectrum of infections, from mild superficial cutaneous infections such as folliculitis to life-threatening necrotizing fasciitis (Palit and Inamadar, 2010). Based on the depth of the skin layer infected, there are massive conditions that define the severity of bacterial skin infections as complicated SSSIs (cSSSIs) and uncomplicated SSSIs (Eisenstein, 2008). In 2008, new terminology was proclaimed by the U.S. Food and Drug Administration (FDA), namely acute bacterial SSSI (ABSSSI), to describe bacterial-aided SSSIs with a characteristic lesion size of at least 75 cm^2

(Pulido-Cejudo et al., 2017). ABSSSIs are a huge encumbrance in clinical settings because the sensitivity pattern of infecting bacteria to current antibiotic treatment is found to be altered (Eisenstein, 2008). Besides, bacteria form biofilm on healthy and diseased epidermal surfaces (Vlassova et al., 2011). Biofilm is a sessile microbial population encased in an extracellular polymeric matrix and is a defensive action presented by microbes to sustain antibiotic treatment and host immune attack (Subramenium et al., 2015a). Skin biofilm formed in diseased state worsens the severity of infection by enduring the antibiotic treatments (Vlassova et al., 2011). The incessant increase in the development of resistance to A recent generation of antibiotics by bacteria impairs the management of ABSSSIs and demands alternate therapies to tackle the situation (Shah and Shah, 2011). This standpoint describes the requisite of deep insights to the clinical presentations, pathogen involved, mode of pathogen entry, severity of infection, and proper choice of treatment for effective management of ABSSSIs (Ki and Rotstein, 2008). This chapter mainly reviews different ABSSSIs, their pathophysiology, clinical presentations, diagnosis, prevailing treatment strategies, and other emerging alternative approaches for efficacious control of ABSSSIs outbreak.

10.2 Skin, an indispensable innate barrier

Skin furnishes the magnificent integumentary system and is the largest organ of higher eukaryotes that makes up to approximately 16% of whole body weight of an individual and covers a surface area of about 1.8 m^2. It is bequeathed with three structural layers, namely epidermis (outermost), dermis (middle), and hypodermis (innermost). Despite the fact that it is reliably structured throughout the body, the age of an individual and the anatomical location greatly influences the thickness of the skin (Kanitakis, 2002).

Skin is an indispensable inborn barrier that constitutes the first layer of defense and concretes protective armor to the inside of an individual against environmental assaults. It plays numerous roles as a barrier (Figure 10.1) such as:

- Physical barrier to shield against mechanical disturbances,
- UV barrier to resist UV mutilations,
- Oxidant barrier to protect the cell membrane and lipids from oxidative stresses,
- Thermal barrier to acclimatize the body from different climatic conditions,
- Permeability barrier to thwart water loss and resist entry of undesirable allergens, and
- Microbial barrier to preclude the colonization of other pathogens on host (Menon and Kilgman, 2009).

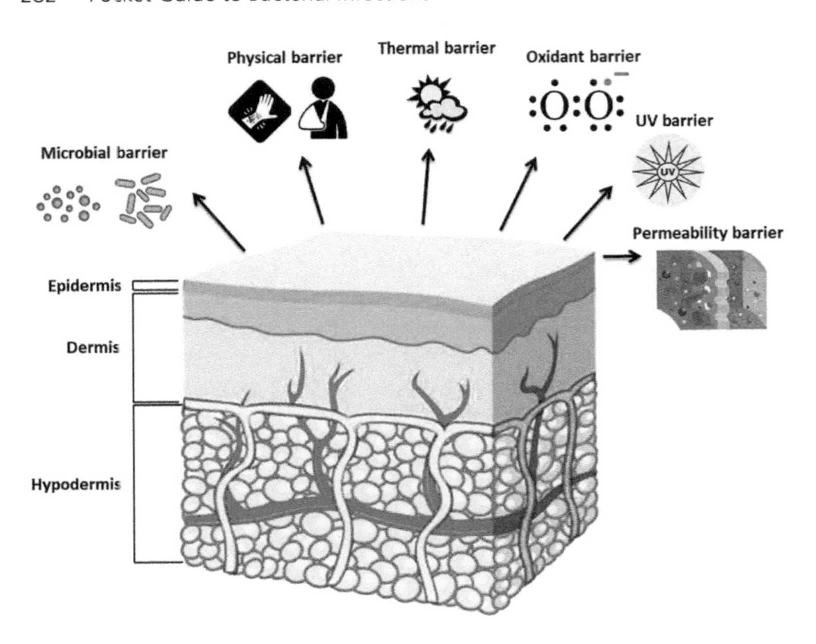

Figure 10.1 The barrier functions of skin. Skin is an indispensable barrier that protects the inside of an individual from mechanical and chemical assaults, oxidative stresses, ultraviolet mutilations, different climatic conditions, water loss, and entry of pathogenic microbes. Disruption of skin integrity paves way for skin infections.

Despite the multiple barrier functions performed by the skin, there are multitude cutaneous infections caused by the mutilation of epidermal integrity and alterations in skin microbial community influenced by environmental factors (Ibrahim et al., 2015).

10.3 Skin microflora

Ever since birth, the skin serves as the host for diversified collection of microbes such as bacteria, fungi, and viruses and affords an apposite environment for microbial growth (Chiller et al., 2001). A handful of researches have used amplicon sequencing and shotgun metagenomics sequencing techniques, which have erected the primary pipelines for understanding the diversity of skin microbial communities (Costello et al., 2009; Grice et al., 2009; Findley et al., 2013; Oh et al., 2014, 2016). The exploration of skin microbial community in healthy individuals has divulged the primary abundance of bacterial kingdom at almost all sites of the skin and fungal species as the least abundant kingdom (Oh et al., 2016). Commonly touted skin residential

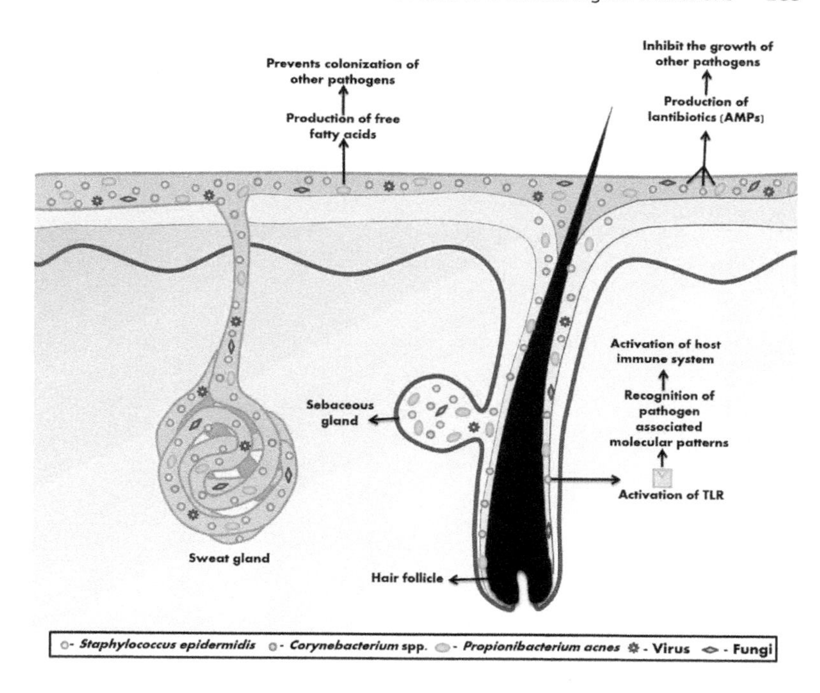

Figure 10.2 Skin microflora. Skin harbors a diversified collection of microbes such as bacteria, fungi, and viruses. The abundant bacterial kingdom majorly constitutes the microbial barrier, which includes *S. epidermidis*, *P. acnes*, and *Corynebacterium* spp. These bacterial species produce lantibiotics and prevent the entry of other pathogens. *S. epidermidis* in a competitive environment triggers toll-like receptor (TLR) pathway, which in turn recognizes the pathogen-associated molecular patterns and activates the host immune system accordingly. The production of free fatty acids by *P. acnes* as a result of triglyceride metabolism prevents the colonization of *S. aureus*, *S. pyogenes*, several gram-negative species, yeasts, molds, and so on.

bacterial species descend as *Staphylococcus* spp., *Corynebacterium* spp., and *Propionibacterium* spp. (Cogen et al., 2008). However, the skin microflora greatly diverges in proportions and in structure among individuals as a consequence of host genetics, sex, age, and regional and local environmental factors (Rosenthal et al., 2011). These commensals prominently play a beneficial role of protecting the skin either directly or indirectly from the attack of other pathogens (Figure 10.2). For instance, *S. epidermidis* armors the skin from colonization of infectious *S. aureus* by binding the keratinocyte receptors on skin. Similarly, *P. acnes*, an indigenous colonizer of sebaceous glands, catabolizes the sebum lipids and releases fatty acids, which in turn, establishes an inapt milieu for surviving virulent *S. pyogenes* (Chiller et al., 2001).

In circumstances, environmental factors such as climatic conditions, individual's lifestyle, choice of clothing, usage of antibiotics, usage of cosmeceutical products such as soaps, cream, lotion, and so on, altogether impacts the normal microflora of skin. Alterations in skin microflora are now deliberated to play essential role in skin infections. Additionally, dermal breaches perturb the barrier function of the skin and permeate the invasion of commensals as opportunistic pathogens and colonization of other pathogens, which eventually culminate in skin infections with various degree of severity (Grice and Serge, 2011).

Though understanding the interaction of skin microbiota with host and with other pathogens remains unclear, it is an important prerequisite to gain deep insights into the severity and underlying risk factors of skin infections, which will help to gain advancement in designing the appropriate treatment strategies to preclude the difficulties in the management of skin infections.

10.4 Acute bacterial skin and skin structured infections

Impairment in barrier activities of skin paves way for the infiltration of skin commensals as pathogens and other invading microbes into the underlying soft tissues and layers of skin, which results in the development of SSSIs (Ki and Rotstein, 2008). Until 2008, SSSIs was stratified based on the severity of the infection into two categories namely, uncomplicated SSSI and cSSSI. Uncomplicated SSSIs generally include mild infections such as impetigo, furunculosis, carbuncles, cellulitis, erysipelas, and other mild abscesses. cSSSIs encompass spreading and deeper tissue infections such as necrotizing cellulitis, infected wound burns, deep-tissue abscesses, and infected ulcers, which require immediate attention (Ramana et al., 2013). A decade ago, the FDA categorized cSSSIs and uncomplicated SSSIs as acute bacterial skin and skin structured infections (ABSSSIs) (Figure 10.3) (Shah and Shah, 2011). According to the guidelines of the FDA, the skin infections with minimum lesion size of 75 cm^2 and characteristic symptoms such as redness, edema, and other local signs have been recognized as ABSSSIs (Moran et al., 2013).

Bacterial skin infections are opportunistic and have a great predilection for patients who are immunocompromised and individuals with other predisposed conditions. Identification of nature of infection such as mild or deep and localized or spreading is quintessential, which lays foundation for determining the urgency and proper strategy of the treatment. It is generally seen as almost curable if prompt diagnosis and immediate and appropriate

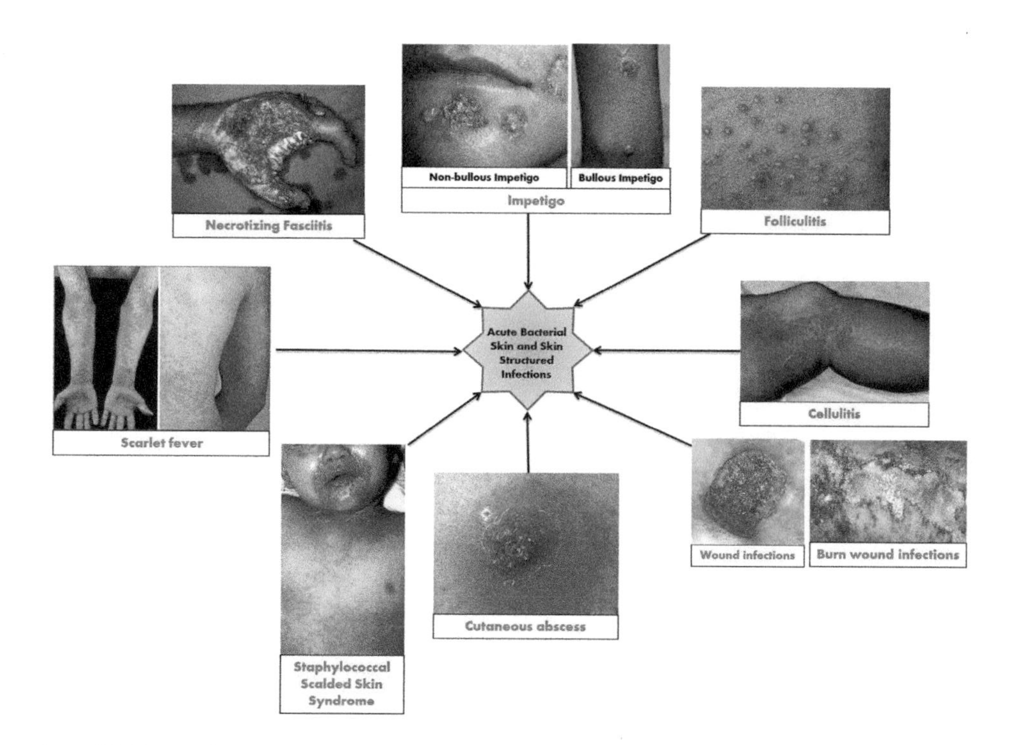

Figure 10.3 Acute bacterial skin and skin structured infections (ABSSSIs). ABSSSIs include a wide spectrum of infections ranging from mild superficial impetigo, folliculitis to deep-seated cellulitis and life-threatening necrotizing fasciitis.

treatment is provided to the patient. However, delayed or improper treatment worsens the scenario, and the resulting secondary effects can be life threatening (Hedrick, 2003). Thus, reviewing the etiologic agents, clinical presentations, diagnosis, and treatment options of common bacterial skin infections is essential.

10.4.1 Impetigo

Impetigo is a common superficial skin infection, which is greatly contagious and usually affects children ages 2 to 5. It is classified into two types, non-bullous impetigo (impetigo contagiosa) and bullous impetigo.

Impetigo contagiosa is slightly pruritic, which is mainly characterized by the development of small pustules that exude at later stages and form honey-colored crust. *S. aureus* is the predominant etiologic agent, whereas group A Streptococcus (GAS) hits several cases either alone or in association with *S. aureus*. The dissemination of nonbullous form to surrounding areas occurs through autoinoculation. Secondary impetigo, a subtype of nonbullous impetigo, increases the severity of systemic infections (such as diabetes mellitus) and also prompts the individual with dermal breaches to develop common impetigo. The recovery rate usually involves 2 to 3 weeks without any treatment, whereas antibiotic treatment speeds up the rate of recovery and lessens discomfort (Parker, 1955; Hedrick, 2003).

Bullous form of impetigo mainly develops due to the exfoliative toxin A (ETA) produced by coagulase positive *S. aureus*. Neonates remain to be the primary choice of bullous impetigo. However, it affects the elderly, too. The clinical presentations of bullous form comprise development of small vesicles that enlarge to become flaccid bullae and ooze out on exudation (Amagai et al., 2000).

The severe and deeper form of impetigo is referred to as *ecthyma*, which spreads to the dermis and is characterized by ulceration under crusted surface. The management of ecthyma remains the same as that of impetigo because the bacterial etiologic agents implicated in the infections are same (Pye, 2010).

10.4.1.1 Epidemiology – Impetigo accounts for 10% of overall skin problems and 30% of it are observed to be bullous form. Staphylococcal scalded skin syndrome (SSSS) is the localized form of bullous impetigo, which causes less than 3% of deaths in children and more than 60% in the elderly population (Hanakawa et al., 2002). Moreover, it has been reported that the global incidence of impetigo in neonates at any particular time point hits more than 162 million, which also heightens the incidence of life-threatening rheumatic heart disease and arthritis due to rheumatic fever (Bowen et al., 2015).

10.4.1.2 Diagnosis – Impetigo is generally identified clinically by culturing the lesion material acquired from infected area. Gram staining is rarely performed to confirm the infection (Cole et al., 2007).

10.4.1.3 Prevailing treatment options – The current treatment options of impetigo include topical and oral antibiotics. Mupirocin (commercially known as Bactroban), a topical antibiotic, is used as the first-line of defense for treating impetigo. However, patients who remain non-responsive to topical treatment are treated with oral antibiotics such as penicillin V, cephalexin, amoxicillin/clavulanate, and dicloxacillin. Antibiotic resistance among bacterial species increases the treatment cost and incapacitates the management of impetigo (Bangert et al., 2012).

10.4.2 Folliculitis

Folliculitis is the inflammation of hair follicle caused due to virus, fungi, or bacteria. Poor hygiene, exposure to certain chemicals, humid environment, maceration, hyperhydration, and occlusion are several predisposing factors that conciliate the development of folliculitis. Folliculitis is characterized by formation of small pustules with erythematous base that later expand into clusters and crusts. Bacterial folliculitis may be superficial or deep that occurs at different sites such as the scalp, buttocks, axillae, medial thigh, and face of children and adults. *S. aureus* is the major cause of bacterial folliculitis; however, other bacterial species such as *Pseudomonas*, *Streptococcus*, *Proteus*, and some Coliform bacteria are also involved in the infection (Luelmo-Aguilar and Santandreu, 2004).

The symptoms usually include boils, scarring, furunculosis, and permanent hair loss at the site of infection. The spread of infection into the tissues of follicles culminates in the development of furuncles and carbuncles and the untreated folliculitis may transpire into life-threatening necrotizing fasciitis (Palit and Inamadar, 2009).

10.4.2.1 Epidemiology – A survey of bacterial skin infections in the northern part of India among schoolchildren had reported that 64.4% cases of skin infections were observed to be folliculitis together with impetigo (Palit and Inamadar, 2009). Also, it accounts a total of 11% of all primary scarring hair loss or cicatricial alopecia cases (Otberg et al., 2008).

10.4.2.2 Diagnosis – The physical examination of patient for clinical signs and studying the patient history for analyzing other predisposing factors helps in preliminary identification of folliculitis. And, the laboratory examination of pustular material collected from the site of folliculitis is usually considered for identification of etiologic agent (i.e., bacteria, fungi, or virus) implicated in the infection. Histopathologic studies confirm and ease the process of diagnosis and proper classification of infection (Luelmo-Aguilar and Santandreu, 2004).

10.4.2.3 Prevailing treatment options – The use of topical antibiotics such as mupirocin, neomycin, and fusidic acid for 7 to 10 days remains to be the first-line of defense for generalized therapy. If treatment fails to work and or the infection is spreading and severe, proper diagnosis of etiologic agent usually soothes the process of identifying specified antibiotic treatment. In the latter case, use of systemic antibiotics such as erythromycin and flucloxacillin for 1 week are the prime choice of treatment. Heavy dose of trimethoprim prevails as the choice of treatment, if the etiologic agent implicated in the infection is identified to be a gram-negative microorganism (Pye, 2010).

10.4.3 Cellulitis

Cellulitis s a common, deeper, and generally nonpurulent inflammatory skin infection caused by bacteria. This disseminating skin infection is usually characterized by edema, pain, tenderness, swelling, warmth, local erythema, lymphangitis and white blood cell infiltration (Hook et al., 1986). The dermal and subcutaneous layers of skin remain inflamed in cellulitis, due to infection caused by *S. pyogenes*, coagulase positive *S. aureus*, *P. aeruginosa*, *Haemophilus influenzae* and other beta hemolytic Streptococcus (BHS) that gain their access through impaired skin barrier, dermatitis, and skin ulceration. However, *S. pyogenes* is the major etiologic agent of cellulitis, which generates inflammation of skin by enzymatic hydrolysis of cellular components (Hedrick, 2003; Chira and Miller, 2010).

Frequently witnessed facial cellulitis is now a rarely observed type, which is recently demarcated as erysipelas (superficial cellulitis) in some literature. The vital complication associated with facial cellulitis comprises odontogenic or orbital infections, which require immediate attention and surgical episodes (Stevens et al., 2014). Perianal cellulitis is witnessed in young children and is characterized by purulent drainage, perianal pruritus, fissures, and rectal bleeding. Periorbital cellulitis is observed in eyelid portion and periorbital tissues. Buccal cellulitis is greatly observed in children before vaccination with conjugated *H. influenzae* type b vaccine and is responsible for 25% of the facial cellulitis cases (Swartz, 2004). Purulent cellulitis may also be observed in several cases, which occur as an extension of initial abscess and culminate in secondary cellulitis followed by purulent drainage and exudation (Ibrahim et al., 2015).

10.4.3.1 Epidemiology – In the United States, 14.5 million cases of cellulitis were observed annually and hospitals specialized for cellulitis were also mounted (Arakaki et al., 2014). Moreover, cellulitis is an important complication associated with HIV, wherein, a retrospective study of epidemiology of cellulitis in patients with HIV has shown that cellulitis accounted for 3.02% of overall admissions (Manfredi et al., 2002)

10.4.3.2 Diagnosis – Analysis of morphological characteristics of infection and patient history for identification of predisposing factors remains as a standard method for diagnosing cellulitis. Identification of bacterial species implicated in infection is quite difficult, unless pus or open wound have developed on the skin (Hedrick, 2003). Elevated white cell count and C-reactive protein (CRP) are several inflammatory markers employed for diagnosis, which supports detection, but cannot be used solely to identify the infection because of non-specificity. Skin biopsies and magnetic resonance imaging (MRI) are engaged for distinguishing the fatal necrotizing fasciitis from cellulitis because misdiagnosis occurs frequently and increases the treatment cost (Schmid et al., 1998; Swartz, 2004).

10.4.3.3 Prevailing treatment options – Management of mild cellulitis involves oral antibiotic treatment such as penicillin and flucloxacillin (500 mg to 1 g for four times a day). The recovery rate usually takes about 1 to 2 weeks. Patients allergic to penicillin are treated with clarithromycin as an alternate. Intravenous antibiotic therapy is suggested for severe cellulitis cases and patients diagnosed with other systemic infections and other complications (Pye, 2010).

10.4.4 Scarlet fever

Scarlet fever, or scarlatina, is the infection caused by GAS, typically as an extension of strep throat and is characterized by fever, erythematous sore throat, sand-paper like rash, and enlarged papillae on tongue (strawberry-like tongue). The papillary eruptions generally initiate from groin, then disseminate jointly to the trunk and axillae, and finally reach extremities in a week's time, followed by desquamation on palms and soles. Though, it affects people of all ages with predominance among males over females, the episodes of scarlet fever in children is still high (Basetti et al., 2017).

The erythrogenic toxins produced by GAS (now classified as pyrogenic exotoxins) are the major etiologic agent of this infection (Wessels, 2016). Recent researchers have divulged a collection of related streptococcal pyrogenic exotins (SPE). Eleven such toxins are involved in infection and no single toxin was found in all the cases because each Streptococcal strain was capable of generating four to six toxins (Spaulding et al., 2013). However, the prime toxins SpeA, SpeC, and SSA are often found in combination in several episodes (Davies et al., 2015). Complications that are possibly associated with scarlet fever include rheumatic fever, more invasive streptococcal toxic shock syndrome, suppurative arthritis, toxic myocarditis, osteomyelitis, allergic glomerulonephritis, and meningitis (Hedrick, 2003).

10.4.4.1 Epidemiology – Previously demoted as less episodic, scarlet fever has surged as a major health issue with massive episodes in recent decades. The outbreaks of scarlet fever were stated to be surged in England, Hong Kong, Vietnam, South Korea, and China. The epidemic of scarlet fever in 2009 was found to be very high, affecting nearly 23,000 individuals in Vietnam and 100,000 individuals in mainland China (Basetti et al., 2017).

10.4.4.2 Diagnosis – Because the severities linked with scarlet fever are extensive, early clinical diagnosis remains difficult. Clinical examination and case history analysis are useful in diagnosis. Centor score is an aiding technique used for the identification of scarlet fever across numerous healthcare settings in various countries. This robust technique includes four clinical signs and symptoms and is used to range the infection as "specific" and "very specific" for GAS pharyngitis, when the clinical condition satisfies more than three or four symptoms, respectively. However, prolonged pyrexia and tachycardia together with dissemination of rashes to trunk ascertains the clinical condition of scarlatina (Aalbers et al., 2011).

10.4.4.3 Prevailing treatment options – Beta-lactam antibiotics remain as the classic choice of treatment. Penicillin is prescribed frequently because of its reduced costs and ability to outstrip cephalosporin and macrolides. Macrolides are reported to favor antibiotic resistance, produce adverse effects, and found to be toxic at effectual doses. For managing scarlet fever, phenoxymethylpenicillin penicillin (V) is prescribed for 10 days (four times a day) and the repercussions are usually lessened using over-the-counter drugs such as paracetamol, ibuprofen, and so on (Basetti et al., 2017).

10.4.5 Staphylococcal scalded skin syndrome

SSSS is blistering skin disorder with preponderance for neonates and children younger than ages 5 or 6 and not in adults. In 1878, Baron Gotfried Ritter von Rittershain witnessed 297 cases of SSSS, and referred to it as "dermatitis exfolitiva neonatorum," which was later demarcated as Ritter's disease (Patel and Finlay, 2003).

SSSS is mediated by epidermolytic exfoliative toxins A and B (ETA and ETB) produced by *S. aureus*. These ETA and ETB exaggerate the epidermal blistering by proteolytic cleavage of desmoglein 1 (a cell-cell adhesion molecule), which is expressed by keratinocytes in epidermal zona granulosa (Ross and Shoff, 2017). SSSS initially commences with fever, pruritus, discomfort, redness of skin, and development of fluid-filled blisters on groin, armpits, and body orifices such as the nose and ears, which disseminates to the trunk and extremities followed by epidermal exudation within 72 hours that leaves marks similar to that of burns. The clinical indications include chillness, fever, and malaise with early signs of conjunctivitis or sore throat. The potential complications comprise electrolytic imbalance, pneumonia, and sepsis (Hedrick, 2003).

In neonates, SSSS is frequently observed in the diaper area and axillary and naval regions. Adults are usually less prone to SSSS because of the presence of antibodies specific to the toxins, which are developed during childhood period. However, adults with a compromised immune systems and renal failure are predisposed to the attack of SSSS (Lamanna et al., 2017).

10.4.5.1 Epidemiology – A retrospective study in France of the Czech Republic estimated the outbreak of SSSS to be 2.53 cases per million per year. In Europe, 0.56 cases per million per year was admitted with SSSS (Lamanna et al., 2017). The mortality rate associated with SSSS was reported to be approximately 4% in children and about 50%–60% in adults (Pye, 2010; Li et al., 2014).

10.4.5.2 Diagnosis – An assessment of case history along with evaluation of clinical features of infection is usually done. Clinical diagnosis of SSSS is performed by skin biopsies, analysis of frozen sections of blisters, and exfoliative cytology. Culturing of samples obtained from blisters in affected area is helpful in identifying the etiologic agent of infection (Hedrick, 2003; Ross and Schoff, 2017). Also distinguishing SSSS from toxic epidermal necrolysis (TEN) is vital because the latter involves high mortality rate.

10.4.5.3 Prevailing treatment options – Intravenous antibiotics are usually administered for management of SSSS. Flucloxacillin remains as the preferred choice of treatment. Substitution of intravenous antibiotics with oral antibiotics can be done after several days based on treatment response. The risk of electrolytic imbalance is resolved by the administration of intravenous fluids. The recovery of skin health and heat loss reduction is carried out using nonadherent dressings and emollients on skin (Pye, 2010; Ross and Schoff, 2017).

10.4.6 Wound infections

The wound is generally abrasions, breaches, or breaks on anatomical structures such as skin that can extend to underlying tissues, dermal layers, and even to the bone as a consequence of trauma or surgery (Boateng and Catanzano, 2015). Wound infections are caused by the bacterial colonization on chronic wounds that prolongs the recovery rate and remains too unfavorable for wound-healing process (Edwards and Harding, 2004). Burn wounds predispose patients for high pathogenic attack and raise the probability of mortality and morbidity by increasing the risks of other complications such as multiple-organ dysfunction syndrome, systemic inflammatory response syndrome, and severe sepsis (Schwacha et al., 2005).

In 2000, a retroactive study conducted by Giacometti et al. (2000) among 676 patients who underwent surgery for the analysis of possible bacterial strains associated with wound infections showed the prominence of aerobic bacteria in almost all cases with different percentage coverage by different bacterial strains such as *S. aureus* (28.2%), *P. aeruginosa* (25.2%), *Escherichia coli* (7.8%), *S. epidermidis* (7.1%), and *Enterococcus faecalis* (5.6%). Polymicrobial infections were also witnessed with more predominance of *P. aeruginosa* and *S. aureus* combination. Another experimental study by Bessa et al. (2015) using 312 wound samples collected from 213 patients at different wound sites have witnessed the prominence of bacterial strains in the order *S. aureus* (37%), *P. aeruginosa* (17%), *Proteus mirabilis* (10%), *E. coli* (6%), and *Corynebacterium* spp. (5%) with 59.2% incidence of polymicrobial infections.

Bacterial biofilms are highly perceived on chronic wounds, including pressure ulcers (PUs), diabetic foot ulcers (DFUs), and venous leg ulcers (VLUs). Biofilms are a great menace to the management of wound infections because most of the antibiotics administered remains ineffective against bacterial biofilms (Malone et al., 2017).

10.4.6.1 Epidemiology – During mid of nineteenth century, surgical site wound infections were predominantly observed that accounted for approximately 70% to 80% of mortality. Surgical site infections ranks third place of all recurrently reported infections in patients who are hospitalized, and it responsible for 12% to 16% of all nosocomial infections (Cooper, 2013).

10.4.6.2 Diagnosis – The conventional methods for the identification of causative agents of wound infections include tissue biopsies, wound swabs, culturing of samples recovered from infected site, and polymerase chain reactions (PCR) (Weinstein and Mayhall, 2003).

10.4.6.3 Prevailing treatment options – As identification of etiologic agent of wound infections is quite time consuming, empirical antibiotic treatments are often preferred, wherein generalized antibiotics that target a broad spectrum of bacteria ire administered (Cooper, 2013). However, proper identification of causative agents eludes the risk of drug resistance and reduces the health expenditure. Also, wound dressings that deliver the actives such as antiseptic agents or antibiotics at the wound sites are used often to reduce the bacterial load and soothe the process of wound healing. Topical antimicrobials in the form of lotions, creams, and liquids are also used in the management of wound infection (Boateng and Catanzano, 2015).

10.4.7 Cutaneous abscess

Cutaneous and other soft skin abscesses are usually trailed by skin trauma and are characterized by pus collection, redness, edema, and swelling of soft tissues accompanied by erythema. Lesions may be encountered on neck, face, extremities, trunk, axillae, and perianal regions. Based on the severity, abscesses may be simple or uncomplicated skin abscesses (superficial) or complicated skin abscesses (disseminating and deeper). The latter type is likely to be associated with complications such as cellulitis, regional lymphadenopathy, septic phlebitis, lymphangitis, and leukocytosis (Hedrick, 2003). It is predominantly observed among injection drug users (IDUs). The use of heroin and cocaine mixtures (also called "speedballs") containing injections is reported to induce soft-tissue ischemia and predispose the individual to develop abscessws (Murphy et al., 2001). The major etiologic agents that spurred the development of infectious abscess include *S. aureus* (especially community acquired methicillin-resistant *S. aureus* [CA-MRSA]), alpha hemolytic and nonhemolytic Streptococcus, *Propionibacterium* spp., and *Bacteriodes* spp. (Hedrick, 2003; Lee et al., 2004).

10.4.7.1 Epidemiology – In urban medical centers, cutaneous abscesses are reported to account for approximately 2% of all cases that visited emergency facilities with predominance for IDUs (Talan et al., 2000). In the United States, the annual epidemic of SSSI has been chronicled to be inclined from 1.2 to 3.4 million between 1993 and 2005, primarily with massive episodes of cutaneous abscesses at emergency departments (Qualls et al., 2012).

10.4.7.2 Diagnosis – The medical history and drug practices of the patient are studied initially. Then, the type of skin abscess such as postoperative wound abscess, perirectal or perianal abscess, abscess developed from infected breast cysts, or other kind of skin and soft tissue abscess, is analyzed (Talan et al., 2000). The pus or draining material collected from major abscess site is cultured to identify the bacteriology of infection (Murphy et al., 2001).

10.4.7.3 Prevailing treatment options – Empirical antibiotic therapy is usually prescribed. For uncomplicated or simple skin abscess, a simple incision followed by drainage of purulent material is the preferred treatment. Doxycycline and minocycline administration together with incision and drainage remains successful in the management of uncomplicated MRSA infections (Stevens et al., 2014). Trimethoprim sulfamethoxazole (TM-Sulpha) is also prescribed as an alternate for uncomplicated MRSA infections (Talan et al., 2016). Oral or intravenous antibiotics are administered to patients with complicated skin abscess to recover from the risk of CA-MRSA infection (Lee et al., 2004).

10.4.8 Necrotizing fasciitis

Necrotizing fasciitis (NF) is a rare, deep, and aggressive soft-tissue infection that involves the deeper layer of skin and spreads across dermal fascia and remains detrimental for surrounding tissues. The physical manifestations include skin lesions with redness, pain, swelling, erythematous, and edema (Giuliano et al., 1977). It usually occurs as an extension of impaired skin integrity (Pye, 2010). NF is classified microbiologically into three types:

Type I NF: It is mostly polymicrobial infection with high incidence of combinations of gram-positive and -negative microorganism along with anaerobes.
Type II NF: It is mainly a monomicrobial infection instigated by GAS, non-GAS, *S. aureus*, and *Clostridia* spp.
Type III NF: It is rarely encountered type of NF wherein, *Vibrio vulnificus* gain access through dermal breaches exposed to seawater and cause NF (Elliott et al., 2000).

The aforesaid microorganisms instigate NF by producing endotoxins or exotoxins and invade subcutaneous layer leading to the ischemic necrosis of tissue (Stevens and Bryant, 2017).

10.4.8.1 Epidemiology – Between 2001 and 2003, a population-based surveillance conducted in Canada by Eneli and Davies (2007) revealed that the epidemic of GAS associated NF was 2.12 cases per million children and that of non-GAS NF was 0.81 cases per million children with mortality rate of 5.4%. In New Zealand, a national study between 1990 and 2006 conducted by Das et al. (2011) showed that the annual outbreak of NF has increased from 0.18 to 1.69 cases per 100,000 persons with mortality rate of 0.3 cases per 100,000 persons. In the United States, a nationwide study conducted by Psoinos et al. (1993) between 1990 and 2010 witnessed an annual incidence of NF ranging from 3,800 to 5,800 cases.

10.4.8.2 Diagnosis – A Laboratory Risk Indicator for NF score (LRINEC) is used for early diagnosis of NF from other soft-tissue infections. LRINEC involves estimation of white cell count, CRP, sodium, creatinine, glucose, and hemoglobin levels. Based on the level of variables, a score is assigned.

Potential NF is feared, when the score is more than 6. Score ≤5 indicates low risk of NF. Computed tomography (CT) and MRI are also performed for the diagnosis of NF, when LRINEC scoring system remains ambiguous (Wong et al., 2004).

10.4.8.3 Prevailing treatment options – Early diagnosis and proper distinction of NF from other soft-tissue infections is quintessential to elude the risk of complications and for proper treatment. Surgical debridement of the affected part is usually suggested in severe NF cases. However, the management of early diagnosed NF can be achieved using administration of broad-spectrum antibiotics (Pye, 2010).

10.5 Predominant bacteria in skin infections

A tremendous uplift of diagnosis of SSSIs in both community and healthcare settings has been observed lately. Over recent decades, visits at ambulatory care and emergency departments and hospitalization of patients are reported to be enormously inclined as a ramification of substantial surge in SSSIs (Esposito et al., 2016).

A bacterial role in dermatological diseases is a well-known fact. However, the understanding of bacterial pathogenesis, the behavioral pattern of bacteria to antibiotic treatments, and the likelihood of pathogenesis and antibiotic resistance is still in its nascent period of development, which plays a potential role in manipulating proper treatment strategies to manage infections.

Recent findings have divulged that bacterial ecology of skin is more multifaceted and harbors a wide spectrum of no-fastidious or noncultivable bacteria in healthy and in diseased states, which posit the possible role of commensals in skin infections (Vlassova et al., 2011). This viewpoint shows a departure from previously established views of beneficial role of skin microflora in maintaining skin health. Thus, reviewing the virulent traits of bacteria involved in pathogenesis, altered behavior of commensal microflora in diseased condition, probable use of recent technological advancements as alternate approaches for managing the global burden of emerging antimicrobial resistance is a dire need.

Frequently addressed gram-positive and -negative opportunistic human pathogens in SSSIs include *S. aureus*, *Streptococcus* spp., and *P. aeruginosa*. Beneath these formidable bacterial species are the profuse and underrated skin commensals such as *S. epidermidis*, *P. acnes*, and *Corynebacterium* spp., which have recently evolved as opportunistic pathogens in SSSIs (Figure 10.4). These commensals are also molecular reservoirs of virulence

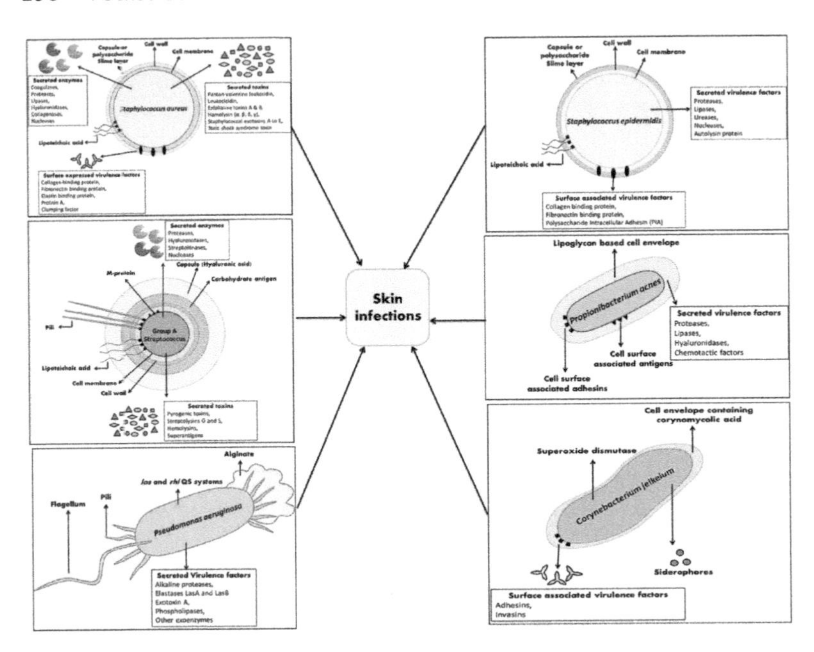

Figure 10.4 The basic cell structure and virulent determinants of predominant bacteria in skin infections. The left column includes predominant pathogens such as *S. aureus*, *S. pyogenes*, and *P. aeruginosa* in skin infections, whereas the right column includes predominant skin commensals as opportunistic pathogens such as *S. epidermidis*, *P. acnes*, and *C. jeikeium* in skin infections. The virulence determinants hosted by these bacterial species in addition to their ability to form biofilm on animate and inanimate surfaces help to elude a multitude of antibiotic treatments and host immune system and pose a great menace to the clinical settings.

and antimicrobial resistance traits for other pathogens, which elicit the severity of infections (Levy and Marshall, 2004). Thus, the subsequent part of this chapter details the essentials of virulence traits of the aforementioned pathogens and their role in SSSIs.

10.5.1 *S. aureus*

The gram-positive and coagulase positive *S. aureus* is a member of Firmicutes and commensal in one-third of the human population (Sethupathy et al., 2017). However, it is also an extensively isolated and leading human opportunistic pathogen (Alegre et al., 2016). The reportage of *S. aureus* contribution in SSSIs hit 80%, of which 63% was observed in cellulitis and cutaneous abscesses (Malachowa et al., 2015). This remarkable escalation of infectious role is the result of the advent of MRSA in community and healthcare settings. This notion affords a hostile situation for medical practitioners to manage SSSIs, especially at emergency departments (Frazee et al., 2005).

Acquisition of type IV staphylococcal cassette chromosome *mec* (SCC*mec*, a transposable element) comprising the *mec*A gene, which codes for penicillin-binding protein (PBP) 2a is the actual cause of the emergence of MRSA. The site-specific recombination of SCC*mec* and genomic DNA makes MRSA to resist beta-lactam antibiotics, which basically targets cell wall PBP to interrupt the synthesis of peptidoglycan layer and ultimately kills the bacterium (Cogen et al., 2008).

An epidemiological population based study of *S. aureus* instigated SSSIs at Georgia had unveiled the occurrence of 72% of MRSA in SSSIs. Among MRSA isolates, CA-MRSA accounted for 87%, of which 99% CA-MRSA were MRSA USA 300 clones. Moreover, notorious role of two types of clones namely, USA 300 and USA 400 in CA-MRSA associated SSSIs in the United States have been recognized by Centers for Disease Control and Prevention (CDC). It has been stated that 65% of preliminary antibiotic treatments carried out for CA-MRSA infections remains inadequate and ineffective than infections associated with methicillin-susceptible *S. aureus* (MSSA). This differential feedback of treatments used for CA-MRSA and MSSA infections is primarily the consequence of presence of virulent factor named Panton–Valentine leucocidin (PVL) in CA-MRSA strains (King et al., 2006).

The prolonged cohabitation of *S. aureus* with the host, together with its ability to express numerous secreted and cell-associated virulence factors has led to its adaption to host immune attack at many stages of SSSIs (Malachowa et al., 2015). The repertoire of virulence determinants of *S. aureus* include biofilm formation, secretion of toxins (exfoliative toxins ETA and ETB), hemolysin (α, β, δ, and γ), staphylococcal enterotoxins (A to E), leukocidin, PVL, toxic shock syndrome toxin, etc.), secretion of enzymes (coagulases, proteases, lipases, collagenases, hyaluronidases and nucleases), and surface expressed virulent factors (collagen-binding protein, elastin-binding protein, fibronectin-binding protein, protein A, and clumping factor) (Sethupathy et al., 2017).

The virulent enzymes sturdily aid survival of *S. aureus* by host tissue damage, whereas opsonization of *S. aureus* by surface expressed protein A and clumping factor helps in the evasion of host phagocytic attack. The production of hemolysins triggers the nuclear factor (NF-κB) inflammatory pathway through pore formation of targeted host cell membranes (Cogen et al., 2008). The exfoliative toxins ETA and ETB disrupt the cell-cell adhesion protein molecule desmoglein-1, which causes cutaneous blistering in SSSS (Hedrick, 2003). PVL in CA-MRSA complicates SSSIs by polymorphonuclear karyorrhexis, capillary dilation, and skin necrosis prompted by the production of severe inflammatory lesions (Dufour et al., 2002). The surface-expressed binding proteins elicit the severity of infections by promoting the invasive lifestyle of *S. aureus* through initial adhesion of *S. aureus* to host cell surface, followed by colonization and invasion (Shinji et al., 2011). Also, a novel class of virulent determinant (observed initially in CA-MRSA USA 300 clone and later in other *S. aureus* lineages) named arginine catabolic mobile element (ACMB) is involved in immune modulating functions such as conferring tolerance to polyamines (a nonspecific immune response), which in turn facilitate the successful survival of *S. aureus* by outnumbering the competitors, host colonization, and invasion (Shore et al., 2011).

In addition, the novel lantibiotics synthesized by commensal coagulase negative *Staphylococcus* spp. were found to synergize with cathilicidin (a cationic antimicrobial peptide), which curtails the growth of *S. aureus* (Nakatsuji et al., 2017). However, *S. aureus* evade the attack of host cationic antimicrobial peptides (CAMP) by modulating its cell surface charge, which is actively accomplished by Dlt protein and MprF enzyme that neutralize the negative charge of cell wall surface by substituting D-alanine in cell wall teichoic and lipoteichoic acids, and adding L-lysine to phosphatidyl glycerol, respectively (Cogen et al., 2008). Besides, staphyloxanthin, a golden-colored carotenoid pigment produced by *S. aureus,* protects it from oxidants and neutrophilic attack by exhibiting antioxidant activity (Sethupathy et al., 2017).

Apart from the aforementioned virulence determinants that provide protection from host immune attack, biofilm formation of *S. aureus* is rated as one of the important virulence factors that support antibiotic tolerance. The biofilm formation of CA-MRSA strain, a leading notorious pathogen in SSSIs, makes it to resist beta-lactam antibiotics and other non-beta-lactam antibiotics such as erythromycin, oxacillin, kanamycin, and tetracycline and confers intermediate resistant to fusidic acid (Dufour et al., 2002; Vanhommerig et al., 2014).

The current treatment options for MSSA associated skin infections include first-generation cephalosporins, clindamycin, oxacillin, and nafcillin (Stevens et al., 2014). On the other hand, the inefficiency of beta-lactam antibiotics to treat MRSA-associated infections was remedied by vancomycin in 1980. However, the emergence of vancomycin-intermediate MRSA (VIMRSA) in late 1990s and subsequent vancomycin-resistant MRSA (VRMRSA) strains momentously incapacitates the previous treatment strategies. Acquisition of *van*A gene complex from vancomycin-resistant enterococci (VRE) has led to the advent of VIMRSA, wherein vancomycin could still bind to thickened cell wall of VIMRSA but efficiently curtails the diffusion of vancomycin. In the current scenario, the inefficiency of vancomycin is balanced by tedizolid, daptomycin, linezolid, oritavancin, dalbavancin telavancin, ceftobiprole, and ceftaroline (Chambers and Leo, 2009). Lately, TM-sulfa is clinically used as efficient antimicrobial agent to tackle MSSA and MRSA infections (Frazee et al., 2005).

Alternative treatment strategies such as formulation of anti-biofilm and anti-virulence regimen, which effectively targets only the virulence of the pathogen rather than its growth, could be used in clinical settings for evading the emergence of drug resistance because the use of antibiotics impose Darwinian pressure on the pathogen (Sivasankar et al., 2016). Also, an experimental study carried out in mouse model by Wang et al. (2016) have posited that a nanoparticle-based anti-virulence vaccine apprehended with staphylococcal α-hemolysin (Hla) could be used in clinical settings for managing MRSA infections, which curtails pathogenesis and invasiveness of MRSA through active elicitation of anti-Hla antibodies. Additionally, an in vitro study by Ramsey et al. (2016) involving commensal *Corynebacterium striatum* as a modulator of pathogenic and invasive lifestyle of *S. aureus* to commensal without affecting the growth has unveiled the possibility of new treatment options. Thus, nonbactericidal anti-virulent strategies are the requirement of current age for combating *S. aureus* infections for eluding antibiotic resistance.

10.5.2 *Streptococcus* spp.

The important bacterial species witnessed frequently after *S. aureus* in both complicated and uncomplicated SSSIs is the gram-positive *Streptococcus* spp., which causes numerous skin infections ranging from mild infections such as impetigo, folliculitis, abscess, and scarlet fever to deep-seated cellulitis and life-threatening invasive necrotizing fasciitis (Abrahamian et al., 2008).

Streptococcus, a member of Firmicutes is a chain-forming, catalase negative, coagulase-negative coccus, and a non-spore-forming facultative anaerobe. Several streptococcal strains are indigenously colonized on human skin and throat and become opportunistic pathogen at suitable predisposal conditions, whereas certain strains remain pathogenic (Ralph and Carapetis, 2013). A backdated grouping of *Streptococcus* and *Enterococcus* under the same genus *Streptococcus* was distinguished as discrete genera after 1984. The grouping of *Streptococcus* was found to be complicated during past. Initially, based on the pattern of hemolytic activity observed on blood agar plates, *Streptococcus* was classified into three groups namely, α-hemolytic, β-hemolytic, and non-hemolytic *Streptococcus*. In 1933, Lancefield performed serological typing of β-hemolytic *Streptococcus* isolated from humans, other animals, cheese, and milk based on anti-C precipitin test. The anti-C precipitin test majorly relied on the carbohydrate content (such as polysaccharide and teichoic acid) of antigens found on bacterial cell wall by classifying *Streptococcus* into groups A, B, C, D, and E, wherein group D and E included *Enterococcus* (Lancefield, 1933; Hardie et al., 1997).

Among three different hemolytic streptococci, β-hemolytic streptococci are extensively associated with skin infections, whereas GAS (*S. pyogenes*) is the predominant etiologic agent followed by group B streptococci (GBS, *Streptococcus agalactiae*) (Bisno and Stevens, 1996; Schuchat, 1998). A retroactive population-based study including adults with purulent SSSIs has unveiled the frequency of streptococci isolates as 7% in abscesses, 9% in infectious wounds, and 13% in purulent cellulitis (Abrahamian et al., 2008). In addition, the advent of highly invasive GAS remains as the predominant etiologic agent of series of severe and life-threatening SSSIs (Bisno and Stevens, 1996).

Similar to *S. aureus*, GAS also presents an arsenal of virulent determinants for eluding host immune attack and antibiotic treatments. These virulence determinants include biofilm formation, surface-associated lipoteichoic acids (LTA), M protein and pili, hyaluronic acid capsule and pyrogenic exotoxins such as leukocidin, streptolysins (O & S), streptokinase, superantigens, hyaluronidase, hemolysins, and proteases. Biofilm formation of GAS is a devastating and multifaceted process, which involves M-proteins, pili, and LTA for initial attachments and includes numerous other virulent factors to build a strong impediment against host immune attack and antibiotic treatments and further capable of complicating the treatment by causing severe invasive diseases using surface-associated invasins (Subramenium et al., 2015a; Ibrahim et al., 2016).

The surface-associated pilus proteins and M-proteins of GAS strains involved in skin and throat infections are found to be different, which are discriminated in clinical settings using T-antigen and M-protein serotyping (Cogen et al., 2008). The T-antigen genes sheltered in FCT region (i.e., loci of Fibronectin, collagen-binding protein, and T-antigen) encode pilus proteins and adhesins, which establish mechanical stabilization of covalent bonds and supports endurance of shearing forces formed during initial attachment of GAS to integrin found on host extracellular matrix (ECM) (Bessen, 2016). Also, the interaction of host integrin α5β1 with bacterial fibronectin-binding protein is reported to promote invasion by eliciting the cellular signaling pathways, which in turn, leads to conformational changes of host cytoskeletal actin (Walker et al., 2014).

Furthermore, the surface associated M-proteins comprising M1, M3, and M6 proteins are helpful in bacterial colonization, aggregation, and invasion of host ECM, wherein the surface-exposed hypervariable N-terminus of M-proteins facilitate attachment. M-proteins also bind and interfere with immunoglobulins and complementary regulatory components, thereby eluding host immune attack. The secreted proteases SpeB curtail the employment of phagocytes at the infection site by cleaving chemokines such as complement component c5a and interleukin 8 (IL8). Similar to *S. aureus*, GAS recruits Dlt proteins to neutralize the negative charge of cell wall LTA, thereby hindering the attachment of CAMP. Hyaluronic acid capsulation of GAS helps in opsonophagocytosis and also to evade the attack of antimicrobial peptides produced by commensals. The cytolysins such as streptolysin O and streptolysin S help in the survival of host immune attack by facilitating the rupture of neutrophils and macrophage (Walker et al., 2014).

Established with multitude of virulent determinants, this host-adapted GAS exhibits an archetypal defense system to escape host immune attack and antibiotic treatment. The antibiotic resistance profile of GAS has been reported to include macrolides, tetracycline, and fluoroquinolone, which pose a great threat to the management of streptococcal infections (Ibrahim et al., 2016). The empiric antibiotic treatment administered in several cases of SSSIs executes a greater risk of emergence of antibiotic resistance.

Despite of less resistance reported for penicillin, it is used currently as the first-line antibiotic for streptococcal-associated SSSIs. For individuals allergic to penicillin, macrolides are administered (Ibrahim et al., 2016). Phenoxymethyl penicillin (V) is used for managing scarlet fever (Basetti et al., 2017). Daptomycin is reported to be safe and efficient in managing *Streptococcus*-associated cSSSIs (Arbeit et al., 2004).

Numerous studies report anti-biofilm and anti-virulent agents from natural sources such as 2, 4-Di-tertbutylphenol from *Bacillus subtilis* (Viszwapriya et al., 2016), limonene found in citrus fruits (Subramenium et al., 2015), cinnamaldehyde and its derivatives from cinnamon (Shafreen et al., 2014), usnic acid from lichen (Nithyanand et al., 2015), and morin from orange and guava (Green et al., 2012) against GAS, which reduces the virulence of the pathogen and helps in natural clearing of bacterial load by host immune system and aids effective penetration of antibiotics. These anti-virulence–based treatment strategies could be established in clinical settings as alternate for evading the risk of advent of antibiotic resistance.

10.5.3 *P. aeruginosa*

P. aeruginosa, a ubiquitous gram-negative, rod-shaped and highly motile aerobic bacterium is the part of normal skin microflora that innocuously colonizes the human skin, mouth, and some nonsterile regions of healthy individuals. But at suitable predisposing conditions, it could effectively infect any region of the body it comes into contact, thus behaving as an opportunistic pathogen (Cogen et al., 2008). It is encountered predominantly in hot-tub folliculitis and acute and chronic wound and burn wound infections, which are reported to be associated with higher risk of morbidity and mortality (Percival et al., 2012; Serra et al., 2015).

P. aeruginosa greatly colonize the moist skin surfaces (such as burn wounds) rather than dry skin and exhibits pathogenic mechanisms with great predilection toward individuals who are immunocompromised and hospitalized (Morrison and Richard, 1984). Moreover, it has been stated that 2.5% of patients with acute thermal injuries or burns are highly prone to develop *Pseudomonas septicemia*, wherein the rate of mortality has been observed to be 76% (Stieritz and Holder, 1975).

Also, *P. aeruginosa* infects a broad area of wound site and complicates the wound-healing process owing to its high rate of acquired antibiotic resistance. The wound-healing process is reported to be still more complex, when *P. aeruginosa* is found to be coinfected on wounds with *S. aureus*. This bacterial interaction modulates the virulence rate, curtails the wound-healing process, and exhibits altered feedback for antibiotic treatments, which mystifies the identification of suitable treatment strategy (Serra et al., 2015). These recalcitrant *P. aeruginosa*–infected chronic and acute wounds and burn wounds stance a great economic as well as medical burden.

Furthermore, *P. aeruginosa*–colonized burn wounds rapidly worsen and leads to complete dissemination followed by death within few weeks, whereas, *P. aeruginosa*–colonized chronic wounds lasts for a long time and shows less probability for mortality. Altogether, all the aforementioned notions necessitate the unpinning of associated virulent determinants and molecular mechanisms underlying the differential response of *P. aeruginosa* infected burn and chronic wounds, which in turn, would facilitate the management of wound infections (Turner et al., 2014).

The secreted (i.e., alkaline protease, elastases LasA and LasB, phospholipases, exoenzyme S and exotoxin A) and cell-associated (i.e., alginate, flagellum, pili and adhesins) virulence determinants afford multiple benefits to *P. aeruginosa* such as evasion from host immune attack, host invasion, and endurance of antibiotic treatments. Additionally, quorum sensing systems *las* and *rhl* of *P. aeruginosa* are reported to be involved in chronic wound infections, which are putatively known to control several virulence factors (Rumbaugh et al., 1999). Quorum sensing (QS) is the cell-to-cell communication process that occurs at the onset of increased signaling molecules called autoinducers (AI) attained during high cell density (HCD) and monitors the level of signaling molecules and modulates gene expression suitably to endure adverse situations. The *las* QS system involves AI synthase LasI to secrete 3-oxo-C12-homoserine lactone (3OC12HSL), which in turn, triggers the transcription activator LasR at high AI level to regulate the expression of target genes encoding exotoxin A, proteases, and elastases. The *rhl* QS system works in series with *las* QS system, wherein the LasR-3OC12HSL complex formed at high AI level targets the AI synthase RhlI and triggers it to produce butanoyl homoserine lactone (C4HSL). This in turn, induces the activation of transcription regulator RhlR at high AI level and regulates the expression of target genes encoding proteases, rhamnolipid, biofilm formation, swarming, elastases, siderophores and pyocyanin (Rutherford and Bassler, 2012).

The type IV pili and flagella are reported to be involved in burn wound infections by mediating the twitching motility of *P. aeruginosa* through liquid interface of wound, thereby helping the establishment of appropriate adhesion with skin surface. The flagellar components also aid dissemination of *P. aeruginosa* from infected wound site. Additionally, the flagellar glycosylation supports the colonization of *P. aeruginosa* and increases the virulence, which leads to death (Arora et al., 2005). Cell-associated alginate mediates attachment and biofilm viscoelasticity and confers resistance toward host immune defenses and antibiotics during chronic infections. Rhamnolipids are supportive in spreading of infection by mediating the dispersal of *P. aeruginosa* from biofilm and helps in reestablishment of biofilm in a new niche (Kostakioti et al., 2013). Elastase, a type of protease, plays and essential role in burn infections by exerting tissue-damaging activity, including degradation of plasma proteins such as complement factors, immunoglobulins, and so on (Wretlind and Pavlovskis, 1983).

Altogether with robust defense strategies, *P. aeruginosa* remains as an intractable multidrug resistant (MDR) pathogen and poses a great menace to clinical settings for managing the infections. The antibiotic resistant profile of *P. aeruginosa* includes a multitude of antibiotics such as gentamicin, ceftizoxime, cephalothin, carbenicillin, ceftazidime, ciprofloxacin, tetracycline, amikacin, and so on. The empiric antibiotic treatment is found to be futile. Early diagnosis, appropriate identification of strain type of wound colonizers, and their antibiotic sensitivity pattern is reported to be appropriate for managing the burn infections (Pruitt et al., 1983).

Alternative approaches, including development of quorum quenchers, anti-biofilm, and anti-virulence agents, could act as efficient therapeutic regimen. For instance, curcumin, a natural compound found in turmeric plants, is reported to act as a potent quorum quencher of *P. aeruginosa* (Sethupathy et al., 2016). Recently, ciprofloxacin-loaded keratin hydrogels developed by Roy et al. (2015) were evidenced to progress the healing process of *P. aeruginosa*–associated wound infections. Thus, these alternate strategies could be adopted in clinical settings after appropriate clinical trials, which are posited to help in eluding the advent of antibiotic resistance.

10.5.4 Predominant skin commensals as opportunistic skin pathogens

10.5.4.1 *S. epidermidis* – *S. epidermidis* is a ubiquitous gram-positive skin and mucosal membrane colonizer, which exerts a mutualistic relationship with host. It forms the major part of microbial barrier that precludes the colonization of other pathogens. In a competitive environment, it secretes lantibiotics (i.e., lanthionine-containing antibacterial peptides) often referred as *bacteriocins*, which prevent the colonization of *S. aureus* and GAS (Sahl, 1994; Cogen et al., 2007). Also, accessory gene regulator (*agr*) locus found in commensal *S. epidermidis* produces peptide pheromones that activate the *agr* QS system of competing bacteria, which in turn, reduces colonization and down-regulates the expression of virulence factors by increasing the production of pheromones such as phenol soluble modulin (Otto, 2001). In addition, *S. epidermidis* boosts the host immune defense by eliciting the signaling of toll like receptor (TLR). The pattern recognition receptors TLRs, in turn, specifically recognize different pathogen-associated molecular patterns and activate the host immune system accordingly.

Despite the aforesaid beneficiary roles of *S. epidermidis*, it has been identified as opportunistic pathogen during past two decades predominantly affecting drug abusers and individuals who are immunocompromised. *S. epidermidis* is reported to be encountered less frequently in abscesses, cellulitis, and several wound infections (Cogen et al., 2008). It is also reported to elicit the severity of miliaria (prickly heat), a skin ailment frequently witnessed in profusely sweating individuals and neonates with undeveloped sweat glands. Miliaria occurs as a result of occlusion of sweat glands by periodic acid sciff (PAS) positive extracellular polysaccharide substance produced by *S. epidermidis* (Mowad et al., 1995).

These detrimental infectious roles of *S. epidermidis* are well allied with its ability to form biofilm and produce virulence factors such as autolysin protein, proteases, lipases, polysaccharide intracellular adhesion (PIA), and surface-associated fibronectin and collagen-binding protein. The surface-associated fibronectin and collagen-binding proteins mediate the attachment of *S. epidermidis* (Williams et al., 2002). Autolysin proteins facilitate autolysis process of competing bacteria and uses the extracellular DNA (eDNA) attained through autolysis for establishment of robust biofilm. Proteases and lipases are involved in the tissue-damage process, which increases the severity of infection. Once the biofilm is established, PIA helps intracellular adhesion and increases the virulence (McKenney et al., 2000).

These virulence factors equip a robust and recalcitrant biofilm and incapacitate the antibiotic treatments at clinical settings. Thus, exploration of novel treatment strategies to combat biofilm-assisted infections are still in progress. Antibodies designed against *S. epidermidis* surface-associated proteins and interferon therapy is reported to possess positive feedbacks against recalcitrant biofilms (Boelens et al., 2000). Alpha-mangostin, an anti-biofilm agent found in mangosteen, has been reported to eradicate *S. epidermidis* biofilms (Sivaranjani et al., 2017). These treatment options could be included in clinical trials for finding the suitability of these strategies in clinical settings.

10.5.4.2 *P. acnes* – *P. acnes* is a gram-positive and anaerobic bacillus, which is an indigenous colonizer of sebaceous glands. It also forms the part of microbial and plays mutualistic role similar to *S. epidermidis*. It catabolizes the sebum lipids of sebaceous glands and release fatty acids, which precludes the colonization of sebaceous gland by other pathogens. *P. acnes* are also known to produce antimicrobial peptides called *bacteriocins*, which include jenseniin, acnecin, and propionicin. These bacteriocins prevent the colonization of other *Propionibacterium* spp., several gram-negative bacteria, yeast, molds, and so on.

P. acnes also exhibits infectious role by causing inflammatory acne vulgaris majorly and synovitis, acne, pustulosis, hyperostosis, and osteitis (SAPHO) syndrome rarely. The free fatty acids produced by *P. acnes* as the result of triglyceride metabolism trigger the inflammatory response. Acne vulgaris is ailment of pilosebaceous follicles caused as result of hormonal imbalance, immune hypersensitivity, *P. acnes* infection, and follicular keratinization. It is frequently witnessed in 80% youngsters of the U.S. population. Several putative predisposal factors such as genetic factors, stress, androgens, follicular pattern of the individual, and use of steroids are known to influence the onset of acne vulgaris.

Excessive sebum production attained as the result of high level androgen mediates the colonization and biofilm formation of *P. acnes*

(Coenye et al., 2008). In addition to biofilm formation, *P. acnes* secretes several virulence determinants such as proteases, hyaluronidase, lipases, and chemotactic factors, which play substantial role in acne vulgaris. Furthermore, it has been reported that the aforementioned virulence factors of *P. acnes* triggers the nonspecific immune response by activating the proinflammatory cytokines and supports differentiation and proliferation of keratinocytes. The matrix metalloproteases (MMP) produced by keratinocytes deteriorates the hair follicles and induces inflammatory acne lesions (Dessinioti et al., 2010; Saising and Voravuthikunchai, 2012).

A number of treatment options are available to manage acne vulgaris, which include topical retinoids, oral antibiotics, oral isotretinoin, and administration of benzyl peroxide (Cogen et al., 2008). However, alternative strategies such as anti-biofilm therapy would be useful in managing recalcitrant biofilm-assisted *P. acne* infections. Sivasankar et al. (2016) investigated the in vitro and in vivo anti-biofilm potential of ellagic acid and tetracycline against *P. acnes* biofilm and other associated virulence factors, which remain a good revelation for managing the infections with combined approaches.

10.5.4.3 *Corynebacterium* spp. – *Corynebacterium* spp. are diphtheroid, gram-positive, and facultative anaerobic mycobacterium, which constitute approximately 50% of skin microflora (Blaise et al., 2008). The skin microflora majorly constitute two species of *Corynebacterium* namely, *C. diphtheriae* and nondiphtheriae corynebacteria (diphtheroids). *C. diphtheriae* are reported to be witnessed in cutaneous ulcers of drug abusers, alcoholics, and individuals exposed to poor hygienic environment. The nondiphtheriae cornynebacteria or diphtheroids include totally 17 species, which are diversely distributed in humans and other animals (Cogen et al., 2008). The diphtheroid majorly found in human epithelium is *C. jeikeium*. It acts as microbial barrier and protects the host from pathogenic attack by the production of bacteroids. The enzyme superoxide dismutase produced by *C. jeikeium* for shielding against the attack of free radicals, confers protection to host against free radicals (Storz and Imlay, 1999).

However, *C. jeikeium* is capable of developing papular eruptions in patients who are immunocompromised and individuals with skin abrasions or traumas. *C. jeikeium* harbors numerous virulence traits such as siderophores, invasins, adhesins, biofilm formation, and superoxide dismutase (Joh et al., 1999; Ton-That and Schneewind, 2004; Blaise et al., 2008). The siderophores mediates iron and manganese sequestration and helps to evade host attack by surviving superoxide radicals. Adhesins and invasins promote the virulence by aiding adhesion to host epithelium and host invasion. Also, the cell envelope of *C. jeikeium* contains corynomycolic acid, which helps it to resist multiple antibiotic treatments (Cogen et al., 2008).

The biofilm formation together with other virulence traits affords resistance toward many antibiotics. However, *C. jeikeium* remains susceptible to glycopeptides group of antibiotics such as vancomycin. Currently, erythromycin is used as first-line treatment for skin infections instigated by *Corynebacterium* spp. and is administered continuously for 3 to 4 weeks. Fusidic acid is also used in the treatment. The emerging bacterial resistance debilitates the conventional antibiotic treatments. Yet, reduction of hyperkeratosis using kerolytic agent and lessening of sweat production using topical aluminum hydroxide could upsurge the efficacy of prevailing strategies (Blaise et al., 2008).

10.6 Conclusion and future prospects

Bacterial role as commensal and pathogen still remains cryptic. Though, skin microflora play numerous protective role and constitute microbial barrier for shielding the host from pathogenic attack, their role in certain infectious diseases still remains ambiguous. Thus, unpinning the role of skin commensals in SSSIs, alterations in physiology of commensals during infections, interactions between commensals and pathogens, factors influencing the severity of infections, molecular mechanisms underlying infectious role of pathogens, as well as commensals and role of commensals in antibiotic resistance of pathogens remains quintessential for safeguarding the mankind from the advent of new MDR strains and infectious pathogens.

Recently, many researches have been dedicated in reconnecting the traditional medical practices to the current scenario through the use of advanced technologies for identifying the specific bioactive molecules present in the natural sources to outstrip the deceiving resistance mechanisms of bacterial pathogens. However, after documenting the in vitro anti-infective potential of active lead(s) against infectious pathogen(s), many researches halt midway and do not prolong the research toward the in vivo analysis of their efficacy. Albeit many researchers take their innovation to next level by deciphering the molecular mechanisms underlying the anti-infective potential of active lead and assessing their activity in various in vivo models, the analysis of aptness of active lead(s) for clinical applications has to be still carried out. The proper formulation of active lead(s) with less cytotoxicity, high stability, and bioavailability, followed by appropriate clinical trials for employing the identified natural-based therapeutic molecules in clinical use for eluding the global burden of antibiotic resistance and betterment of human life will serve as the future endeavors.

Acknowledgments

The authors thankfully acknowledge the support extended by Department of Science and Technology, Government of India through PURSE [Grant No. SR/S9Z-23/2010/42 (G)] & FIST Level I, Phase II (Grant No. SR/FST/LSI-639/2015(C)), and University Grants Commission (UGC), New Delhi, through SAPDRS1 [Grant No. F.3-28/2011 (SAP-II)]. The authors also gratefully thank the Bioinformatics Infrastructure Facility (BIF) funded by Department of Biotechnology, Government of India [Grant no. BT/BI/25/015/2012].

References

Aalbers, J., K. K. O'Brien, W. S. Chan, G. A. Falk, C. Teljeur, B. D. Dimitrov, and T. Fahey. 2011. Predicting streptococcal pharyngitis in adults in primary care: A systematic review of the diagnostic accuracy of symptoms and signs and validation of the centor score. *BMC Medicine* 9:67.

Abrahamian, F. M., D. A. Talan, and G. J. Moran. 2008. Management of skin and soft-tissue infections in the emergency department. *Infectious Disease Clinics* 22:89–116.

Alegre, M. L., L. Chen, M. Z. David, C. Bartman, S. Boyle-Vavra, N. Kumar, A. S. Chong, and R. S. Daum. 2016. Impact of *Staphylococcus aureus* USA300 colonization and skin infections on systemic immune responses in humans. *Journal of Immunology* 197:1118–1126.

Amagai, M., N. Matsuyoshi, Z. H. Wang, C. Andl and J. R. Stanley. 2000. Toxin in bullous impetigo and staphylococcal scalded-skin syndrome targets desmoglein 1. *Nature Medicine* 6:1275.

Arakaki, R. Y., L. Strazzula, E. Woo and D. Kroshinsky. 2014. The impact of dermatology consultation on diagnostic accuracy and antibiotic use among patients with suspected cellulitis seen at outpatient internal medicine offices: A randomized clinical trial. *JAMA Dermatology* 150:1056–1061.

Arbeit, R. D., D. Maki, F. P. Tally, E. Campanaro, B. I. 2004. Eisenstein and Daptomycin 98-01 and 99-01 Investigators. The safety and efficacy of daptomycin for the treatment of complicated skin and skin-structure infections. *Clinical Infectious Diseases* 38:1673–1681.

Arora, S. K., A. N. Neely, B. Blair, S. Lory and R. Ramphal. 2005. Role of motility and flagellin glycosylation in the pathogenesis of *Pseudomonas aeruginosa* burn wound infections. *Infection and Immunity* 73:4395–4398.

Bangert, S., M. Levy and A. A. 2012. Hebert bacterial resistance and impetigo treatment trends: A review. *Pediatric Dermatology* 29:243–248.

Basetti, S., J. Hodgson, T. M. 2017. Rawson and A. Majeed. Scarlet fever: A guide for general practitioners. *London Journal of Primary Care* 9:77–79.

Bessa, L. J., P. Fazii, Di M. Giulio and L. Cellini. 2015. Bacterial isolates from infected wounds and their antibiotic susceptibility pattern: Some remarks about wound infection. *International Wound Journal* 12:47–52.

Bessen, D. E. 2016. Tissue tropisms in group A Streptococcus: What virulence factors distinguish pharyngitis from impetigo strains? *Current Opinion in Infectious Diseases* 29:295.

Bisno, A. L. and D. L. Stevens. 1996. Streptococcal infections of skin and soft tissues. *The New England Journal of Medicine* 334:240–246.

Blaise, G., A. F. Nikkels, T. Hermanns-Lê, N. Nikkels-Tassoudji, and G. E. Piérard. 2008. Corynebacterium-associated skin infections. *International Journal of Dermatology* 47:884–890.

Boateng, J. and O. Catanzano. 2015. Advanced therapeutic dressings for effective wound healing—A review. *Journal of Pharmaceutical Sciences* 104:3653–3680.

Boelens, J. J., T. Van der Poll, J. Dankert and S. A. Zaat. 2000. Interferon-γ protects against biomaterial-associated *Staphylococcus epidermidis* infection in mice. *The Journal of Infectious Diseases* 181:1167–1171.

Bowen, A. C., A. Mahé, R. J. Hay, R. M. Andrews, A. C. Steer, S. Y. Tong and J. R. Carapetis. 2015. The global epidemiology of impetigo: A systematic review of the population prevalence of impetigo and pyoderma. *PLoS One* 10:e0136789.

Chambers, H. F. and F. R. DeLeo. 2009. Waves of resistance: *Staphylococcus aureus* in the antibiotic era. *Nature Reviews Microbiology* 7:629.

Chiller, K., B. A. Selkin and G. J. Murakawa. 2001. Skin microflora and bacterial infections of the skin. *Journal of Investigative Dermatology Symposium Proceedings* 6:170–174.

Chira, S. and L. G. Miller. 2010. *Staphylococcus aureus* is the most common identified cause of cellulitis: A systematic review. *Epidemiology Infection* 138, 313–317.

Coenye, T., K. Honraet, B. Rossel and H. J. Nelis. 2008. Biofilms in skin infections: *Propionibacterium acnes* and acne vulgaris. *Infectious Disorders Drug Targets* 8:156–159.

Cogen, A. L., V. Nizet and R. L. Gallo. 2008. Skin microbiota: A source of disease or defence? *British Journal of Dermatology* 158:442–455.

Cogen, A. L., V. Nizet and R. L. Gallo. 2007. *Staphylococcus epidermidis* functions as a component of the skin innate immune system by inhibiting the pathogen Group A Streptococcus. *Journal of Investigative Dermatology* 127:S131.

Cole, C. and J. Gazewood. 2007. Diagnosis and treatment of impetigo. *American Family Physician* 75:859–864.

Cooper, R. A. 2013. Surgical site infections: Epidemiology and microbiological aspects in trauma and orthopaedicsurgery. *International Wound Journal* 10:3–8.

Costello, E. K., C. L. Lauber, M. Hamady, N. Fierer, J. I. Gordon and R. Knight. 2009. Bacterial community variation in human body habitats across space and time. *Science* 326:1694–1697.

Das, D. K., M. G. Baker and K. Venugopal. 2011. Increasing incidence of necrotizing fasciitis in New Zealand: A nationwide study over the period 1990 to 2006. *Journal Infection* 63:429–433.

Davies, M. R., M. T. Holden, P. Coupland, J. H. Chen, C. Venturini, T. C. Barnett, N. L. B. Zakour, H. Tse, G. Dougan, K. Y. Yuen and M. J. Walker. 2015. Emergence of scarlet fever Streptococcus pyogenes emm12 clones in Hong Kong is associated with toxin acquisition and multidrug resistance. *Nature Genetics* 47:84.

Dessinioti, C. and A. D. Katsambas. 2010. The role of *Propionibacterium acnes* in acne pathogenesis: Facts and controversies. *Clinics Dermatology* 28:2–7.

Dufour, P., Y. Gillet, M. Bes, G. Lina, F. Vandenesch, D. Floret, J. Etienne and H. Richet. 2002. Community-acquired methicillin-resistant *Staphylococcus aureus* infections in France: Emergence of a single clone that produces Panton-Valentine leukocidin. *Clinical Infectious Diseases* 35:819–824.

Edwards, R. and K. G. Harding. 2004. Bacteria and wound healing. *Current Opinion in Infectious Diseases* 17:91–96.

Eisenstein, B. I. 2008. Treatment challenges in the management of complicated skin and soft-tissue infections. *Clinical Microbiology Infection* 14:17–25.

Elliott, D., J. A. Kufera and R. A. Myers. 2000. The microbiology of necrotizing soft tissue infections. *American Journal of Surgery* 179:361–366.

Eneli, I. and HD. Davies. Epidemiology and outcome of necrotizing fasciitis in children: An active surveillance study of the Canadian Paediatric Surveillance Program. *Journal Pediatrics* 151, 79–84 (2007).

Esposito, S., S. Noviello and S. Leone. 2016. Epidemiology and microbiology of skin and soft tissue infections. *Current Opinion in Infectious Diseases* 29:109–115.

Findley, K., J. Oh, J. Yang, S. Conlan, C. Deming, J. A. Meyer, D. Schoenfeld, E. Nomicos, M. Park, Sequencing N. I. S. C. C. and J. Becker. 2013. Topographic diversity of fungal and bacterial communities in human skin. *Nature* 498:367–370.

Frazee, B. W., J. Lynn, E. D. Charlebois, L. Lambert, D. Lowery and F. Perdreau-Remington. 2005. High prevalence of methicillin-resistant *Staphylococcus aureus* in emergency department skin and soft tissue infections. *Annals Emergency Medicine* 45:311–320.

Giacometti, A., O. Cirioni, A. M. Schimizzi, Del M. S. Prete, F. Barchiesi, M. M. D'errico, Petrelli and G. Scalise. 2000. Epidemiology and microbiology of surgical wound. *Journal Clinical Microbiology Infection* 38:918–922.

Giuliano, A., F. Lewis Jr, K. Hadley and F. W. Blaisdell. 1997. Bacteriology of necrotizing fasciitis. *American Journal of Surgery* 134:52–57.

Green, A. E., R. S. Rowlands, R. A. Cooper and S. E. Maddocks. 2012. The effect of the flavonol morin on adhesion and aggregation of *Streptococcus pyogenes*. *FEMS Microbiology Letter* 333:54–58.

Grice, E. A. and J. A. Segre. 2011. The skin microbiome. *Nature Reviews Microbiology* 9:244–253.

Grice, E. A., H. H. Kong, S. Conlan, C. B. Deming, J. Davis, A. C. Young, G. G. Bouffard, R. W. Blakesley, P. R. Murray, E. D. Green and M. L. Turner. 2009. Topographical and temporal diversity of the human skin microbiome. *Science* 324, 1190–1192.

Hanakawa, Y., N. M. Schechter, C. Lin, L. Garza, H. Li, T. Yamaguchi, Y. Fudaba et al., 2002. Molecular mechanisms of blister formation in bullous impetigo and staphylococcal scalded skin syndrome. *The Journal Clinical Investigation* 110:53–60.

Hardie, J. M. and R. A. Whiley. 1997. Classification and overview of the genera Streptococcus and Enterococcus. *Journal Applied Microbiology* 83:1–11.

Hedrick, J. 2003. Acute bacterial skin infections in pediatric medicine. *Paediatric Drugs* 5:35–46.

Hook, E. W., T. M. Hooton, C. A. Horton, M. B. Coyle, P. G. Ramsey and M. Turck. 1986. Microbiologic evaluation of cutaneous cellulitis in adults. *Archives of Internal Medicine* 146, 295–7.

Ibrahim, F., T. Khan and G. G. Pujalte. 2015. Bacterial skin infections. *Prim Care: Clinics in Office Practice* 42, 485–99.

Ibrahim, J., J. A. Eisen, G. Jospin, D. A. Coil, G. Khazen and S. Tokajian. 2016. Genome analysis of *Streptococcus pyogenes* associated with pharyngitis and skin infections. *PLoS One* 11:e0168177.

Joh, D., E. R. Wann, B. Kreikemeyer, P. Speziale and M. Höök. 1999. Role of fibronectin-binding MSCRAMMs in bacterial adherence and entry into mammalian cells. *Matrix Biology* 18, 211–223.

Kanitakis, J. 2002. Anatomy, histology and immunohistochemistry of normal human skin. *European Journal Dermatology* 12, 390–399.

Ki, V. and C. Rotstein. 2008. Bacterial skin and soft tissue infections in adults: A review of their epidemiology, pathogenesis, diagnosis, treatment and site of care. *Can Journal Infectious Diseases and Medical Microbiology* 19, 173–184.

King, M. D., B. J. Humphrey, Y. F. Wang, E. V. Kourbatova, S. M. Ray and H. M. Blumberg. 2006. Emergence of community-acquired methicillin-resistant *Staphylococcus aureus* USA 300 clone as the predominant cause of skin and soft-tissue infections. *Annals of Internal Medicine* 144:309–317.

Kostakioti, M., M. Hadjifrangiskou and S. J. Hultgren. 2013. Bacterial biofilms: Development, dispersal, and therapeutic strategies in the dawn of the postantibiotic era. *Cold Spring Harbor Perspectives in Medicine* 3:p.a010306.

Lamanna, O., D. Bongiorno, L. Bertoncello, S. Grandesso, S. Mazzucato, G. B. Pozzan, M. Cutrone, M. Chirico, F. Baesso, P. Brugnaro and V. R. Cafisoapid. 2017. containment of nosocomial transmission of a rare community-acquired methicillin-resistant *Staphylococcus aureus* (CA-MRSA) clone, responsible for the Staphylococcal Scalded Skin Syndrome (SSSS). *Italian Journal Pediatrics* 43:5.

Lancefield, R. C. 1933. A serological differentiation of human and other groups of hemolytic streptococci. *Journal Experimental Medicine* 57:571–595.

Lee, M. C., A. M. Rios, M. F. Aten, A. Mejias, D. Cavuoti, G. H. Mccracken JR, and R. D. Hardy. 2004. Management and outcome of children with skin and soft tissue abscesses caused by community-acquired methicillin-resistant *Staphylococcus aureus*. *Pediatric Infectious Disease Journal* 23:123–127.

Levy, S. B. and B. Marshall. 2004. Antibacterial resistance worldwide: Causes, challenges and responses. *Nature Medicine* 10:S122.

Li, M. Y., Y. Hua, G. H. Wei and L. Qiu. 2014. Staphylococcal Scalded Skin Syndrome in Neonates: An 8-Year Retrospective Study in a Single Institution. *Pediatric Dermatology* 31:43–47.

Luelmos-Aguilar, J. and M. S. Santandreu. 2004. Folliculitis recognition and management. *American Journal Clinical Dermatol* 5:301–310.

Malachowa, N., S. D. Kobayashi, D. E. Sturdevant, D. P. Scott and F. R. De Leo. 2015. Insights into the *Staphylococcus aureus*-host interface: Global changes in host and pathogen gene expression in a rabbit skin infection model. *PLoS One* 10:e0117713.

Malone, M., T. Bjarnsholt, A. J. McBain, G. A. James, P. Stoodley, D. Leaper, M. Tachi, G. Schultz, T. Swanson and R. D. Wolcott. 2017. The prevalence of biofilms in chronic wounds: A systematic review and meta-analysis of published data. *Journal Wound Care* 26:20–25.

Manfredi, R., L. Calza and F. Chiodo. 2002. Epidemiology and microbiology of cellulitis and bacterial soft tissue infection during HIV disease: A 10-year survey. *Journal of Cutaneous Pathology* 29:168–172.

Mckenney, D., K. Pouliot, Y. Wang, V. Murthy, M. Ulrich, G. Döring, J. C. Lee, D. A. Goldmann and G. B. Pier. 2000. Vaccine potential of poly-1-6 β-DN-succinylglucosamine, an immunoprotective surface polysaccharide of *Staphylococcus aureus* and *Staphylococcus epidermidis*. *Journal Biotechnology* 83:37–44.

Menon, G. K. and A. M. Kligman. 2009. Barrier functions of human skin: A holistic view. *Skin Pharmacology and Physiology* 22:178–189.

Moran, G. J., F. M. Abrahamian, F. LoVecchio and D. A. Talan. 2013. Acute bacterial skin infections: Developments since the 2005 Infectious Diseases Society of America (IDSA) guidelines. *Journal Emergency Medicine* 44:e397–412.

Morrison Jr, A. J. and R. P. Wenzel. 1984. Epidemiology of infections due to *Pseudomonas aeruginosa*. *Reviews of Infection Diseases* 6:S627–S642.

Mowad, C. M., K. J. McGinley, A. Foglia and J. J. Leyden. 1995. The role of extracellular polysaccharide substance produced by *Staphylococcus epidermidis* in miliaria. *Journal American Academy of Dermatology* 33:729–733.

Murphy, E. L., D. DeVita, H. Liu, E. Vittinghoff, P. Leung, D. H. Ciccarone and B. R. Edlin. 2001. Risk factors for skin and soft-tissue abscesses among injection drug users: A case-control study. *Clinical Infectious Diseases* 33:35–40.

Nakatsuji, T., T. H. Chen, S. Narala, K. A. Chun, T. Yun, F. Shafiq, P. F. Kotol et al., 2017. Antimicrobials from human skin commensal bacteria protect against *Staphylococcus aureus* and are deficient in atopic dermatitis. *Science Translation Medicine* 9:eaah4680.

Nithyanand, P., R. M. B. Shafreen, S. Muthamil and S. K. Pandian. 2015. Usnic acid, a lichen secondary metabolite inhibits Group A Streptococcus biofilms. *Antonie van Leeuwenhoek* 107:263–272.

Oh, J., A. L. Byrd, C. Deming, S. Conlan, B. Barnabas, R. Blakesley, G. Bouffard et al., 2014. Biogeography and individuality shape function in the human skin metagenome. *Nature*, 514:59–64.

Oh, J., A. L. Byrd, M. Park, H. H. Kong, J. A. Segre and NISC. 2016. Comparative sequencing program. Temporal stability of the human skin microbiome. *Cell* 165:854–866.

Otberg, N., H. Kang, A. A. Alzolibani and J. Shapiro. 2008. Folliculitis decalvans. *Dermatologic Therapy* 21:238–244.

Otto, M. 2001. *Staphylococcus aureus* and *Staphylococcus epidermidis* peptide pheromones produced by the accessory gene regulator *agr* system. *Peptides* 22:1603–1608.

Palit, A., and A. C. Inamadar. 2009. Childhood cutaneous vasculitis: A comprehensive appraisal. *Indian Journal of Dermatology* 54:110–117.

Palit, A. and A. C. Inamadar. 2010. Current concepts in the management of bacterial skin infections in children. *Indian Journal Dermatology Venereology Leprology* 76:476.

Parker, M. T., A. J. H. Tomlinson and R. E. O. Williams. 1955. Impetigo contagiosa. The association of certain types of *Staphylococcus aureus* and of *Streptococcus pyogenes* with superficial skin infections. *Epidemiology and Infection* 53:458–473.

Patel, G. K. and A. Y. Finlay. 2003. Staphylococcal scalded skin syndrome. *American Journal Clinical Dermatology* 4:165–175.

Percival, S. L., C. Emanuel, K. F. Cutting and D. W. Williams. 2012. Microbiology of the skin and the role of biofilms in infection. *International Wound Journal* 9:14–32.

Proksch, E., J. M. Brandner and J. M. Jensen. 2008. The skin: An indispensable barrier. *Experimental Dermatology* 17:1063–1072.

Pruitt Jr, B. A., R. B. Lindberg, W. F. McManus, and A. D. Mason Jr, 1983. Current approach to prevention and treatment of Pseudomonas aeruginosa infections in burned patients. *Reviews of Infectious Diseases* 5:S889–S897.

Psoinos, C. M., J. M. Flahive, J. J. Shaw, Y. Li, S. C. Ng, J. F. Tseng and H. P. Santry. 2013. Contemporary trends in necrotizing soft-tissue infections in the United States. *Surgery* 153:819–827.

Pulido-Cejudo, A., M. Guzmán-Gutierrez, A. Jalife-Montaño, A. Ortiz-Covarrubias, J. L. Martínez-Ordaz, H. F. Noyola-Villalobos and L. M. Hurtado-López. 2017. Management of acute bacterial skin and skin structure infections with a focus on patients at high risk of treatment failure. *Therapeutic Advances in Infectious Disease* 4:143–161.

Pye, L. 2010. Bacterial skin infections. *InnovAiT* 3:388–395.

Qualls, M. L., M. M. Mooney, C. A. Camargo Jr, T. Zucconi, D. C. Hooper and D. J. Pallin. 2012. Emergency department visit rates for abscess versus other skin infections during the emergence of community-associated methicillin-resistant Staphylococcus aureus, 1997–2007. *Clinical Infectious Diseases* 55:103–105.

Ralph A. P., Carapetis J. R. (2013) Group a streptococcal diseases and their global burden. In. *Host-Pathogen Interactions in Streptococcal Diseases. Current Topics in Microbiology and Immunology.* Edited by.G. Chhatwal. Springer, Berlin, Germany.

Ramana, K. V., V. B. Pinnelli, B. Prakash, S. Kandi, C. H. V. Sharada, A. Kalaskar, S. D. Rao, R. Mani and R. Rao. 2013. Complicated skin and skin structure infections (cSSSI's): A comprehensive review. *American Journal Medical and Biological Research* 1:159–164.

Ramsey, M. M., M. O. Freire, R. A. Gabrilska, K. P. Rumbaugh and K. P. 2016. Lemon *Staphylococcus aureus* shifts toward commensalism in response to *Corynebacterium* species. *Frontiers in Microbiology* 7, 1230.

Rosenthal, M., D. Goldberg, A. Aiello, E. Larson and B. Foxman. 2011. Skin microbiota: Microbial community structure and its potential association with health and disease. *Infection, Genetics and Evolution* 11, 839–848.

Ross, A. and H. W. Shoff. 2017. Staphylococcal scalded skin syndrome. In *StatPearls Treasure Island* [Internet]. R. Miller, T. Sneden, E. Hughes, B. Beatty, and G. Rubio, editors. Louisville, KY: University of Louisville.

Roy, D. C., S. Tomblyn, D. M. Burmeister, N. L. Wrice, S. C. Becerra, L. R. Burnett, J. M. Saul and R. J. Christy. 2015. Ciprofloxacin-loaded keratin hydrogels prevent *Pseudomonas aeruginosa* infection and support healing in a porcine full-thickness excisional wound. *Advances in Wound Care* 4:457–468.

Rumbaugh, K. P., J. A. Griswold, B. H. Iglewski and A. N. Hamood. 1999. Contribution of quorum sensing to the virulence of *Pseudomonas aeruginosa* in burn wound infections. *Infection and Immunity* 67:5854–5862.

Rutherford, S. T. and B. L. Bassler. 2012. Bacterial quorum sensing: Its role in virulence and possibilities for its control. *Cold Spring Harbor Perspectives Medicine* 2:a012427.

Sahl, H. G. 1994. Staphylococcin 1580 is identical to the lantibiotic epidermin: Implications for the nature of bacteriocins from Gram-positive bacteria. *Applied and Environmental Microbiology* 60:752–755.

Saising, J. and S. P. Voravuthikunchai. 2012. Anti *Propionibacterium acnes* activity of rhodomyrtone, an effective compound from *Rhodomyrtus tomentosa* (Aiton) Hassk. leaves. *Anaerobe* 18:400–404.

Schmid, M. R., T. Kossmann and S. Duewell. 1998. Differentiation of necrotizing fasciitis and cellulitis using MR imaging. *American Journal Roentgenology* 170:615–620.

Schuchat, A. 1998. Epidemiology of group B streptococcal disease in the United States: Shifting paradigms. *Clinical Microbiology Reviews* 11:497–513.

Schwacha, M. G., L. T. Holland, I. H. Chaudry and J. L. Messina. 2005. Genetic variability in the immune-inflammatory response after major burn injury. *Shock* 23:123–128.

Serra, R., R. Grande, L. Butrico, A. Rossi, U. F. Settimio, B. Caroleo, B. Amato, L. Gallelli and S. de Franciscis. 2015. Chronic wound infections: The role of *Pseudomonas aeruginosa* and *Staphylococcus aureus*. *Expert Review of Anti-infective Therapy* 13:605–613.

Sethupathy, S., K. G. Prasath, S. Ananthi, S. Mahalingam, S. Y. Balan and S. K. Pandian. 2016. Proteomic analysis reveals modulation of iron homeostasis and oxidative stress response in *Pseudomonas aeruginosa* PAO1 by curcumin inhibiting quorum sensing regulated virulence factors and biofilm production. *Journal of Proteomics* 145:112–126.

Sethupathy, S., L. Vigneshwari, A. Valliammai, K. Balamurugan and S. K. Pandian. 2017. L-Ascorbyl 2, 6-dipalmitate inhibits biofilm formation and virulence in methicillin-resistant *Staphylococcus aureus* and prevents triacylglyceride accumulation in *Caenorhabditis elegans*. *RSC Advances* 7:23392–406.

Shafreen, B., R. Mohmed, C. Selvaraj, S. K. Singh and S. Karutha Pandian. 2014. *In silico* and *in vitro* studies of cinnamaldehyde and their derivatives against LuxS in *Streptococcus pyogenes*: Effects on biofilm and virulence genes. *Journal of Molecular Recognition* 27:106–116.

Shah, M. and H. D. Shah. 2011. Acute bacterial skin and skin structure infections: Current perspective. *Indian Journal of Dermatology* 56:510.

Shinji, H., Y. Yosizawa, A. Tajima, T. Iwase, S. Sugimoto, K. Seki and Mizunoe, Y. 2011. Role of fibronectin-binding proteins A and B in *in vitro* cellular infections and *in vivo* septic infections by *Staphylococcus aureus*. *Infection and Immunity* 79:2215–223.

Shore, A. C., A. S. Rossney, O. M. Brennan, P. M. Kinnevey, H. Humphreys, D. J. Sullivan, R. V. Goering, R. Ehricht, S. Monecke and D. C. Coleman. 2011. Characterization of a novel arginine catabolic mobile element (ACME) and staphylococcal chromosomal cassette mec composite island with significant homology to *Staphylococcus epidermidis* ACME type II in Methicillin-resistant *Staphylococcus aureus* genotype ST22-MRSA-IV. *Antimicrobial Agents and Chemotherapy* 55:1896–1905.

Sivaranjani, M., M. Prakash, S. Gowrishankar, J. Rathna, S. K. Pandian and A. V. Ravi. 2017. *In vitro* activity of alpha-mangostin in killing and eradicating *Staphylococcus epidermidis* RP62A biofilms. *Applied Microbiology and Biotechnology* 101:3349–3359.

Sivasankar, C., S. Maruthupandiyan, K. Balamurugan, P. B. James, V. Krishnan and S. K. Pandian. 2016. A combination of ellagic acid and tetracycline inhibits biofilm formation and the associated virulence of *Propionibacterium acnes in vitro* and *in vivo*. *Biofouling* 32:397–410.

Spaulding, A. R., W. Salgado-Pabón, P. L. Kohler, A. R. Horswill, D. Y. Leung and P. M. Schlievert. 2013. Staphylococcal and streptococcal superantigen exotoxins. *Clinical Microbiology Reviews* 26:422–447.

Stevens, D. L. 1997. Necrotizing clostridial soft tissue infections. In *The Clostridia, Molecular Biology and Pathogenesis [Internet]*. The J. I. Rood, B. A. McClane, J. G. Songer and R. W. Titball editors. Seattle, WA: University of Washington School of Medicine.

Stevens, D. L., A. L. Bisno, H. F. Chambers, E. P. Dellinger, E. J. Goldstein, S. L. Gorbach, J. V. Hirschmann, S. L. Kaplan, J. G. Montoya and J. C. Wade. 2014. Practice guidelines for the diagnosis and management of skin and soft tissue infections: 2014 update by the Infectious diseases society of america. *Clinical Infectious Diseases* 59:e10–e52.

Stieritz, D. D. and I. A. Holder. 1975. Experimental studies of the pathogenesis of infections due to *Pseudomonas aeruginosa*: Description of a burned mouse model. *Journal of Infectious Diseases* 131:688–691.

Storz, G. and J. A. Imlayt. 1999. Oxidative stress. *Current Opinion in Microbiology* 2, 188–194.

Subramenium, G. A., K. Vijayakumar and S. K. Pandian. 2015. Limonene inhibits streptococcal biofilm formation by targeting surface-associated virulence factors. *Journal of Medical Microbiology* 64:879–890.

Subramenium, G. A., D. Viszwapriya, P. M. Iyer, K. Balamurugan and S. K. Pandian. 2015a. CovR mediated antibiofilm activity of 3-furancarboxaldehyde increases the virulence of Group A Streptococcus. *PLoS One* 10:p.e0127210.

Swartz, M. N. 2004. Cellulitis. *New England Journal of Medicine* 350:904–912.

Talan, D. A., W. R. Mower, A. Krishnadasan, F. M. Abrahamian, F. Lovecchio, D. J. Karras, M. T. Steele, R. E. Rothman, R. Hoagland and G. J. Moran. 2016. Trimethoprim–sulfamethoxazole versus placebo for uncomplicated skin abscess. *New England Journal Medicine* 374:823–832.

Talan, D. A., P. H. Summanen and S. M. Finegold. 2000. Ampicillin/sulbactam and cefoxitin in the treatment of cutaneous and other soft-tissue abscesses in patients with or without histories of injection drug abuse. *Clinical Infectious Diseases* 31:464–471.

Ton-That, H. and O. Schneewind. 2004. Assembly of pili in Gram-positive bacteria. *Trends Microbiology* 12:228–234.

Turner, K. H., J. Everett, U. Trivedi, K. P. Rumbaugh and M. Whiteley. 2014. Requirements for *Pseudomonas aeruginosa* acute burn and chronic surgical wound infection. *PLoS Genet* 10:p.e1004518.

Vanhommerig, E., P. Moons, D. Pirici, C. Lammens, J. P. Hernalsteens, H. De Greve, S. Kumar-Singh, H. Goossens and S. Malhotra-Kumar. 2014. Comparison of biofilm formation between major clonal lineages of Methicillin-resistant *Staphylococcus aureus*. *PLoS One* 9:p. e104561.

Viszwapriya, D., U. Prithika, S. Deebika, K. Balamurugan and S. K. Pandian. 2016. *In vitro* and *in vivo* antibiofilm potential of 2, 4-Di-tert-butylphenol from seaweed surface associated bacterium *Bacillus subtilis* against Group A Streptococcus. *Microbiological Research* 191:19–31.

Vlassova, N., A. Han, J. M. Zenilman, G. James and G. S. Lazarus. 2011. New horizons for cutaneous microbiology: The role of biofilms in dermatological disease. *British Journal of Dermatology* 165:751–759.

Walker, M. J., T. C. Barnett, J. D. McArthur, J. N. Cole, C. M. Gillen, A. Henningham, K. S. Sriprakash, M. L. Sanderson-Smith and V. Nizet. 2014. Disease manifestations and pathogenic mechanisms of group A Streptococcus. *Clinical Microbiology Reviews* 27:264–301.

Wang, F., R. H. Fang, B. T. Luk, C. M. J. Hu, S. Thamphiwatana, D. Dehaini, P. Angsantikul, A. V. Kroll, Z. Pang, W. Gao and W. Lu. 2016. Nanoparticle-Based antivirulence vaccine for the management of methicillin-resistant *Staphylococcus aureus* skin infection. *Advanced Functional Materials* 26:1628–1635.

Weinstein, R. A. and C. G. Mayhall. 2003. The epidemiology of burn wound infections: Then and now. *Clinical Infectious Diseases* 37:543–550.

Wessels, M. R. 2016. Pharyngitis and scarlet fever. In *Streptococcus pyogenes: Basic Biology to Clinical Manifestations [Internet]*. Ferretti, J. J., Stevens, D. L., Fischetti V. A., editors. Oklahoma City, OK: University of Oklahoma Health Sciences Center.

Williams, R. J., B. Henderson, L. J. Sharp and S. P. Nair. 2002 Identification of a fibronectin-binding protein from *Staphylococcus epidermidis*. *Infection and Immunity* 70:6805–6810.

Wong, C H., L. W. Khin, K. S. Heng, K. C. Tan and C. O. Low. 2004. The LRINEC (Laboratory Risk Indicator for Necrotizing Fasciitis) score: A tool for distinguishing necrotizing fasciitis from other soft tissue infections. *Critical Care Medicine* 32:1535–1541.

Wretlind, B. and O. R. Pavlovskis. 1983. *Pseudomonas aeruginosa* elastase and its role in pseudomonas infections. *Reviews of Infectious Diseases* 5:S998–S1004.

11 Bacteriology of Ophthalmic Infections

Arumugam Priya and Shunmugiah Karutha Pandian

Contents

11.1 Introduction

Despite the apparent sturdiness of the ocular surface and structure, the eye is constantly exposed to the external microbial communities. The existence of ocular microflora divulges the interplay of microbes in the ocular health and disease. The crosstalk between the ocular microbial flora and the ocular defense system coordinates the furtherance of ocular surface homeostasis and health. The disparities in the host factors, environmental stipulations, nutritional, and disease status of an individual may arbitrate ocular microfloral shift (Miller and Iovieno, 2009). Pathological shift in the indigenous microbial community can cause hostile immune reactions or ocular surface cellular damage and exert a pathogenic effect. Through numerous earlier reports, native ocular flora has been shown to be predominantly *Staphylococcus epidermidis*, *Staphylococcus aureus*, and *Corynebacterium* spp. along with other less common organisms such as *Propionibacterium*, *Haemophilus*, *Pseudomonas*, and so on (Capriotti et al., 2009). Bacterial ophthalmic infections vary greatly in severity from common conjunctivitis to endophthalmitis, an adverse infectious postoperative complication. Instantaneous detection and appropriate medication for ocular infections will lessen the incidence of visual impairment and ocular morbidity (Muthiah and Radhakrishnan, 2017). Hence, deciphering the relationship between the normal microflora and etiology of ocular infections is imperative.

This chapter is aimed at describing the characteristics and clinical manifestations of common bacterial intra- and extraocular infections and the role of diverse microflora in infectious diseases associated with ocular surfaces.

11.2 Eye—The photoreceptive and the infection prone organ

The eye, the organ of visual perception, is structurally comparted into three layers or tunics, organized as internal segment that is composed of anterior and posterior chambers, the iris, the lens, the vitreous cavity, the retina, the ciliary body, the choroid and the intrinsic ocular muscles, and enclosed by external segments (i.e., the conjunctiva, the cornea, the sclera, and the tear film). The internal segment of the eye presents a sterile environment by efficient blood-retinal barrier (BRB), whereas the external segment, which is exposed to the environment, is subjected to numerous microbial challenges (Lu and Liu, 2016). Infections may occur in almost any part or tissue of the ocular surface, orbit, and adnexa. The transference of the infection from a site to the other may be ensued by the direct contact or indirectly through blood vessels and nerves (Rumelt, 2016). The conjunctiva, the eyelid, and the cornea are the frequently infected sites of eye because they act as a first line preventive barrier against foreign bodies (Alfonso and Miller, 1990). Prolonged infection or infection to delicate tissue like the cornea may lead to the impairment of normal vision and can extend to loss of sight.

11.3 Ocular microbiota

Perceptive features of the ocular microflora are fundamental in understanding the ocular diseases and infections. Axenfeld (1908) stated that the microbiota of eyelid and conjunctiva are similar to that of the skin and upper respiratory tract. Since then, microbial flora of the ocular surface has been subjected to numerous studies to investigate the indigenous flora of the healthy eyes, as a comparative analysis to interpret the microbial shift during diseased state, to assess the microbial community before intraocular surgeries, or to review the prophylactic strategies in postoperative infections. Axenfeld founds that *Staphylococcus albus* and *Corynebacterium* were frequently isolated organisms, whereas *Staphylococcus aureus*, *Streptococcus* spp., and few other gram-negative bacteria were found with least incidence. The classification of ocular microbiota based on the culture-dependent methods was alleged to be predominantly conquered by gram-positive species such as *Staphylococcus*, *Streptococcus*, *Propionibacterium*, and *Corynebacterium*; gram-negative species such as *Neisseria, Haemophilus*, and few fungal species (Miller and Iovieno, 2009). Culture-based characterization significantly surpassed cultivable and fastidious growing organisms. With the advent of molecular techniques, (Dong et al., 2011) instigated the genome based detection of ocular microbiota and revealed diverse microbial community including commensal, environmental, and opportunistic pathogens (Dong et al., 2011). The 12 genera, *Pseudomonas*, *Propionibacterium*, *Bradyrhizobium*, *Corynebacterium*, *Acinetobacter*, *Brevundimonas*, *Staphylococci*, *Aquabacterium*, *Sphingomonas*, *Streptococcus*, *Streptophyta*, and *Methylobacterium*, were represented as core microbiome of the conjunctiva. Based on the sequencing of 16S rDNA V3–V4 hypervariable segments of bacteria from conjunctival swab, Huang et al. (2016) linked additional genera such as *Millisia*, *Anaerococcus*, *Finegoldia*, *Simonsellia*, and *Veillonella* to the core conjunctival microbiota. Numerous studies have evidenced that the use of contact lenses (Hovding, 1981; Larkin and Leeming, 1991; Fleiszig and Efron, 1992; Iskeleli et al., 2005; Shin et al., 2016), the eyes that endured surgeries (Jabbarvand et al., 2016) and patients with prolonged hospital stays (Sahin et al., 2017) presented variations in the microbial diversity and abundance. Moreover, variation in the ocular microbiota between eyes of an individual and between individuals has also been affirmed (Hovding, 1981).

11.4 Antibacterial protections in ocular surface

The association of indigenous microflora and the ocular mucosal and immune epithelial cells maintains the ocular surface homeostasis through

- barrier preservation,
- inhibition of apoptosis and inflammation,
- producing inhibitory substances such as bacteriocins,
- eliminating harmful pathogens,
- accelerating wound healing and tissue regeneration,
- maintenance of immune tolerance, and
- linkage to adaptive immunity (Miller and Iovieno, 2009).

11.5 Ocular infections

Infections in the eye can be broadly classified into intra- and extraocular infection based on the site of infection origin. Intraocular infections may involve different intraocular structures such as uveal tissues (e.g., choroid, ciliary body, and iris), the retina, and the vitreous chamber. The extraocular infections may emerge in the following surfaces: eyelids, lacrimal sac, conjunctiva, cornea, and the adnexal structures (Table 11.1) (Figure 11.1a and b).

11.5.1 Extraocular infections

11.5.1.1 Blepharitis – The infectious and inflammatory conditions of the lid margin, including the eyelash follicles and sebaceous and apocrine glands are generally described as blepharitis, the most encountered eye infection. It typically occurs bilaterally and exists as a recurrent chronic condition. Blepharitis is a multifactorial complex disease, which institutes several overlapping signs and symptoms (Jackson, 2008). Meibomian gland

Table 11.1 Diagnosis and Treatment Strategies for Intra- and Extraocular Infections

Infection	Diagnosis	Treatment Strategies
Blepharitis	• Clinical diagnosis with presenting signs and symptoms	• Topical antibiotics • Warm compress
Dacryocystisis	• Clinical diagnosis • Differential diagnosis with sinusitis, punctual ectropion, sebaceous cyst, cellulitis	• Local massage over lacrimal sac • Probing and syringing • Dacryocystorhinostomy • Warm compress • Systemic broad-spectrum antibiotics
Bacterial Conjunctivitis	• Clinical diagnosis with presenting signs and symptoms • Conjunctival scraping • PCR • Differential diagnosis including viral, chlamydial and allergic conjunctivitis, superficial keratitis, blepharitis, acute angle closure glaucoma	• Topical application of broad spectrum antibiotics – aminoglycosides, sulfacetamide solution, fluoroquinolones, tetracycline, and chlroamphenicol
Hordeolum	• Typically, no diagnostic testing • Imaging required in case of further complications like periorbital cellulitis	• Lesions drain spontaneously • Warm compress can reduce the abscess • Lid massage/lid scrub with saline or mild shampoo For persistent and large lesions • Erythromycin ophthalmic ointment • Incision and drainage under local anesthesia
Chalazion	• Ophthalmic examination • Clinical diagnosis with the presenting signs and symptoms • Biopsy • CT scan of the face and orbits	• Warm compress • Topical antibiotic ointment • Systemic tetracycline • Subcutaneous injection of steroid triamcinolone acetonide • Incision and curettage

(*Continued*)

Table 11.1 (*Continued*) Diagnosis and Treatment Strategies for Intra- and Extraocular Infections

Infection	Diagnosis	Treatment Strategies
Bacterial Keratitis	• Complete ophthalmic examination (visual acuity, slit-lamp examination, intraocular pressure) • Corneal scraping and pathological examination	• Application of topical antibiotics/corticosteroids • Corneal transplantation
Preseptal and orbital cellulitis	• CT • Ultrasound	• Warm compress • Intramuscular cephalosporin injections • Endoscopical drainage of abscess • Surgical interventions
Endophthalmitis	• Pathological examination from intraocular specimen (vitreous or aqueous humor), blood culture, lumbar puncture • Diagnostic PCR	• Intravitreal antibiotics • Intravitreal steroids • Vitrectomy
Uveitis	• Clinical examination • Laser Flare Cell Meter • Ultrasound biomicroscopy • PCR-based molecular detection of etiological agents • HLA-B27 typing (for nongranulomatous cases) • Local antibody titre determination For Uveitis associated with systemic diseases • Serum lysozyme test • Chest X-ray • Gallium Scintigraphy • Treponemal hemagglutination and Nontreponemal test	• Systemic steroid/ immunosuppressive/ anti-inflammatory regimen • Intravenous polyclonal immunoglobulin treatment

CT, computed tomography; PCR, polymerase chain reaction.

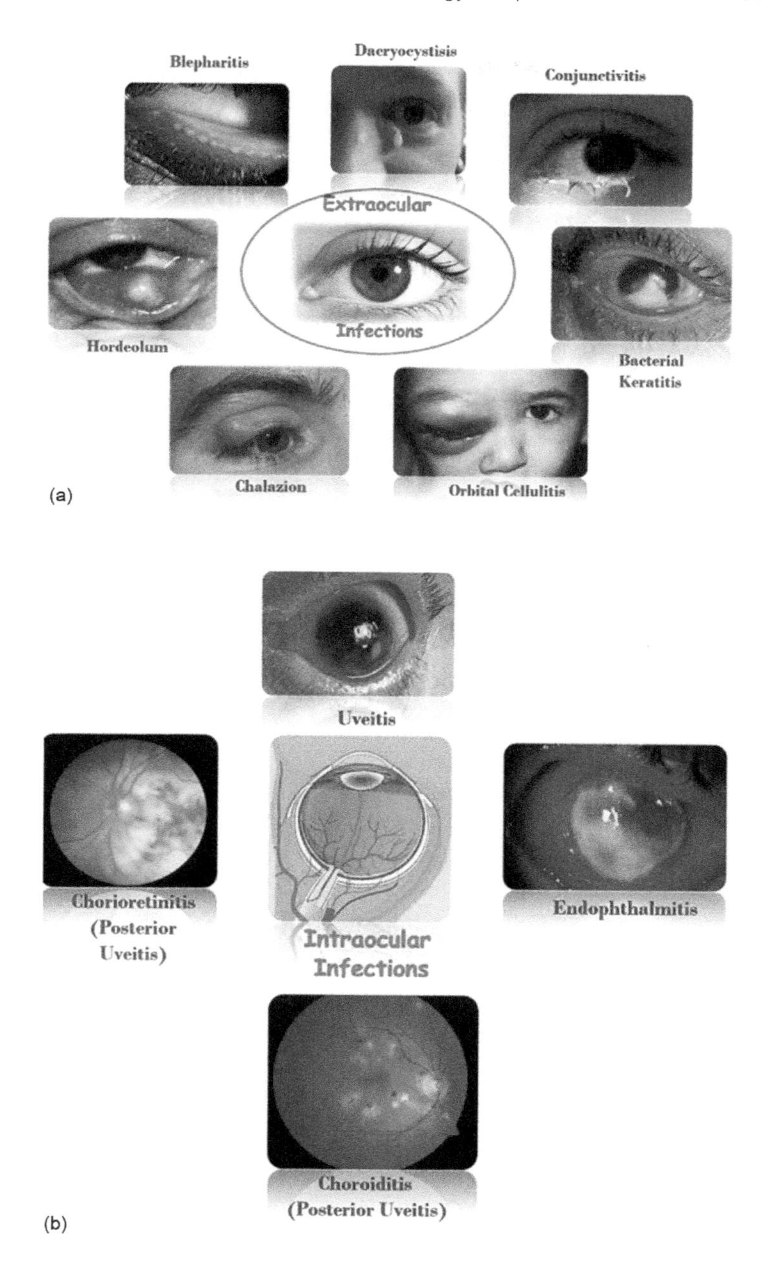

Figure 11.1 (a) Extraocular infections and (b) Intraocular infections.

dysfunction, conjunctival redness, crusting, hyperkeratinization and redness of the eyelid, ocular itching, burning and irritation, dry or watery eyes, and photophobia are typical symptoms of blepharitis (McCulley and Shine, 2000; Favetta, 2015). Surplus colonization of lid-margin microbes, abnormal lid-margin secretion, or dysfunctional tear film will prompt the infection.

11.5.1.1.1 Classification and etiological agents – Anatomically, blepharitis can be classified as anterior and posterior forms. Infections affecting the anterior lid margin and eyelashes result in anterior blepharitis and the infections of Meibomian gland and proximal tissue lead to posterior blepharitis. Apart from the anatomical site-based classification, blepharitis can also be categorized as *Staphylococcal*-mediated blepharitis, seborrheic blepharitis, mixed *Staphylococcal*, and seborrheic blepharitis (McCulley et al., 1981). The frequent organisms associated with blepharitis infection are *S. epidermidis*, *S. aureus*, *P. acnes*, and *Corynebacterium* (Dougherty and McCulley, 1984; Groden et al., 1991). Infection is frequently observed in the contact lens wearers and patients who underwent refractive, cataract, and other ocular surgeries.

11.5.1.1.2 Pathophysiology – The exoenzymes produced by the causative organisms, in particular *S. epidermidis*, initiates irritation in the eyelid and surrounding ocular surface that recruits the inflammatory mediators to the site of infection (Lemp and Nichols, 2009). Lipolytic exoenzymes such as triglyceride lipase, cholesterol esterase, wax esterase, and sterol esters with the deliverance of irritating fatty acids will result in the disruption of tear film integrity (Bron and Tiffany, 2004). Alternatively, the variations in the secretion of Meibomian gland with increased lipid secretion may offer provision for the proliferation of microbes (Nichols et al., 2011). The implication of quorum sensing in the bacterial load rise can increase the signs and symptoms of blepharitis (Haque et al., 2010).

11.5.1.2 Dacryocystitis – Dacryocystitis is a painful inflammatory disorder of the lacrimal sac which occurs because of obstruction in the nasolacrimal duct. The obstruction may be due to primary acquired obstruction or secondary to the trauma, infection or inflammation, or mechanical obstruction. The blockage in the nasolacrimal duct causing stillness in the tears within pathologically congested lacrimal duct will lead to dacryocystitis (Iliff, 1996). It is prevalent among the infants and the middle-aged women. The clinical presentation may be mild as irritation and discomfort to sight threatening. The signs and symptoms vary greatly according to the etiology of the infection. Dacryocystitis was most frequently observed as a unilateral infection.

11.5.1.2.1 Classification and etiological agents – Dacryocystitis may either present in acute or chronic form and congenital in rare instances. Acute dacryocystitis is an acute inflammation of the lacrimal sac due to infection by *S. aureus* or beta hemolytic *Streptococcus*, which results in tenderness and erythema and one fourth of the eye may be present with the lacrimal abscess. If untreated, the infection may propagate to surrounding tissues and cause periorbital or orbital cellulitis. The chronic dacryocystitis exhibits epiphora, mucoid discharge, conjunctival hyperemia, and chronic conjunctivitis (Ali et al., 2013, 2015). The congenital dacryocystitis will be seen since infancy and the presentation will be usually epiphora with mucopurulent discharge. The complications may further lead to orbital cellulitis or even brain abscess and meningitis (Babar et al., 2011). The severity of the infection may range from partial to total obstruction of the nasolacrimal duct. No proper medication or therapy exists. Even the systemic antibiotic therapy will diminish the condition in much slower pace (Cahill and Burns, 1993). The obstruction in the nasolacrimal sac can be relieved by external or endonasal dacryocystorhinostomy (DCR) but leaves a potential risk for endophthalmitis. Hence, chronic dacryocystitis should be cured before intraocular surgeries.

The bacteriology of acute and chronic dacryocystitis has a vast difference. Gram-negative rods dominate in the acute infections, whereas, both gram-negative and -positive isolates were found in the chronic infections. *Staphylococcus* spp., *S. pneumoniae*, *P. aeruginosa*, *E. coli*, (Hartikainen et al., 1997; Iliff, 1996), *Peptostreptococcus*, *Propionibacterium*, *Prevotella*, *Fusobacterium* (Brook and Frazier, 1998), *Streptococcus pyogenes*, and *Streptococcus viridans* (Sarkar et al., 2015) are the predominant causative organisms reported till date.

11.5.1.2.2 Pathophysiology – Under normal conditions, the lacrimal sac is resistant to the infectious organisms. The lacrimal sac usually drains the tear from eye into the nasal cavity. The obstruction in the nasolacrimal duct may result in improper drainage or accumulation of the tears, desquamated cells, and mucoidal secretions. This provides residence for secondary bacterial infection (Iliff, 1996). Bacterial overload at the lacrimal sac will recruit the anti-infective response which leads to acute or chronic dacryocystitis infection (Pinar-Sueiro et al., 2012).

11.5.1.3 Bacterial conjunctivitis – The inflammation of the conjunctiva, the transparent membrane that covers the sclera is termed as conjunctivitis. Conjunctivitis forms the most common cause of red eye and most frequently observed eye infection worldwide. Conjunctivitis can be a cause of bacterial or viral origin. Ocular allergy, other extraocular infections such as blepharitis, dacryocystitis, dry eye, use of contact lenses, ophthalmic solutions, and medications are stated as frequent causes of conjunctivitis. The symptoms include tearing, burning, or stinging sensation, sticky eyelids in the morning, mucopurulent secretions with distinct or severe pain.

The infection begins unilaterally. However, within 1–2 days the fellow eye becomes infected. The condition is also found to be contagious, caused by one or more bacterial species (Bartlett and Jannus, 2008).

11.5.1.3.1 Classification and etiological agents – Conjunctivitis can be classified as hyperacute, acute, and chronic (Morrow and Abbott, 1998). *Neisseria gonorrhoeae* is the common cause of hyperacute conjunctivitis with the characteristic symptoms of abrupt onset, yellow-green purulent secretion, redness, and irritation of the conjunctiva, tenderness and palpation, conjunctival chemosis, conjunctival injection, lid swelling and so on. If hyperacute gonococcal conjunctivitis is left untreated, the corneal involvement and endophthalmitis is inevitable. *N. meningitis* is the second-most common cause for hyperacute bacterial conjunctivitis (Hovding, 2008). Acute bacterial conjunctivitis lasts for 3–4 weeks with burning and irritating sensation obviously followed by purulent discharge. The common bacterial pathogens involved in acute conjunctivitis are *S. aureus*, *S. pneumoniae*, and *Haemophilus influenza*. Prompt medication will lower the contagiousness and reduces further complications like corneal ulceration. The chronic conjunctivitis last for at least 4 weeks with similar signs of itching and burning sensation along with flaky debris, erythema, and bulbar conjunctival injection. The chronic conjunctivitis is commonly caused by *Staphylococcus* species and *Moraxella lacunata* with occasional involvement of other bacteria (Thielen et al., 2000).

11.5.1.3.2 Pathophysiology – Pathophysiology of the conjunctivitis is not well documented. The inflammation caused in the conjunctiva may be induced due to the exogenous and endogenous infections and toxic agents produced by the pathogens (Thielen et al., 2000).

11.5.1.4 Hordeolum – Hordeolum is an acute bacterial infection causing inflammation on the eyelid margin. The infection presents the painful, erythematous, and swollen furuncle. The onset of infection is spontaneous and is dependent on the influence of lid hygiene. It is one of the most common infections of the eye. Hordeola may be associated with various complications such as diabetes, blepharitis, seborrheic dermatitis, and individuals with high levels of lipid secretion. Chalazion and hordeolum frequently presents similar signs and are often misdiagnosed. The hordeolum affects the oil glands of the eye either internally (inside the eyelids) or externally (on the eyelid, near eye lashes) (Lindsley et al., 2017).

11.5.1.4.1 Classification and etiological agents – The internal inflammation affects the Meibomian gland, whereas the external inflammation affects the Zeis or Moll gland (Wald, 2007). The external hordeola is more common and are referred to as styes. In most instances, the lump resolves spontaneously over a period even left untreated. However, the inflammation might spread to other ocular glands and cause secondary infections like cellulitis. The incomplete elimination of bacteria may result in recurrent hordeolum. Internal hordeola tend to be more painful than the external hordeolum (Lindsley et al., 2013). Internal hordeolum if untreated may lead to development of chalazion. In both the forms, the size of the lesion is directly proportionate to the severity of the infection (Lebensohn, 1950). The infection is usually caused by the *Staphylococcus* spp. that infects the eyelash hair follicles (Mueller and Mcstay, 2008).

11.5.1.4.2 Pathophysiology – The deceptive secretory functions of the ocular glands such as Zeis or Moll and Meibomian gland may result in hordeolum. The Zeis gland secrets sebum with antiseptic properties whereas Moll gland secrets IgA, mucin and lysosomes, which act as immune barrier against the pathogenic organisms. The obstructions in these glands lead to impaired defense system that is prone to infectious organisms. Further bacterial infection with *S. aureus*, the most common pathogen will activate the immune system and elicit immediate localized inflammatory response followed by purulent or abscess development (Bragg, 2017).

11.5.1.5 Chalazion – Chalazion, also known as a Meibomian cyst, is a common eyelid disorder of all age groups. It is lipogranulomatous inflammation of the ocular glands caused due to retention of the Meibomian secretion in the sebaceous gland. Inflammation and irritation of the eyelid and ocular surface with the formation of cyst are the common clinical presentations. The cyst formation usually does not affect the normal visual perception. However, the size of the cyst may have an impact. Larger chalazion cyst may interrupt the normal vision or induce astigmatism which can lead to eye morbidity (Park and Lee, 2014). The predisposing factors associated with chalazion include, Meibomian gland dysfunction, chronic blepharitis, dry eye, seborrheic dermatitis, gastritis, and smoking (Nemet et al., 2011). Other factors such as exposure to ultraviolet (UV) light, poor lid hygiene, use of cosmetic products, and stress also contribute to cyst development, but their role in disease is poorly understood.

11.5.1.5.1 Etiological agents – Bacteria, especially *S. aureus*, is known to be the primary causative agent of chalazion. However, the severity and eye morbidity due to chalazion is dictated by the secondary bacterial superimposed infections (Otulana et al., 2008). Incidence of other bacterial agents has not been encountered in years.

11.5.1.5.2 Pathophysiology – Retention of the Meibomian gland secretion leads to the accumulation of lipid which further develops into a lipogranulomatous form. Histological sections of chalazion have shown to be composed of histiocytes, mononuclear granulocyte cells, lymphocytes, plasma cells, polymorphonuclear cells, and eosinophils (Mustafa and Oriafage, 2001). Presence of a pseudo-capsule connective tissue was often observed around the lesion (Ozdal et al., 2004). Liberation of lipids provide hosting environment for the infectious organisms, which worsens the case.

11.5.1.6 Bacterial keratitis – Keratitis is a complicated ocular infectious disease of the cornea that can potentiate unilateral or bilateral ocular morbidity. It is of microbial origin and predominantly due to bacteria. The infection of the cornea will lead to defect in the corneal epithelium, eventual corneal scaring and perforation followed by stromal inflammation. The common risk factors include contact lens wear, previous ocular surgeries, presence of surgical sutures, ocular trauma, persistent topical corticosteroid usage, and certain ocular surface diseases like blepharitis. Several systemic diseases such as diabetes mellitus, rheumatoid arthritis, immunodeficiency, and smoking can also contribute to keratitis (Bourcier et al., 2003; Keay et al., 2006)

11.5.1.6.1 Etiological agents – *P. aeruginosa, S. aureus, coagulase negative Staphylococcus*, and *S. pneumoniae* are considered prime causative organisms (Green et al., 2008). Few other organisms including *S. epidermidis, Moraxella, S. marcescens, Bacillus, Corynebacterium, H. influenza*, and *Alcaligenes xyloxidans* are reported for their association with keratitis (Dart et al., 2008).

11.5.1.6.2 Pathophysiology – Cornea acts as a preventive barrier against the invading pathogen. In addition to this, host defensive mechanism will prevent the corneal tissue from bacterial infections. Failure of the host defense mechanism or corneal breaching will allow bacterial invasion (Andrew et al., 2003). Inflammatory mediators infiltrate rapidly at the site of corneal damage and results in corneal cloudiness around the infected tissue. Various immune secretions will further damage the corneal tissue for instance, IL-8 secretion will lead to neovascularization. Corneal damage and blood vessel formation will interfere the normal vision and if left untreated fallouts in ocular morbidity (Schaefer et al., 2001).

11.5.1.7 Cellulitis – Cellulitis is generally described as the connective tissue or subcutaneous tissue inflammation caused majorly due to infectious pathogens. Cellulitis in the orbital surface is the infection in the soft tissues of the orbit which predominantly affects the children. It is potentially a sight-threatening infection, which can also cause several systemic and life-threatening illnesses such as cavernous sinus thrombosis, meningitis, cerebritis, endophthalmitis, and brain abscess (Donahue and Schwartz, 1998; Georgakopoulos et al., 2010). Common signs include cutaneous tenderness, erythema, severe pain, leukocytosis, and so on. Several pediatric orbital cellulitis has been reported as a secondary infection of paranasal sinus infection.

11.5.1.7.1 Classification and etiological agents – Partition of the soft tissues of eyelid and orbit by orbital septum creates preseptal and postseptal space. Infections at the preseptal space which is anterior to the orbital septum is defined as the preseptal or periorbital cellulitis (Nageswaran et al., 2006). In orbital cellulitis, the infection is restricted to the posterior of orbital septum (Mawn et al., 2000). Trauma, sinusitis, and bacteremia are the major routes of orbital infections. Clinical manifestations of periorbital cellulitis include erythema, induration, tenderness, chemosis, proptosis, limited ocular motility, optic neuritis, hypesthesia, sensory distribution, and so on. Infection of the periorbital cellulitis is restricted to the preseptal eyelid tissue. *H. influenza*, beta hemolytic *Streptococcus* spp., *S. aureus*, *S. epidermidis*, and *S. pyogenes* are major etiological agents. Other rare causative agents are *P. aeruginosa*, *N. gonorrhoeae*, *Treponema pallidum*, and *Mycobacterium tuberculosis* among others (Ambati et al., 2000; Carlisle and Fredrick, 2006).

11.5.1.7.2 Pathophysiology – Paranasal sinuses are the cavities that surround the orbit of the eye. Infections in the sinuses will also contribute to the pathophysiology of orbital cellulitis. In the thin paranasal sinuses, several natural perforations occur to pass the valveless blood vessels and nerves. These perforations are the major route of infections. Additionally, due to thin architect of the orbital bones, formation of abscess from the adjacent sinusitis is probably high (Lee and Yen, 2011).

11.5.2 Intraocular infections

11.5.2.1 Endophthalmitis – Endophthalmitis is an inflammatory disease of the posterior eye segment (vitreous/aqueous humor) due to intraocular bacterial or fungal infections (Callegan et al., 1999; Durand, 2017). It occurs infrequently, but the infection results in devastated eye state and irreversible vision loss. The clinical presentation usually initiates with the blurred vision and causes mild to severe pain, redness, absence of fundus, hypopyon, vitritis, and inflammation in the anterior chamber (Jackson et al., 2014).

11.5.2.1.1 Classification and etiological agents – Based on the route of infection, endophthalmitis is classified into endogenous bacterial or exogenous postoperative endophthalmitis. Exogenous form occurs following the introduction of pathogens directly after ocular surgeries such as cataract surgery, penetrating keratoplasty, or placement of keratoprosthetics. Infection for endogenous endophthalmitis occurs through the bloodstream after crossing the blood retinal barrier (Jackson et al., 2014). In addition to the previously specified clinical presentations, endogenous endophthalmitis will present systemic infections such as fever or influenza-like symptoms. The clinical features and bacteriology of these categories vary relatively. In recent years, the postoperative endophthalmitis has been reported in almost every type of ocular surgery, but predominantly following cataract surgery. The major etiological agents causing endophthalmitis includes *S. aureus*, *S. epidermidis*, *Bacillus cereus*, *Enterococcus faecalis*, *K. pneumoniae*, *P. aeruginosa*, *N. meningitides*, *S. pneumoniae*, Group B streptococci, and *Nocardia* spp. (Callegan et al., 1999; Jackson et al., 2014).

11.5.2.1.2 Pathophysiology – In endogenous endophthalmitis, the organisms from the bloodstream enter the ocular surface either through retina or uvea and invades the tissue after transecting through the blood ocular barrier. With subsequent establishment, the pathogens reside in the aqueous or vitreous humor. Bacterial load and the infiltration of inflammatory cells will initiate destruction of the tissue and physiological imbalance resulting in the loss of function in the anterior segment. If the pathogen enters via retinal artery, the dissemination of bacteria along the retinal vessels will cause irreversible tissue damage due to toxins produced by the pathogens and by the activity of inflammatory cells (Greenwald et al., 1986).

11.5.2.2 Uveitis – Uveitis, the inflammation of the uveal tissue that encompasses iris, choroid, and ciliary body is a major blinding disorder (Biziorek et al., 2001). It is usually found in all age groups, but the severity of the disease is much higher in pediatrics than adult disease. Pediatric uveitis is usually asymptomatic, which results in inability to detect the disease in earlier stage subsequently and leads to permanent vision loss (Curragh et al., 2017). Neither simple clinical examination nor the noninvasive investigation clearly state the causation of disease and the etiology remains unknown in maximum cases. The association of trauma, infection, systemic diseases such as tuberculosis, sarcoidosis, spondylarthropathies, Bechet's disease,

Whipple's disease, Koyanagi-Harada syndrome, and inflammation can critically lead to uveitis (Biziorek et al., 2001).

Though usually non-symptomatic, certain types of uveitis may cause discomfort, pain, photophobia accompanied with lacrimation, congestion, and iridocyclitis which often leads to keratitis or keratopathy (Bartlett and Jannus, 2008). As the disease is related with the vision accomplishing pathway, severe inflammation of the uveal tissue often leads to unilateral or bilateral ocular morbidity based on the origin of the disease (Hogan et al., 1959).

11.5.2.2.1 Classification and etiological agents – Based on the ocular site of inflammation, uveitis can be classified into four major types as anterior uveitis, intermediate uveitis, posterior uveitis, and panuveitis. The inflammation of anterior chamber or the iris lesion or keratic precipitates are usually demarcated as anterior uveitis. Intermediate uveitis can be defined as the inflammation of the vitreous chamber with or without the involvement of peripheral retina. Inflammation affecting retina, choroid, retinal vessels, or posterior vitreous humor is defined as the posterior uveitis. Combination of inflammation in all three described sites is collectively termed *panuveitis* (Bodaghi et al., 2001).

Predominant infectious agents of uveitis include certain parasites such as *Toxoplasma gondii*, *Streptococcus* spp., and fastidious bacteria such as spirochetes, intracellular bacteria such as *Chlamydia* spp., *Rickettsia* spp., *Coxiella burnetiid*, and the gram-positive pathogen *Tropheryma whipplei*. *Bartonella henselae* and *B. grahamii* are also reported as causative organisms of uveitis (Drancourt et al., 2008; Terrada et al., 2009).

11.5.2.2.2 Pathophysiology – Bacterial products such as cell wall components, proteins, endo- and exotoxins are suspected to be the triggers of uveitis. The ocular surface synthesis proteins such as toll-like receptors (TLR) and nod-like receptors (NLR) which respond well to the bacterial products and initiate inflammation on the infected site. NLRs are reported to be closely associated with uveitis than TLRs. Though the association of these proteins are known to cause the infection, the exact mechanism is not well understood, which is essential in understanding the disease state and to develop the treatment strategies for uveitis (Rosenbaum et al., 2008).

11.6 Infectious bacteria in ocular diseases

Microorganisms, specifically bacteria, are the leading causal mediators of numerous infectious diseases, among which ophthalmic infections cannot be ruled out certainly. Though, fungal (e.g., keratitis and endophthalmitis) and other parasitic (e.g., *Acanthamoeba* keratitis, Chagas disease, giardiasis, Leishmaniasis, and Toxoplasmosis) infections are common, the prevalence of bacterial ocular infections are exponential (Nimir et al., 2012; Squissato et al., 2015). The organisms that cause ocular infections are predominantly exogenous that find route into eye during surgeries through contaminations from instruments and infusing fluids. Bacterial infections of the ocular surface can be of mono- or polymicrobial origin. Several predisposing factors such as contact lens wear, ocular surgeries, poor hygiene of the eye, obstruction of nasolacrimal duct, diminished immune status, previous ocular infections, and certain environmental factors make the eye susceptible to bacterial infections. Both gram-positive and -negative bacteria can cause ocular infections. Gram-positive pathogens are more predominant in causing eye infections than the gram-negative organisms. *S. aureus*, Coagulase negative *Staphylococcus*, *S. pneumoniae*, and *P. aeruginosa* are the predominant bacterial isolates found in ocular infections. *S. epidermidis*, *N. gonorrhoeae*, *K. pneumoniae* and *H. influenza* are less common isolates from infected eyes. Several reports evidence the occurrence of *Enterobacter*, *Corynebacterium*, *Acinetobacter*, *Propionibacterium*, and *B. subtilis* in various ocular infections. The types of bacteria, their distribution, load, and the site of infection determine the severity of the ocular disease. Initial treatment strategy include course of topical or systemic antibiotic regimen, which upon regular use can promote antibiotic resistance mechanisms. Emergence of antibiotic resistance may have serious consequences such as development of sight-threatening complications (e.g., keratitis, endophthalmitis, orbital cellulitis, retinitis, and dissemination of infections to other major organs like brain). Moreover, quorum sensing and bacterial biofilm play a critical role in recurrent infections.

11.6.1 *Staphylococcus* spp.

S. aureus and coagulase negative staphylococci (CoNS) are the most predominantly isolated pathogens of infected eye and a major cause of nosocomial eye infections. Among the heterogenous group of CoNS, *S. epidermidis*, and *S. saprophyticus* are the frequently encountered organisms (Mshangila et al., 2013). Despite their normal existence as commensal ocular microbiota, *S. aureus* and CoNS are frequent causative agents of most of the eye infections with increasing frequencies over the course of time. The incidence of *Staphylococcus* species in postoperative cases are even higher, especially in cataract surgeries. In a clinical study, conducted with the conjunctival culture of the patients who underwent cataract surgery, around 80% of the cultures were found to be *Staphylococcus* spp., 45.2% and 35% of CoNS and *Staphylococcus* spp., respectively (Lin et al., 2017).

A 20-year retrospective study of posttraumatic endophthalmitis showed the predominance of *S. epidermidis* (21.8%) and *S. saprophyticus* (12%) (Long et al., 2014). Other than the aforementioned CoNS, *S. cohnii*, and *S. haemolyticus* were observed with lesser incidence (Sherwal and Verma, 2008). In other clinical cases, methicillin-sensitive *Staphylococcus aureus* (MSSA) and methicillin-resistant *Staphylococcus aureus* (MRSA) were found to be associated with ocular infections. MRSA isolates were principally found in lid and lacrimal disorders (Chuang et al., 2012). Community-associated MRSA (CA-MRSA) and healthcare associated MRSA (HA-MRSA) are emerging healthcare risk factors of various ocular infections (Hsiao et al., 2012; Wong et al., 2017). Yet another report states that 50%–66% of hospital workers are found to be infected with *Staphylococcus* (Rashid et al., 2012).

Staphylococcus infections can range from mild infections such as blephero conjunctivitis, corneal ulcer to sight-threatening diseases such as keratoconus, orbital cellulitis, endophthalmitis, keratitis, dacryocystitis, chorioretinitis, scleritis, and so on.

11.6.1.1 Virulence factors – *S. aureus* secretes numerous virulence factors, most of which are part of defense mechanism against the host immunity. The secretory proteins of *S. aureus* include alpha-toxin, beta-toxin, gamma-toxin, panto-valentine leucocidin, and so on. In addition, *S. aureus* also produces proteases, lipases, leucocidin, and exfolatin (O'Callaghan et al., 1997).

The two major barriers of the cornea are the mucin layer and the intracellular tight junction of corneal epithelium, which make the binding and penetrance of infectious agents an impossible event. But the disruption of these barriers increases the susceptibility of cornea to staphylococcal infection, which may lead to keratitis. The binding of *S. aureus* to the corneal cell is achieved through two bacterial proteins, fibronectin-binding protein and collagen-binding adhesin, which leads to significant tissue damage by penetration of bacteria to the corneal epithelium (Rhem et al., 2000; Jett and Gilmore, 2002).

The alpha-toxin binds to the protease receptor ADAM 10; thereby it can form pores on the cell membrane, allowing small molecules to pass through. The binding of alpha-toxin with the immune cells such as neutrophils, monocytes, T-cells, and platelets allow calcium ions to pass through the pores and thus causing cellular dysregulation. It can also cleave the E-cadherin molecules, which are associated with the attachment of cells. Thus, the alpha-toxin binding to the corneal epithelium will eventually disrupt the epithelial layer and initiate the corneal ulceration. In endophthalmitis, the alpha-toxin can cause inflammation in the retina, which can terminate in loss of retinal function (Berube and Juliance, 2013; Kumar and Kumar, 2015).

Beta-toxin, a sphingomyelinase, acts on the scleral epithelium and not the cornea. The F and S component of the gamma toxin, when bound to the cell membrane, can penetrate and lyse the cell, which subsequently leads to infiltration of neutrophils to cornea and iris, conjunctival reddening, chemosis, and fibrin accumulation in the anterior chamber. The gamma-toxin is also toxic to the vitreous chamber (Callegan et al., 1994).

The bacterial surface antigens such as poly-N-acetylglucosamine (PNAG) and lipoproteins, which act on TLRs can mediate release of proinflammatory cytokines such as IL6 and IL8 and other antibacterial molecules, hBD-2, LL-37, and iNOS that can trigger local inflammation and tissue destruction (Li et al., 2008).

The secreted proteases of *S. aureus* can inactivate beta-crystalline protein, which prevents apoptosis in retina. Hence destruction of beta-crystalline often leads to loss of retinal cells (Whiston et al., 2008).

PVL toxin is produced mainly by CA-MRSA strains. It is a two-component toxin system composed of F and S protein, which are likely to produce more severe form of keratitis than non-PVL producing strains. This toxin significantly contributes to corneal virulence but the mechanism is poorly understood (Sueke et al., 2013).

In conjunctivitis, the goblet cells of the conjunctiva when exposed to the *S. aureus* toxin will activate the caspase 1 resulting in production of cytokine IL-1 β which is an efficient inducer of inflammation (McGilligan et al., 2013).

In blepharitis, the lipid accumulation on eyelid forms a cyst. The lipid accumulated is the cleavage product of cholesterol and fatty acids by the action of enzyme lipase. The growth of *S. aureus* is stimulated by the presence of cholesterol, which results in excessive colonization of bacteria on the eyelid. On the other hand, the overgrowth of *S. aureus* and *S. epidermidis* on the nasolacrimal duct potentiate the blockage of duct. This infection can spread to the cornea, resulting in a corneal ulceration (Shine et al., 1993).

In addition to these virulence factors, *S. aureus* produces elastase for its defense against the host immune system, which can form corneal ulcers ultimately leading to keratitis (Wu et al., 1999).

Though the ocular immune system produces multiple components to protect from *S. aureus* infections, the corneal or retinal involvement will result in adverse effects like corneal scarring, reduced visual acuity, or even loss of vision.

11.6.1.2 Antibiotic resistance – The emerging antibiotic resistance and production of numerous virulence factors potentiates the infectious nature of *Staphylococcus* spp. in ocular infections. *Staphylococcus* spp. can evolve very rapidly against a range of antibiotics. *S. aureus* isolates, which are known as MRSA that are resistant to methicillin, are commonly resistant to other beta-lactam antibiotics such as penicillin, carbapenem, cephalosporin, and monobactam (Rayner and Munckhof, 2005). The antibiotic susceptibility of CoNS is unpredictable. Penicillin resistance is extremely common worldwide. Hence, penicillinase-resistant beta-lactam antibiotics such as flucloxacillin and oxacillin are commonly used for first-line therapy. Most of the strains of CoNS, CA-MRSA, and HA-MRSA were found to be multiresistant. In an antibiotic susceptibility study conducted in the year 2012, MRSA strains were found to be susceptible to vancomycin, teicoplanin, and gentamicin (Kotlus et al., 2006; Hsiao et al., 2012). Ocular MRSA strains were highly resistant to fluoroquinolones including the fourth generation (Kotlus et al., 2006). CA-MRSA and HA-MRSA strains were found to be highly sensitive to cotrimoxazole, rifampicin, fusidic acid, and minocycline. The frequently used topical antibiotic chloramphenicol was relatively found to be sensitive to almost all *Staphylococcus* spp. (Wong et al., 2017). *S. aureus* develops resistance against antibiotic with various mechanisms including enzymatic inactivation of antibiotics with penicillinase and aminoglycoside-modifying enzymes, altering the target and thereby decreasing the affinity of antibiotics, trapping of antibiotics and efflux pumps. Horizontal gene transfer, spontaneous mutations, and positive selection can also confer antibiotic resistance (Pantosti et al., 2007) (Figure 11.2).

11.6.1.3 Therapeutic interventions – The use of antibiotics results in resistance mechanisms over a period. The combinations of different antibiotics, for which resistance has not been evolved or use of natural antibacterial agents will potentiate the therapeutic strategies. Dajcs et al. (2001) reported the effectiveness of lysostaphin in treating MRSA-mediated endophthalmitis and keratitis. Lysostaphin is a zinc metalloproteinase enzyme isolated from *S. simulans*, which can potentially slay down *S. aureus* by rapid cell wall digestion. Hence, topical application or intravitreal injection of lysostaphin will be an effectual therapy for ocular staphylococcal infections.

Bacteriophage therapy, antibody therapy targeting the virulence factors of *Staphylococcus*, will be the effective alternates to antibiotics. Caballero et al. (2015) produced a high-affinity monoclonal antibody against *S. aureus* alpha-toxin and proved its effectiveness as a therapy for *S. aureus*–mediated keratitis (Caballero et al., 2015). Bacteriophage-derived lytic proteins such as

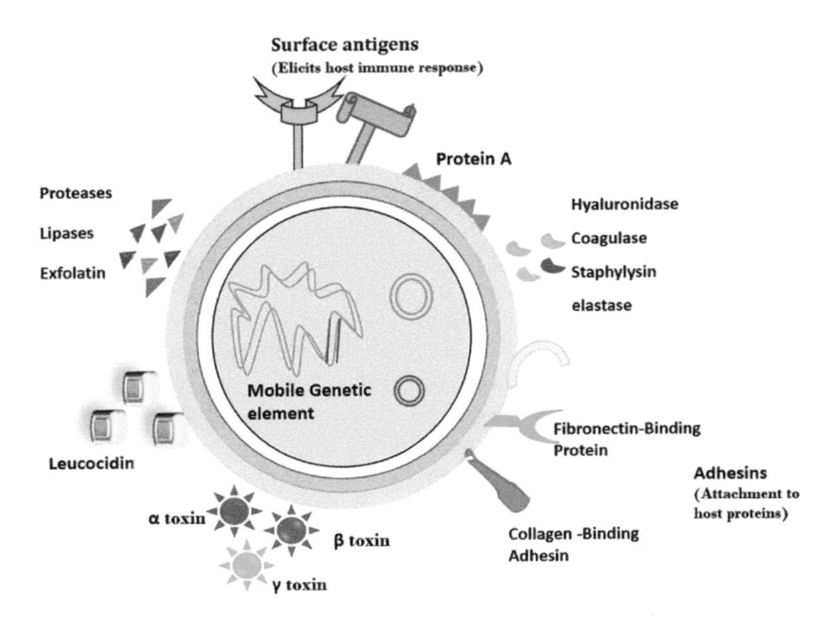

Figure 11.2 Virulence factors of *Staphylococcus* spp.

endolysins have been investigated as antimicrobials. Hence, endolysin technology for killing of *S. aureus* has been reported in the recent past (Schmelcher et al., 2012; Roach and Donovan, 2015). In addition to this, antimicrobial blue light therapy has been reported as a potential alternative or a combinatorial therapy for treatment of keratitis-mediated by *S. aureus* (Zhu et al., 2017).

With these emerging therapeutic procedures, the staphylococcal infections of the eye could be minimized in the near future.

11.6.2 *Streptococcus* spp.

Streptococci are gram-positive bacteria belonging to the phylum Firmicutes, which encompass numerous species that either exist as a commensal or human pathogen. Based on the surface antigens, streptococcal species are classified into various groups. Group A streptococcus (GAS)–*S. pyogenes* and Group B streptococcus–*S. agalactiae* (GBS) are beta hemolytic, whereas *S. pneumoniae* are alpha-hemolytic. GAS and GBS can cause invasive diseases. Though infrequently observed, contributions of *Streptococcus* spp. in various ocular infections have been reported extensively. *Streptococcus* spp. were found in both intra- and extraocular infections including non-sight-threatening conditions such as conjunctivitis, blepharitis, dacryocystitis to sight-threatening preseptal and orbital cellulitis, keratitis, and endophthalmitis. *S. pneumoniae* was reported to be the frequent pathogenic isolate from pediatric conjunctivitis (Friedlaender, 1995; Buznach et al., 2005; Patel et al., 2007). In some bacterial conjunctivitis cases, *S. viridians* were observed as the sole positive culture (Cavuoto et al., 2008). *S. pneumoniae* is considered one among the three-most common pathogens of bacterial conjunctivitis. Along with *S. aureus, S. pneumoniae* forms the most common gram-positive isolate of dacryocystitis followed by GAS and viridian streptococcus (Pinar-Sueiro et al., 2012; Sarkar et al., 2015). *Streptococcus* spp. are predominant in preseptal and orbital cellulitis (Donahue and Schwartz, 1998; Georgakopoulos et al., 2010). Among them, the adult with periorbital cellulitis are more likely to be infected with GAS, whereas, younger children are likely to be infected with *S. pneumoniae* (Schwartz and Wright, 1996). After *S. pneumoniae, S. pyogenes* is frequently observed in ocular cellulitis and in rare instances with other *Streptococcal* spp. such as *S. sanguinis* and *S. milleri* (Ambati et al., 2000; Howe et al., 2004). Endophthalmitic infections were found to be caused by *Streptococcus* spp., namely *S. pneumoniae, S. dysgalactiae,* and GBS (Jackson et al., 2014). Due to routine use of pneumococcal vaccine against pneumoniae, Hauser et al. (2010) reported that the incidence of *S. pneumoniae* in ocular infections are on decline. Yet, among other *Streptococcus* spp., *S. pneumoniae* is frequently being reported in various eye infections.

11.6.2.1 Virulence factors – *Streptococcus* spp. remains a major pathogen of man despite advent antibiotic therapy. The pathogenicity of *Streptococcus* has been conferred by numerous virulence traits. Bacterial components such as capsule, cell wall components, secretary proteins, and enzymes are thought to be the key virulent factors of *Streptococcus* spp. Major virulence factors and their mechanisms are now detailed.

Pneumolysin (PLY) is a potent 53 KDa pore-forming cytotoxin synthesized and located in the cytoplasm of pneumococci. Release of this toxin occurs immediately after spontaneous autolysis of the bacterial cell. Upon release, they interact with the cholesterol and bind the lipid bilayer followed by transmembrane pore formation and lysis of the cell. The pore-forming ability of the PLY can potentially disrupt the corneal epithelium upon infection. In addition to the cytotoxic property, it can directly activate the classical complement system. Activation of complement can either result in production of chemotactic molecules or direct the complement-mediated membrane attack on the host cell. The initiation of proinflammatory cytokines will mediate inflammation and tissue damage. Hence, the release of PLY on the corneal surface eventually leads to corneal ulceration followed by keratitis (Paton, 1996).

Though the capsular polysaccharide of *S. pneumoniae*, also known as pneumococcal PS capsule, does not essentially involve in the inflammatory response but contributes to the disease progression by effectively defending the complement mediated phagocytic attack (Mitchell et al., 1997). In addition to this, secretory proteases cleave the capsule-specific human antibodies, immune components, and blocks the inflammatory responses induced by the cell wall components (Rubins and Janoff, 1998). Surface protein A and adhesins mediate the attachment of bacterial cell surface components to the host cells. Cell wall components can also generate an array of inflammatory mediators, which are more likely to initiate tissue damage.

Autolysin is a cell wall degrading enzyme found on the cell envelop. It is activated under conditions such as nutrient starvation and blockage of cell wall synthesis and autolyzes the bacterial cell to release the cytoplasmic content into the surrounding. The cell wall degradation products and the virulent components from cytoplasmic content will exert inflammatory and toxic effect on the host tissue (Mitchell et al., 1997).

The enzyme neuraminidase assists in bacterial colonization and facilitate adherence on the populated surface. This is achieved by the cleavage of sialic acid residues on the epithelial layer. During conjunctivitis, the production of neuraminidase by ocular surface colonization of *S. pneumoniae* will damage mucins, which are made up of sialic acid residues and considered a major component of the mucus layer of cell surface epithelia, which leads to further complications (Williamson et al., 2008).

The enzyme hyaluronidase is produced by wide range of gram-positive organisms including most of the *Streptococcus* spp. This enzyme degrades the hyaluronic acid, a major component of connective tissue. Cornea and sclera, which are connective tissues of eye, are prone to tissue damage by hyaluronidase enzyme. Hence, the damaged cornea will lead to invasion of pathogens and mark in infectious conditions such as keratitis, endophthalmitis, and so on (Mitchell et al., 1997).

Superoxide dismutases (SOD) are metalloenzymes that act as an oxidative stress defense mechanism. There are two types of SOD in *S. pneumoniae*, MnSOD and FeSOD. MnSOD has been reported to play a major role in virulence of pneumococcus (Yesilkaya et al., 2000; Mitchell, 2000).

11.6.2.2 Antibiotic resistance – *S. pneumoniae* and *S. pyogenes* have been reported to show resistant against a wide range of antibiotics including penicillin, cephalosporin, macrolide, and lincosamide. The alteration of penicillin-binding protein (PBP) confers resistance to penicillin. The efficiency of other antibiotics such as beta-lactam, cephalosporin, and carbapenem are also reliant on PBP. Hence, the activity of these antibiotics is reduced in penicillin resistant *Streptococcus* spp. (Baquero et al., 1991; Doern et al., 1996). Most of the clinical isolates are now being found resistance to treatment with single or combination of antibiotics, an indication of multidrug resistance.

Macrolide resistance has been reported as early as in 1990s. The macrolide resistance is conferred by efflux genes (*mef, mel*) and target-modifying methylase genes (*erm*). The macrolide efflux is mediated by the gene product of *mef* (A) in both *S. pneumoniae* and *S. pyogenes*. The most common form of target site modification is mediated by rRNA methylase, di-methylating a specific adenine residue on the 23S rRNA (Hyde et al., 2001; Farrell et al., 2002). In addition to this, secretary enzymes confer resistance through inactivating the drug molecule (Leclercq et al., 2002). The dissemination of macrolide resistance genes through a mobile genetic element has also been reported. In addition to transformation and recombination, conjugative mobile elements transmit the resistance genes among the pneumococcal and other *Streptococcus* spp. (Chancey et al., 2015) (Figure 11.3).

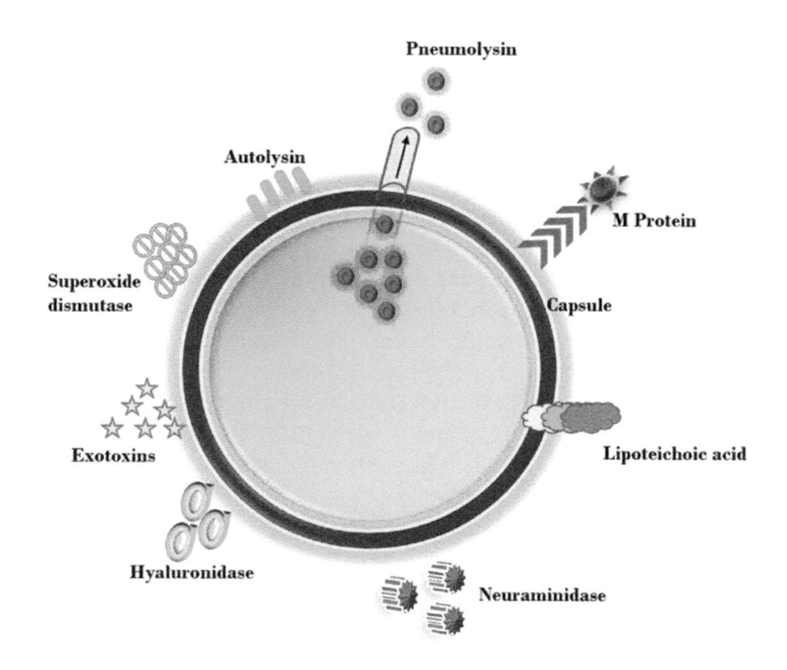

Figure 11.3 Virulence factors of *Streptococcus* spp.

11.6.2.3 Therapeutic interventions – Though antibiotic resistance is widely reported, antibiotics remain the first-line therapy for infections mediated by *Streptococcus* spp. For patients allergic to penicillin, macrolides are the suggested antibiotics. Macrolides and lincosamides are recommended in combinations with beta-lactams for invasive infections (Silva-costa et al., 2015). From various case studies, it is evident that macrolides and lincosamides are the better alternatives to penicillin. In a clinical study, Gregori et al. (2015) used intravitreal injection of antivascular endothelial growth factor (anti-VEGF) for the treatment of infectious endophthalmitis caused by *Streptococcus* and *Staphylococcus* spp.; it lessened the recurrent infection and was found to be a better treatment in preventing vision loss. The treatment of streptococcal infection remains challenging due to their potent virulence factor and antibiotic resistance mechanisms. Hence in recent years, natural sources have been widely explored as an alternate to the existing antibiotics. Viszwapriya et al. (2016; Viszwapriya and Subramenium, 2016) reported betulin, a naturally available triterpenoid and 2,4-Di-tert-butylphenol from seaweed surface-associated bacterium *Bacillus subtilis* as effective anti-infective agents against *S. pyogenes.* Therefore, natural biomolecules alone or in combination with antibiotics will be an effective therapeutic strategy for *Streptococcus* infections.

11.6.3 *Pseudomonas aeruginosa*

P. aeruginosa is a gram-negative, opportunistic bacterial pathogen, which is a major etiological agent of numerous infectious diseases affecting the eye and surrounding tissues. *P. aeruginosa* ocular infections are most frequently observed as from an exogenous origin from an environmental source. However, endogenous infection as a result of metastasis has also been reported. The predisposing factors associated with *P. aeruginosa*–mediated ocular infections includes (1) corneal trauma or ulceration, (2) preexisting ocular infections or diseases, (3) immunosuppressive chemotherapy, (4) immunodeficiency in infants, (5) corneal aberrations caused by contact lens wear, (6) application of contaminated eye area cosmetics, and (7) systemic infections (Wilson and Ahearn, 1977; Reid and Wood, 1979; Kreger, 1983; Wu et al., 2015). In addition to this, implantation of intraocular lens with *Pseudomonas* contamination often leads to endophthalmitis. *P. aeruginosa* can affect almost every part of the eye, namely, diseases affecting the surrounding tissue of the eye, such as dacryocystitis, blepharitis, orbital cellulitis; extraocular infections including conjunctivitis, keratitis, and corneoscleral ulceration; intraocular infection such as endophthalmitis (John et al., 1996; Brito et al., 2003; Delia et al., 2008; Bhattacharjee et al., 2016). *P. aeruginosa* is a common cause of postoperative endophthalmitis.

Corneal damage and invasion of the bacteria to intraocular regions are threatening to sight by development of secondary glaucoma or cataract. Though *P. aeruginosa* is the most recurrently reported *Pseudomonas* species in ophthalmic infections, *P. acidovorans, P. stutzeri*, and *P. fluorescens* also have been less frequently reported as causative organisms of human ocular infections. Dissemination of bacteria from ocular site to bloodstream has not been reported in adults. However, in premature infants, fatal septicemia can develop from the spread of *Pseudomonas* from the eye (Burns and Rhodes, 1961). The presence of various cell-associated and secreted extracellular virulence factors expands the pathogenicity of *P. aeruginosa* in various infections.

11.6.3.1 Virulence factors – Virulence factors of *P. aeruginosa* and their associated mechanisms in ophthalmic infections are detailed next.

Sialic acid specific adhesin based N-acetylneuraminic acid receptor on the cell surface of *P. aeruginosa* mediates the binding to the ocular epithelium which is subsequently followed by colonization and infection (Hazlett et al., 1986). The establishment of adherence to the host surface progresses the infection by damaging the underlying tissue with the extracellular toxic substances and evokes the host immune system. This sequential process will aggravate the tissue damage.

Among all other virulence factors, the ability to invade the host cell is considered the most crucial virulent trait in the pathogenicity of organism. Invasion of *P. aeruginosa* inside the corneal epithelium is reported with two suggestive mechanisms (i.e., transcellular migration and invasion through destruction of corneal epithelium). Fleiszig et al. (1996) reported the invasion and intracellular survival of *P. aeruginosa* within corneal epithelial cells. Because intracellular bacteria can evade the host immune system and antibiotic treatment, *P. aeruginosa* infection can lead to serious consequences like ocular morbidity (Fleiszig et al., 1994). In addition to invasion, where both the bacterial and host cell is viable up to 24 hours, some ocular clinical strains of *P. aeruginosa* are reported as exerting cytotoxic effect on the host cell (Fleiszig et al., 1996).

Exotoxin A (ExoA), the most toxic extracellular product of *P. aeruginosa*, contributes to the corneal damage on infection. It can directly act on epithelial, endothelial, and stromal cells of the cornea and cause necrosis by inhibiting protein synthesis. Exo A is nonproteolytic in nature. However, it inhibits the protein synthesis by transferring adenosine 5'diphosphate ribose (ADPR) of nicotinamide adenine dinucleotide to mammalian elongation factor 2 (Ohman et al., 1980; Lyczak, 2000).

Protease enzymes of *P. aeruginosa* acts as a crucial pathogenic trait in ocular infections. Proteoglycan matrix of the cornea is degraded by this enzyme. *P. aeruginosa* strains usually produce three proteases: protease I, elastase (protease II), and alkaline protease (protease III) (Wretlind and Pavlovskis, 1983). Secretion of elastase (LasB) results in the cleavage of elastin, fibrin, and collagen, which provides mechanical strength to the connective tissue such as cornea. The integrity of the corneal matrix is impaired by the action of elastase. Elastase interferes with the host defense mechanism by cleaving immunological factors such as IgG, IgA, IFNγ, TNF-α, and so on (Willcox, 2007).

The flagella and pili of *Pseudomonas* acts as adhesins in corneal infection mediating cell invasion and cytotoxicity. These extracellular appendages bind specifically to the glycosphingolipidasialo-GM1 on the host cell. This binding event eventually contributes to the corneal damage (Lyczak, 2000).

11.6.3.2 Antibiotic resistance – Extensive treatment of *P. aeruginosa* with antibiotics generates selective pressure that initiates antibiotic resistance. *P. aeruginosa* confers resistance against wide range of antibiotics such as aminoglycosides (i.e., gentamicin, tobramycin, and amikacin) that inhibits the protein synthesis; beta-lactams (i.e., piperacillin and ceftazidime) that inhibit the peptidoglycan layer; and polymyxins (i.e., colomycin and colistin) that bind with the phospholipids of cytoplasmic membrane. Antibiotic resistance in *P. aeruginosa* is imparted through following factors: It is intrinsically resistant to antibiotics because of low permeability of the cell wall, expresses a wide group of resistance mechanisms, attains resistance through mutations in genes that regulate resistant genes, and acquires additional resistance genes from plasmids, transposons, and bacteriophages (Lambert, 2002).

To date, four different efflux systems have been discovered in *P. aeruginosa* mexAB-oprM, mexXY-oprM, mexCD-oprJ, and mexEF-oprN. Extrusion of beta-lactams and quinolones are mediated by mexAB-oprM, aminoglycosides are extruded by mexXY-oprM, and mexEF-oprN extrudes carbapenems and quinolones (Ziha-zarifi et al., 1999).

Inactivation or modification of antibiotics by secretory enzymes and changing the target by spontaneous mutations also contribute to antibiotic resistance. In the current scenario, to combat multidrug resistant *Pseudomonas* infections, effective natural bioactive molecules such as curcumin from *Curcuma longa* will effectively attenuate the virulent traits and lessens the infection without conferring antibiotic resistance (Sethupathy et al., 2016) (Figure 11.4).

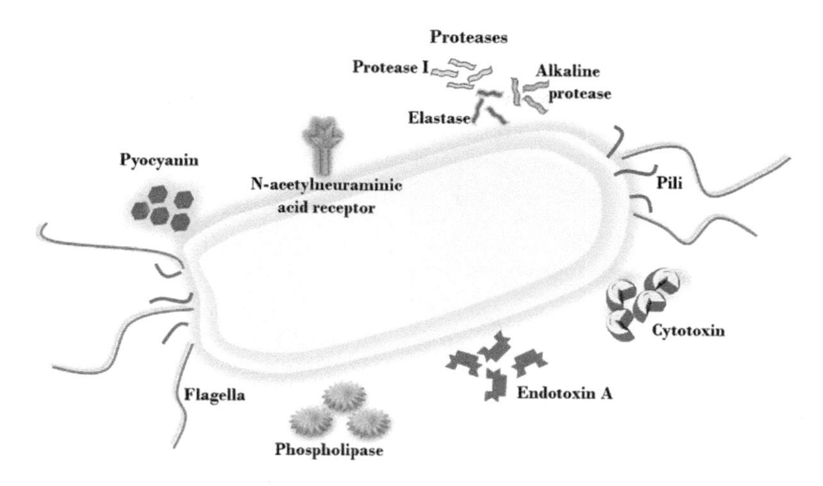

Figure 11.4 Virulence factors of *Pseudomonas aeruginosa*.

11.6.3.3 Therapeutic interventions – Multidrug efflux pump is a major mechanism by which pathogenic organisms exhibit antibiotic resistance. Hence, inhibition of such efflux pumps appears as a promising therapeutic strategy. In search of efflux pump inhibitors, numerous natural sources, synthetic molecules, and existing antibiotics have been screened in recent years. The family of efflux pump that mediates significant antibiotic resistance in *P. aeruginosa* belongs to resistance nodulation division (RND) family. Adamson et al. (2015) demonstrated that the putative efflux pump inhibitors trimethoprim and sertraline alone or in combination with antibiotic, namely levofloxacin, resulted in enhanced therapeutic assistance. Phe-Arg-β-naphthylamide (PAβN) has been reported to be the most effective and broad-spectrum inhibitor of *P. aeruginosa* RND efflux pump. PAβN is now shown to encompass impact on the virulence of *P. aeruginosa.* It is evident that efflux pump inhibitors can have potential anti-virulent property (Rampioni et al., 2017).

Martins et al. (2008) used a unique therapeutic strategy for the treatment of infectious keratitis with the combination of riboflavin and ultraviolet-A (UVA; 365 nm). The UVA-induced riboflavin, efficiently exhibited antimicrobial property against bacterial and fungal isolates of keratitis, including the multidrug resistant *P. aeruginosa*.

Blue light treatment of *P. aeruginosa* exhibited strong antimicrobial activity with inactivation of an array of *Pseudomonas* virulence factors. Blue light treatment in synergy with antibiotics also efficiently decreased the pathogenicity of *P. aeruginosa* (Fila et al., 2017).

Phage therapy is one more potential therapeutic remedy for *P. aeruginosa* infections. Currently, there are 137 completely sequenced *Pseudomonas* phage genomes available in the databases (Pires et al., 2015). The treatment of *Pseudomonas* infections with the combination of phage and other antimicrobial agents as an effectual therapy has also been reported (Torres-Barcelo et al., 2014).

With the advent of these therapeutic approaches, there are high possibilities that infections caused by *Pseudomonas* will decline over time.

11.6.4 Coryneform bacteria

Coryneform bacteria comprises a group of aerobic, asperogenous, gram-positive, rod-shaped bacteria, which includes the following genera: *Corynebacterium, Turicella, Arthrobacter, Brevibacterium, Dermabacter, Propionibacterium, Rothia, Exiguobacterium, Oerskovia, Cellulomonas, Sanguibacter, Microbacterium, Aureobacterium, Arcanobacterium*, and *Actinomyces* (Funke et al., 1997). Though less frequently encountered, *Corynebacterium* and *Propionibacterium* are the most-common coryneform bacteria associated with several systemic and ocular infections. The species *C. macginleyi*, isolated from an eye specimen, was the first lipophilic coryneform bacteria reported (Riegel et al., 1995). *C. macginleyi* resides as an ocular flora but acts as an opportunistic pathogen in conjunctivitis and other ocular infections. It also has been described as a conjunctiva-specific pathogen because it is predominantly isolated from conjunctivitis eyes and infrequently found in sight-threatening infections such as keratitis and endophthalmitis (Joussen et al., 2000; Ly et al., 2006; Suzuki et al., 2007). Ruoff et al. (2010) reported corneal ulceration and scaring caused by *C. macginleyi* as a sole causative agent, suggesting that *C. macginleyi* can cause ocular infection even when it is not associated with other opportunistic or pathogenic organisms. Detection and identification of *C. macginleyi* species is problematic as it is fastidious, requires enriched media, and being sequestered by other organisms. Culture-independent techniques like PCR will enumerate the presence of *C. macginleyi*. In a case report, Ferrer et al. (2004) encountered sterile endophthalmitis followed by cataract surgery. However, the sequence analysis showed *C. macginleyi* as the causative agent of endophthalmitis. *Mycobacterium* keratitis is often misdiagnosed with *Corynebacterium* keratitis (Garg et al., 1998). *C. pseudodipthericum* was found in conjunctivitis in extremely rare instances (Joussen et al., 2000).

P. acnes and other *Propionibacterium* spp. normally inhabit the conjunctiva. Postoperative *Propionibacterium* endophthalmitis has been reported scarcely. *P. acnes* are commonly associated with chronic blepharitis, chalazion, dacryocystitis, endophthalmitis, and keratitis. Predisposing factors for

P. acnes infections include existence of foreign bodies, systemic infections, and diseases such as diabetes, previous ocular surgery, immunodeficiency, and steroid therapy. *P. acnes* infection can appear shortly after surgery or after longer period because they can reside intracellularly and remain dormant for prolonged duration.

11.6.4.1 Virulence factors – Low oxidation-reduction potential within eye provides suitable environment for the growth of anaerobes especially *P. acnes*. Persistent intracellular localization of *P. acnes* and its secretory metabolites and enzymes will result in tissue damage (Csukas et al., 2004). Hemolytic and cytotoxic activity of *P. acnes* has been reported. Various enzymes such as Chondroitin sulfatase, hyaluronidase, gelatinase, phosphatase, lecithinase, and hemolysin production are found in *P. acnes*, which potentiate its virulence (Hoeffler, 1977). The enzyme chondroitin sulfatase initiates hydrolysis of sulfate groups from N-acetyl-D-galactosamine 6-sulfate and keratan sulfate, which are structural carbohydrates of corneal tissue. Hyaluronidase hydrolyses the nonsulfated glycosaminoglycan, the hyaluronic acid of connective tissues. The enzyme lecithinase is a type of phospholipase, which can act on lecithin and cause hemolysis (Figure 11.5).

11.6.4.2 Antibiotic resistance – *Corynebacterium* spp. have developed resistance against antibiotics such as fluoroquinolones, chloramphenicol, cefazolin, vancomycin, sulbenicillin etc., *C. macginleyi* are extremely resistant to fluoroquinolones and macrolides (Joussen et al., 2000; Eguchi et al., 2008),

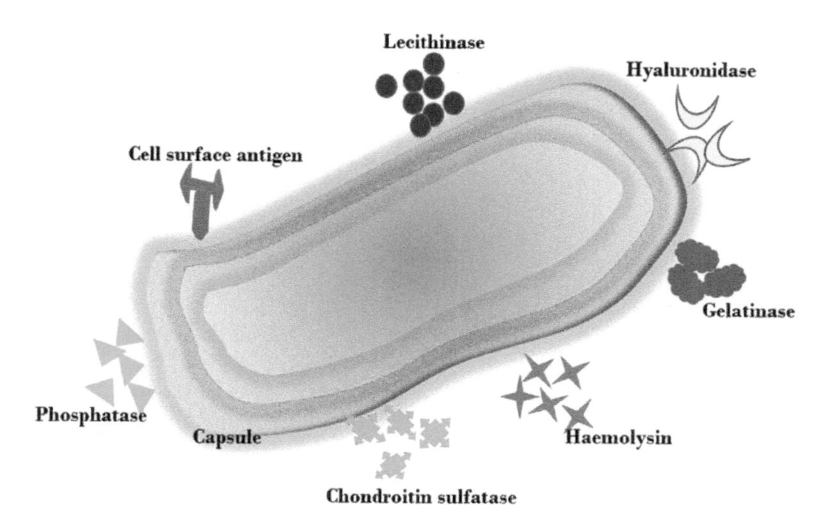

Figure 11.5 Virulence factors of *Coryneform bacteria*.

whereas *P. acnes* are highly resistant to macrolides and tetracyclines (Aubin et al., 2014). *Corynebacterium* spp. attain antibiotic resistance through general resistance mechanisms such as drug efflux, target modification, inactivation, or modification of drug molecule.

11.6.4.3 Therapeutic interventions – As *C. macginleyi* and *P. acnes* are resistant to commonly available topical antibiotics in ophthalmology, use of natural sources offers a potential therapeutic value. Koday et al. (2010) investigated the antimicrobial efficacy of various medicinal plants against *C. macginleyi* and reported the bioactive potential of methanolic extracts of 36 medicinal plants. *Terminalia catappa, Terminalia chebula, Rosa indica, Albizia lebbeck*, and *Butea monosperma* were reported to have significant bactericidal activity.

11.7 Role of bacterial biofilm in ocular infections

Biofilm is a consortium of microorganisms which are structurally enclosed within the exopolymeric conglomerate matrix that generally encompass extracellular DNA, protein, and polysaccharide of both bacterial and host components. Biofilm thus forms a three-dimensional complex architect of microbial community achieved through cell-to-cell communication (Flemming et al., 2016). Biofilm can form on either natural surfaces such as teeth, heart valve, or on abiotic surfaces such as medical implants, catheters, and contact lenses, among others. Cells within biofilm are physiologically heterogenous and that results in decreased susceptibility to antibiotics. Bacteria residing on biofilm matrix are 1000-fold less sensitive to antibiotics than their planktonic counterparts. Biofilm formation in ocular infections is achieved by persistent colonization of bacteria either on abiotic surfaces that are in contact with the ocular tissue or implanted within ocular surface or formation of biofilm on biotic surfaces of the eye. Prosthetic devices of eye range from contact lens to lens implant, glaucoma tubes, punctual plugs, stents, corneal sutures, and scleral buckles (Hou et al., 2012). Among this array of prosthesis, contact lens present the most frequent and severe form of biofilm-related ocular infections. Improper handling of contact lens attributes to contamination. Around 80% of contact lens wearers are reported for bacterial contamination either without presenting signs and symptoms or bacterial keratitis (McLaughlin-Borlace et al., 1998). Extended contact lens wear is a major predisposing factor for keratitis. During surgical cataract removal, bacteria can enter the eye surface and reside on the implanted intraocular lens, leading to endophthalmitis. Biofilm formation of punctal plug, a prosthetic for the treatment of tear-deficient dry eye, has been reported for acute conjunctivitis (Yokoi et al., 2000). In addition to direct effect of biofilm, the secretory components such as endotoxins or lipopolysaccharides from organisms residing in the biofilm matrix has a major role in pathogenicity of the biofilm-associated infections. Endotoxins can exert numerous effect on host tissue by eliciting cytokines, inflammatory components, and complement cascade of host immune system. Diffuse lamellar keratitis is a major complication of corrective surgery laser assisted in situ keratomileusis (LASIK), which is initiated due to the contamination of sterilizer reservoirs with gram-negative pathogens. The use of contaminated instruments recruits several inflammatory mediators and the endotoxin produced by gram-negative organisms from bacterial biofilm will worsen the condition (Holland et al., 2000). Bacteria residing in biofilm matrix of chronic keratitis condition have been implicated in infectious crystalline keratopathy (ICK). Biofilm formation on biotic ocular surface is extremely difficult and infrequent. In certain cases of ICK, biofilm formation in the absence of prosthetic materials has been noted. Mihara et al. (2004) reported a case with movable mass of bacterial biofilm on the ocular surface, which was initially considered calcification. Yellowish white calcification was present on the nasal sclera and corneal region, which moved on blinking. Pathological examination of the calcified mass revealed the formation of biofilm by numerous gram-positive bacilli with neutrophils. Hence the authors suggested that unusual mass on ocular surface without the involvement of biomaterial should not be deliberated merely as calcification. Possibilities of infection, involvement of infectious bacteria, and biofilm formation should be considered before recommending treatment strategies. Biofilm formation by *S. aureus*, *S. epidermidis*, *P. acnes*, and *P. aeruginosa* on ocular surfaces with or without the contribution of biomaterial have been profusely reported (Leid et al., 2002; Hou et al., 2012).

Infections caused by biofilm are massively resistant to antimicrobial agents. Hence, the treatment of biofilm-related infections are extremely difficult to eradicate. Pathogenic bacteria residing on the biofilm matrix employ various tolerance and resistance mechanisms to withstand antibiotics. Production of extracellular DNA, exopolysaccharide matrices, stress response, presence of various biofilm-specific genetic determinants, efflux pumps, intracellular communications such as quorum sensing ,and horizontal gene transfer afford the survival of biofilm on treatment with antimicrobial agents (Hall and Mah, 2017).

Detailed research on biofilm-specific antibiotic resistance and tolerance will help in improvising treatment for biofilm-related infectious diseases. Moreover, due to increased resistance to antimicrobial agents, research on alternates to antibiotics are in surge. Augmentation of antibiotic efficacy with combinations of antibiotics or natural biomolecules, phage therapy,

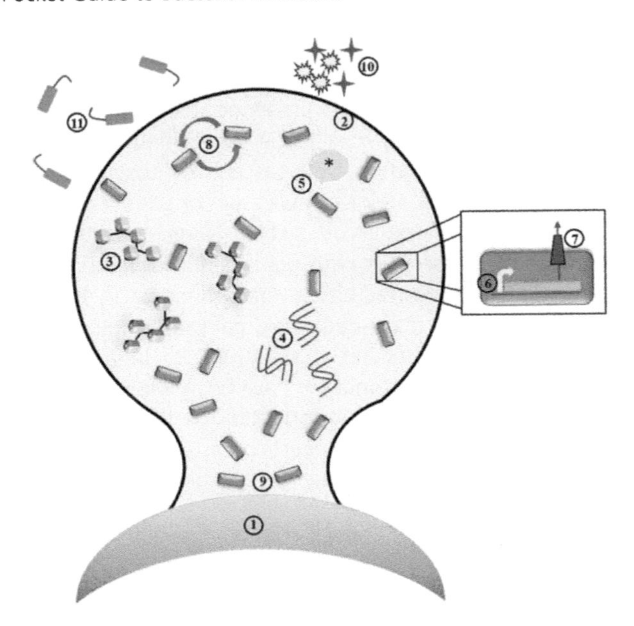

Figure 11.6 Biofilm formation on abiotic surface (i.e., contact lens) and its associated antibiotic resistance and tolerance mechanisms. (1) Abiotic surface (i.e., contact lens); (2) biofilm cells embedded in a matrix; (3) exopolysaccharides; (4) extracellular DNA; (5) stress responses; (6) genetic determinants that are specifically expressed in biofilm cells; (7) multidrug efflux pumps; (8) intracellular interactions (horizontal gene transfer); (9) persister cells; (10) biologically active molecules; and (11) bacteria shed to the environment.

use of bacteriocins, photodynamic therapy, and antibiofilm and antiquorum sensing agents are widely discussed and reported as convincing therapeutic strategies at this time (Allen et al., 2014) (Figure 11.6).

11.8 Conclusion and future perspectives

The bacterial pathogens causing acute or chronic infections on either extraocular or intraocular surfaces increase the incidence of ocular morbidity and are often sight threatening. Proper maintenance of ocular hygiene is the initial precautionary measure for the management of ocular microbial infections. Bacteria of both endogenous and exogenous origin can cause mild to severe forms of ophthalmic infections. Antibiotic treatment and surgery are prevailing therapeutic regimens. Persistent bacterial infections are a result of antibiotic resistance that are caused by long-term topical or systemic antibiotic therapy and the ability of the bacteria to form biofilm on biotic or abiotic surfaces. In addition to this, complications of postoperative infections further deteriorate the condition and fail to provide absolute remedy from ocular infections. Even though there are lot of emerging therapeutic interventions such as phage therapy, photodynamic therapy, bacteriocins, and antibiofilm agents, permutation with phytomolecules will intensify the treatment procedures. Further insight on predisposing factors and appropriate prognosis is imperative and should be explored in detail.

Acknowledgments

The authors thankfully acknowledge the support extended by Department of Science and Technology, Government of India through PURSE [Grant No. SR/S9Z-23/2010/42 (G)] & FIST (Grant No. SR-FST/LSI-087/2008), and University Grants Commission (UGC), New Delhi, through SAP-DRS1 [Grant No. F.3-28/2011 (SAP-II)]. The authors also thankfully acknowledge the Bioinformatics Infrastructure Facility funded by Department of Biotechnology, Government of India [Grant No. BT/BI/25/015/2012 (BIF)].

References

Adamson, D. H., V. Krikstopaityte and P. J. E. 2015. Cootenhanced efficacy of putative efflux pump inhibitor/antibiotic combination treatments versus MDR strains of *Pseudomonas aeruginosa* in a *Galleria mellonella* in vivo infection model. *Journal of Antimicrobial Chemotherapy* 70:2271–2278.

Alfonso, E. C. and D. Miller. 1990. Detection of ocular infections. In *Clinical Applications of the Limulus Amoebocyte Lysate Test*. R. B. Prior, editor. Columbus, OH: The Ohio State University.

Ali, M. J., S. D. Joshi, M. N. Naik and S. G. Honavar 2015. Honavarlinical profile & management outcome of acute dacryocystitis: Two decades of experience in a tertiary eye care center. In *Seminars in Ophthalmology* 30:118–123.

Ali, M. J., S. R. Motukupally, S. D. Joshi and M. N. T Naikhe. 2013. Microbiological profile of lacrimal abscess: Two decades of experience from a tertiary eye care center. *Journal of Ophthalmic Inflammation and Infection* 3:57–61.

Allen, H. K., J. Trachsel, T. Looft and T. A Casey. 2014. Finding alternatives to antibiotics. *Annals of the New York Academy of Sciences* 1323:91–100.

Ambati, B. K., J. Ambati, N. Azar, L. Stratton and E. V. Schmidt. 2000. Periorbital & orbital cellulitis before and after the advent of haemophilus influenzae type B vaccination1. *Ophthalmology* 107:1450–1453.

Andrew, S. E., A. Nguyen, G. L. Jones and D. E. Brooks. 2003. Seasonal effects on the aerobic bacterial & fungal conjunctival flora of normal thoroughbred brood mares in Florida. *Veterinary Ophthalmology* 6:45–50.

Aubin, G. G., M. E. A. Portillo, A. Trampuz and S. Corvec. 2014. *Propionibacterium acnes*, an emerging pathogen: From acne to implant-infections, from phylotype to resistance. *Médecine et Maladies Infectieuses* 44:241–250.

Axenfeld, T. 1908. *Bacteriology of the Eye*. In Macnab, editor. West Chester, PA: William Wood & Company.

Babar, T. F., Z. Masud, N. Saeed and M. D. Khan. 2011. An analysis of patients with chronic dacryocystitis. *Journal of Postgraduate Medical Institute* 18:424–431.

Baquero, F., J. M. Beltren and E. Loza. 1991. A review of antibiotic resistance patterns of *Streptococcus pneumoniae* in Europe. *Journal of Antimicrobial Chemotherapy* 28:31–38.

Bartlett, J. D. and S. D. Jaanus. 2008. Clinical ocular pharmacology. In *Clinical and Experimental Optometry*. A. C. Nathan Efron editor. Melbourne, Australia: University of Melbourne.

Berube, B. J. and J. B. Wardenburg. 2013. *Staphylococcus aureus* α-toxin: Nearly a century of intrigue. *Toxins* 5:1140–1166.

Bhattacharjee, H., K. Bhattacharjee, K. Gogoi, M. Singh, B. G. Singla and A. Yadav. 2016. Microbial profile of the vitreous aspirates in culture proven exogenous endophthalmitis: A 10-year retrospective study. *Indian Journal of Medical Microbiology* 34:153–158.

Biziorek, B., J. Mackiewicz, Z. Zagorski, L. Krwawicz and D. Haszcz. 2001. Etiology of uveitis in rural & urban areas of mid-eastern Pol. *Annals of Agricultural and Environmental Medicine* 8:241–244.

Bodaghi, B. et al. 2001. Chronic severe uveitis: Etiology & visual outcome in 927 patients from a single center. *Medicine* 80:263–270.

Bourcier, T., F. Thomas, V. Borderie, C. Chaumeil and Laroche, L. 300. Bacterial keratitis: Predisposing factors, clinical & microbiological review of 300 cases. *British Journal of Ophthalmology* 87:834–838.

Bragg, K. J. and J. K. Le. 2017. Hordeolum. *In StatPearls* [Internet]. Treasure Island (FL): StatPearls.

Brito, D. V. D., E. J. Oliveira, C. Matos, V. O. S. Abdallah and P. P. Gontijo Filho. 2003. An outbreak of conjunctivitis caused by multiresistant *Pseudomonas aeruginosa* in a Brazilian newborn intensive care unit. *Brazilian Journal of Infectious Diseases* 7:34–35.

Bron, A. J. and J. M. Tiffany. 2004. The contribution of meibomian disease to dry eye. *The Ocular Surface* 2:149–164.

Brook, I. and E. H. Frazier. 1998. Aerobic and anaerobic microbiology of dacryocystitis. *American Journal of Ophthalmology* 125:552–554.

Burns, R. P. and D. H. Rhodes. 1961. *Pseudomonas* eye infection as a cause of death in premature infants. *Archives of Ophthalmology* 65:517–525.

Buznach, N., R. Dagan and Greenberg, D. 2005. Clinical and bacterial characteristics of acute bacterial conjunctivitis in children in the antibiotic resistance era. *Pediatric Infectious Disease* 24:823–828.

Caballero, A. R. et al. 2015. Effectiveness of alpha-toxin Fab monoclonal antibody therapy in limiting the pathology of Staphylococcus aureus keratitis. *Ocular Immunology and Inflammation* 23:297–303.

Cahill, K. V. and J. A. Burns. 1993. Management of acute dacryocystitis in adults. *Ophthalmic Plastic & Reconstructive Surgery* 9:38–41.

Callegan, M. C., M. C. Booth, B. D. Jett and M. S. Gilmore. 1999. Pathogenesis of Gram-positive bacterial endophthalmitis. *Infection and Immunity* 67:3348–3356.

Callegan, M. C., L. S. Engel, J. M. Hill and R. J. O'Callaghan. 1994. Corneal virulence of *Staphylococcus aureus*: Roles of alpha-toxin & protein A in pathogenesis. *Infection and Immunity* 62:2478–2482.

Capriotti, J. A., J. S. Pelletier, M. Shah, D. M. Caivano and D. C. Ritterband. 2009. Normal ocular flora in healthy eyes from a rural population in Sierra Leone. *International Journal of Ophthalmology* 29:81–84.

Carlisle, R. T. and G. T. Fredrick. 2006. Preseptal & orbital cellulitis. *Hospital Physician* 42:15–19.

Cavuoto, K., D. Zutshi, C. L. Karp, D. 2008. Miller and W. Feuer. Update on bacterial conjunctivitis in South Florida. *Ophthalmology* 115:51–56.

Chancey, S. T., S. Agrawal, M. R. Schroeder, M. M. Farley, H. Tettelin and D. S. Stephens. 2015. Composite mobile genetic elements disseminating macrolide resistance in *Streptococcus pneumoniae. Frontiers in Microbiology* 6:26–40.

Chuang, C. C. et al. 2012. *Staphylococcus aureus* ocular infection: Methicillin-resistance, clinical features & antibiotic susceptibilities. *PLoS One* 7:e42437.

Csukas, Z., B. Banizs and F. Rozgonyi. 2004. Studies on the cytotoxic effects of *Propionibacterium acnes* strains isolated from cornea. *Microbial Pathogenesis* 36:171–174.

Curragh, D. S., M. Rooney, M. O'Neill, C. E. McAvoy and E. McLoone. 2017. Paediatric uveitis in a Northern Ireland population–an overview. *Rheumatology 56*:1–8.

Dajcs, J. J., B. A. Thibodeaux, E. B. Hume, X. Zheng, G. D. Sloop and R. J. O'Callaghan. 2001. Lysostaphin is effective in treating methicillin-resistant *Staphylococcus aureus* endophthalmitis in the rabbit. *Current Eye Research* 22:451–457.

Dart, J. K. G., C. F. Radford, D. Minassian, S. Verma and F. Stapleton. 2008. Risk factors for microbial keratitis with contemporary contact lenses: A case-control study. *Ophthalmology* 115:1647–1654.

Delia, A. C., G. C. Uuri, K. Battacharjee, D. Das and U. Gogoi. 2008. Bacteriology of chronic dacryocystitis in adult population of northeast India. *Orbit* 27:243–247.

Doern, G. V., A. Brueggemann, H. P. Holley and A. M. Rauch. 1996. Antimicrobial resistance of *Streptococcus pneumoniae* recovered from outpatients in the United States during the winter months of 1994 to 1995: Results of a 30-center national surveillance study. *Antimicrobial Agents and Chemotherapy* 40:1208–1213.

Donahue, S. P. and G. Schwartz. 1998. Preseptal & orbital cellulitis in childhood: A changing microbiologic spectrum. *Ophthalmology* 105:1902–1906.

Dong, Q., M. Jennifer et al. 2011. Diversity of bacteria at healthy human conjunctiva. *Investigative Ophthalmology & Visual Science* 52:5408–5413.

Dougherty, J. M. and J. P. McCulley. 1984. Comparative bacteriology of chronic blepharitis. *British Journal of Ophthalmology* 68:524–528.

Drancourt, M. et al. 2008. High prevalence of fastidious bacteria in 1520 cases of uveitis of unknown etiology. *Medicine* 87:167–176.

Durand, M. L. 2017. Bacterial and fungal endophthalmitis. *Clinical Microbiology Reviews* 30:597–613.

Eguchi, H. et al. 2008. High-level fluoroquinolone resistance in ophthalmic clinical isolates belonging to the species *Corynebacterium macginleyi*. *Journal of Clinical Microbiology* 46:527–532.

Farrell, D. J., I. Morrissey, S. Bakker and D. Felmingham. 2002. Molecular characterization of macrolide resistance mechanisms among *Streptococcus pneumoniae* and *Streptococcus pyogenes* isolated from the PROTEKT 1999–2000 study. *Journal of Antimicrobial Chemotherapy* 50:39–47.

Favetta, J. R. 2015. Blepharitis management: A clinical approach. *Ophthalmology Times* 40:1–4.

Ferrer, C., J. M. Ruiz-Moreno, A. Rodriguez, J. Montero and J. L. Alio. 2004. Postoperative *Corynebacterium macginleyi* endophthalmitis. *Journal of Cataract & Refractive Surgery* 30:2441–2444.

Fila, G., A. Kawiak and M. S. Grinholc. 2017. Blue light treatment of *Pseudomonas aeruginosa*: Strong bactericidal activity, synergism with antibiotics and inactivation of virulence factors. *Virulence* 8:938–958.

Fleiszig, S. M. and N. Efron. 1992 Microbial flora in eyes of current & former contact lens wearers. *Journal of Clinical Microbiology* 30:1156–1161.

Fleiszig, S. M., T. S. Zaidi, M. J. Preston, M. Grout, D. J. Evans and G. B. Pier. 1996. Relationship between cytotoxicity and corneal epithelial cell invasion by clinical isolates of *Pseudomonas aeruginosa. Infection and Immunity* 64:2288–2294.

Fleiszig, S. M. J., T. S. Zaidi, E. L. Fletcher, M. J. Preston and G. B. Pier. 1994. *Pseudomonas aeruginosa* invades corneal epithelial cells during experimental infection. *Infect Immun* 62:3485–3493.

Flemming, H. C., J. Wingender, U. Szewzyk, P. Steinberg, S. A. Rice and S. Kjelleberg. 2016. Biofilms: An emergent form of bacterial life. *Nature Reviews Microbiology* 14:563–568.

Friedlaender, M. H. 1995. A review of the causes & treatment of bacterial & allergic conjunctivitis. *Clinical Therapeutics* 17:800–810.

Funke, G., A. Von Graevenitz, J. Clarridge and K. A. Bernard. 1997. Clinical microbiology of coryneform bacteria. *Clinical Microbiology Reviews* 10:125–159.

Garg, P., S. Athmanathan and G. N. Rao. 1998. *Mycobacterium chelonei* masquerading as *Corynebacterium* in a case of infectious keratitis: A diagnostic dilemma. *Cornea* 17:230–232.

Georgakopoulos, C. D., M. I. Eliopoulou, S. Stasinos, A. Exarchou, N. Pharmakakis and A. Varvarigou. 2010. Periorbital & orbital cellulitis: A 10-year review of hospitalized children. *European Journal of Ophthalmology* 20:1066–1072.

Green, M., A. Apel and F. Stapleton. 2008. Risk factors & causative organisms in microbial keratitis. *Cornea* 27:22–27.

Greenwald, M. J., L. G. Wohl and C. H. Sell. 1986. Metastatic bacterial endophthalmitis: A contemporary reappraisal. *Survey of Ophthalmology* 31:81–101.

Gregori, N. Z. et al. 2015. Current infectious endophthalmitis rates after intravitreal injections of anti-vascular endothelial growth factor agents and outcomes of treatment. *Ophthalmic Surgery, Lasers and Imaging* 46:643–648.

Groden, L. R., B. Murphy, J. Rodnite and G. I. Genvert. Lid flora in blepharitis. *Cornea* 10:50–53.

Hall, C. W. and T. F. Mah. 2017. Molecular mechanisms of biofilm-based antibiotic resistance and tolerance in pathogenic bacteria. *FEMS Microbiol Rev* 41:276–301.

Haque, R. M. et al. 2010. Multicenter open-label study evaluating the efficacy of azithromycin ophthalmic solution 1% on the signs & symptoms of subjects with blepharitis. *Cornea* 29:871–877.

Hartikainen, J., O. P. Lehtonen and K. M. Saari. 1997. Bacteriology of lacrimal duct obstruction in adults. *British Journal of Ophthalmology* 81:37–40.

Hauser, A. and S. Fogarasi. 2010. Periorbital & orbital cellulitis. *Pediatrics in Review* 31:242–249.

Hazlett, L. D., M. Moon and R. S. Berk. 1986. In vivo identification of sialic acid as the ocular receptor for *Pseudomonas aeruginosa. Infect Immun* 51:687–689.

Hoeffler, U. 1997. Enzymatic and hemolytic properties of *Propionibacterium acnes* and related bacteria. *Journal of Clinical Microbiology* 6:555–558.

Hogan, M. J., S. J. Kimura and P. Thygeson. 1959. Signs and Symptoms of Uveitis: I. Anterior Uveitis. *American Journal of Ophthalmology* 47:155–170.

Holland, S. P., R. G. Mathias, D. W. Morck, J. Chiu and S. G. Slade. 2000. Diffuse lamellar keratitis related to endotoxins released from sterilizer reservoir biofilms. *Ophthalmology* 107:1227–1233.

Hou, W., X. Sun, Z. Wang and Zhang, Y. 2012. Biofilm-forming capacity of *Staphylococcus epidermidis*, *Staphylococcus aureus*, and *Pseudomonas aeruginosa* from ocular infections. *Investigative Ophthalmology & Visual Science* 53:5624–5631.

Hovding, G. 2008. Acute bacterial conjunctivitis. *Acta Ophthalmol* 86:5–17.

Hovding, G. 1981. The conjunctival & contact lens bacterial flora during lens wear. *Acta Ophthalmol* 59:387–301.

Hsiao, C. H. et al. 2012. Methicillin-resistant *Staphylococcus aureus* ocular infection: A 10-year hospital-based study. *Ophthalmology* 119:522–527.

Huang, Y., B. Yang and W. Li. 2016. Defining the normal core microbiome of conjunctival microbial communities. *Clinical Microbiology and Infection* 22:643, e7–e12.

Hyde, T. B. et al. 2001. Macrolide resistance among invasive *Streptococcus pneumoniae* isolates. *JAMA* 286:1857–1862.

Iliff, N. T. 1996. Infections of the lacrimal drainage system. *Ocular Immunology and Inflammation* 2:346–355.

Iskeleli, G., H. Bahar, E. Eroglu, M. M. Torun and Ozkan, S. 2005. Microbial changes in conjunctival flora with 30-day continuous-wear silicone hydrogel contact lenses. *Eye Contact Lens* 31:124–126.

Jabbarvand, M., H. Hashemian, M. Khodaparast, M. Jouhari, A. Tabatabaei and S. Rezaei. 2016. Endophthalmitis occurring after cataract surgery: Outcomes of more than 480 000 cataract surgeries, epidemiologic features, & risk factors. *Ophthalmology* 123:295–301.

Jackson, T. L., T. Paraskevopoulos and I. Georgalas. 2014. Systematic review of 342 cases of endogenous bacterial endophthalmitis. *Survey of Ophthalmology* 59:627–635.

Jackson, W. B. 2008. Blepharitis: Current strategies for diagnosis & management. *Canadian Journal of Ophthalmology* 43:170–179.

Jett, B. D. and M. S. Gilmore. 2002. Internalization of *Staphylococcus aureus* by human corneal epithelial cells: Role of bacterial fibronectin-binding protein & host cell factors. *Infection and Immunity* 70:4697–700.

John, M. A., T. W. Austin and A. M. Bombassaro. 1996. *Pseudomonas aeruginosa* blepharitis in a patient with vancomycin induced neutropenia. *Canadian Journal of Infectious Diseases and Medical Microbiology* 7:63–65.

Joussen, A. M., G. Funke, F. Joussen and G. Herbertz. 2000. *Corynebacterium macginleyi*: A conjunctiva specific pathogen. *British Journal of Ophthalmology* 84:1420–1422.

Keay, L. et al. 2006. Microbial keratitis: Predisposing factors & morbidity. *Ophthalmology* 113:109–116.

Koday, N. K., G. S. Rangaiah, V. Bobbarala and S. Cherukuri. 2010. Bactericidal activities of different medicinal plants extracts against ocular pathogen viz *Corynebacterium macginleyi. Drug Invent Today* 2:5–8.

Kotlus, B. S., R. A. Wymbs, E. M. Vellozzi and I. J. Udell. 2006. In vitro activity of fluoroquinolones, vancomycin, & gentamicin against methicillin-resistant *Staphylococcus aureus* ocular isolates. *American Journal of Ophthalmology* 142:726–729.

Kreger, A. S. 1983. Pathogenesis of *Pseudomonas aeruginosa* ocular diseases. *The Journal of Infectious Diseases* 5:S931–S935.

Kumar, A. and A. Kumar. 2015. Role of *Staphylococcus aureus* virulence factors in inducing inflammation and vascular permeability in a mouse model of bacterial endophthalmitis. *PLoS One* 10:e0128423.

Lambert, P. A. 2002. Mechanisms of antibiotic resistance in *Pseudomonas aeruginosa. Journal of the Royal Society of Medicine* 95:22–26.

Larkin, D. F. P. and J. P. Leeming. 1991. Quantitative alterations of the commensal eye bacteria in contact lens wear. *Eye* 5:70–74.

Lebensohn, J. E. 1950. Treatment of hordeola. *Postgraduate Medical Journal* 7:133–133.

Leclercq, R. 2002. Mechanisms of resistance to macrolides and lincosamides: Nature of the resistance elements and their clinical implications. *Clinical Infectious Diseases* 34:482–492.

Lee, S. and M. T. Yen. 2011. Management of preseptal & orbital cellulitis. *Saudi Journal of Ophthalmology* 25:21–29.

Leid, J. G., J. W. Costerton, M. E. Shirtliff, M. S. Gilmore and M. Engelbert. 2002. Immunology of *Staphylococcal biofilm* infections in the eye: New tools to study biofilm endophthalmitis. *DNA and Cell Biology* 21:405–413.

Lemp, M. A. and K. K. Nichols. 2009. Blepharitis in the United States 2009: A survey-based perspective on prevalence & treatment. *The Ocular Surface* 7:S1–S14.

Li, Q., A. Kumar, J. F. Gui and X. Y. Fu-Shin. 2008. *Staphylococcus aureus* lipoproteins trigger human corneal epithelial innate response through toll-like receptor-2. *Microbial Pathogenesis* 44:426–434.

Lin, Y. H. et al. 2017. Antibiotic susceptibility profiles of ocular & nasal flora in patients undergoing cataract surgery in Taiwan: An observational & cross-sectional study. *BMJ Open* 7:e017352.

Lindsley, K., J. J. Nichols and K. Dickersin. 2013. Interventions for acute internal hordeolum. *Cochrane Database of Systematic Reviews 9:*1–12.

Lindsley, K., J. J. Nichols and K. Dickersin. 2017. Non-surgical interventions for acute internal hordeolum. *The Cochrane Library* 1–14.

Long, C., B. Liu, C. Xu, Y. Jing, Z. Yuan and X. Lin. 2014. Causative organisms of post-traumatic endophthalmitis: A 20-year retrospective study. *BMC Ophthalmology* 14:34–40.

Lu, L. J. and J. Liu. 2016. Focus: Microbiome: Human Microbiota & Ophthalmic Disease. *Yale Journal of Biology and Medicine* 89:325–330.

Ly, C. N., J. N. Pham, P. R. Badenoch, S. M. Bell, G. Hawkins, D. L. Rafferty and K. A. McClellan. 2006. Bacteria commonly isolated from keratitis specimens retain antibiotic susceptibility to fluoroquinolones and gentamicin plus cephalothin. *Clinical & Experimental Ophthalmology* 34:44–50.

Lyczak, J. B., C. L. Cannon and G. B. Pier 2000. Establishment of Pseudomonas aeruginosa infection: Lessons from a versatile opportunist1. *Microbes Infect* 2:1051–1060.

Martins, S. A. R. et al. 2008. Antimicrobial efficacy of riboflavin/UVA combination (365 nm) in vitro for bacterial and fungal isolates: A potential new treatment for infectious keratitis. *Investigative Ophthalmology & Visual Science* 49:3402–3408.

Mawn, L. A., D. R. Jordan and S. P. Donahue. 2000. Preseptal & orbital cellulitis. *Ophthalmology Clinics* 13:633–641.

McCulley, J. P. and W. E. Shine. 2000. Changing concepts in the diagnosis & management of blepharitis. *Cornea* 19:650–658.

McCulley, J. P., J. M. Dougherty and D. G. Deneau. 1981. Chronic blepharitis: Classification & mechanisms. *Immunology Abstract* 24:55–72.

McGilligan, V. E., M. S. Gregory-Kser, D. Li, J. E. Moore, R. R. Hodges, M. S. Gilmore, T. C. Moore and D. A. Dartt. 2013. *Staphylococcus aureus* activates the NLRP3 inflammasome in human & rat conjunctival goblet cells. *PLoS One* 8, e74010.

McLaughlin-Borlace, L., F. Stapleton, M. Matheson, and J. K. G. Dart. 1998. Bacterial biofilm on contact lenses and lens storage cases in wearers with microbial keratitis. *Journal of Applied Microbiology* 84:827–838.

Mihara, E., M. Shimizu, C. Touge and Y. Inoue. 2004. Case of a large, movable bacterial concretion with biofilm formation on the ocular surface. *Cornea* 23:513–515.

Miller, D. and A. Iovieno. 2009. The role of microbial flora on the ocular surface. *Current Opinion in Allergy and Clinical Immunology* 9:466–470.

Mitchell, T. J. 2000. Virulence factors and the pathogenesis of disease caused by *Streptococcus pneumoniae*. *Research in Microbiology* 151:413–419.

Mitchell, T. J., J. E. Alexander, P. J. Morgan and P. W. Andrew. 1997. Molecular analysis of virulence factors of *Streptococcus pneumoniae*. *Journal of Applied Microbiology* 83:62S–71S.

Morrow, G. L. and R. L. Abbott. 1998. Conjunctivitis. *American Family Physician* 57:735–746.

Mshangila, B., M. Paddy, H. Kajumbula, C. Ateenyi-Agaba, B. Kahwa and J. Seni. 2013. External ocular surface bacterial isolates & their antimicrobial susceptibility patterns among pre-operative cataract patients at Mulago National Hospital in Kampala, Uganda. *BMC Ophthalmology* 13:71–76.

Mueller, J. B. and C. M. McStay. 2008. Ocular infection & inflammation. *Emergency Medicine Clinics* 26:57–72.

Mustafa, T. A. and I. H. Oriafage. 2001. Three methods of treatment of chalazia in children. *Saudi Medical Journal* 22:968–972.

Muthiah, S. and N. Radhakrishnan. 2017. Management of extraocular infections. *The Indian Journal of Pediatrics* 84:945–952.

Nageswaran, S., C. R. Woods, D. K. Benjamin Jr, L. B. Givner and A. K. Shetty. 2006. Orbital cellulitis in children. *The Pediatric Infectious Disease Journal* 25:695–699.

Nemet, A. Y., S. Vinker and I. Kaiserman. 2011. Associated morbidity of chalazia. *Cornea* 30:1376–1381.

Nichols, K. K. et al. 2011. The international workshop on meibomian gland dysfunction: Executive summary. *Investigative Ophthalmology & Visual Science* 52:1922–1929.

Nimir, A. R., A. Saliem and I. A. A. Ibrahim. 2012. Ophthalmic parasitosis: A review article. *Interdisciplinary Perspectives on Infectious Diseases* 2012:1–12.

O'Callaghan, R. J. et al. 1997. Specific roles of alpha-toxin & beta-toxin during *Staphylococcus aureus* corneal infection. *Infection and Immunity* 65:1571–1578.

Ohman, D. E., R. P. Burns and B. H. Iglewski. 1980. Corneal infections in mice with toxin A and elastase mutants of *Pseudomonas aeruginosa*. *The Journal of Infectious Diseases* 142:547–555.

Otulana, T. O., O. T. Bodunde and H. A. Ajibode. 2008. Chalazion, a benign eyelid tumour–the sagamu experience. *Nigerian Journal of Ophthalmology* 16:33–35.

Ozdal, P. C., F. Codere, S. Callejo, A. L. Caissie and M. N. Burnier. 2004. Accuracy of the clinical diagnosis of chalazion. *Eye* 18:135–138.

Pantosti, A., A. Sanchini and M. Monaco. 2007. Mechanisms of antibiotic resistance in *Staphylococcus aureus. Future Microbiology* 2:323–334.

Park, Y. M. and J. S. Lee. 2014. The effects of chalazion excision on corneal surface aberrations. *Contact Lens Anterior Eye* 37:342–345.

Patel, P. B., M. C. G. Diaz, J. E. Bennett and M. W. Attia. 2007. Clinical features of bacterial conjunctivitis in children. *Academic Emergency Medicine* 14:1–5.

Paton, J. C. 1996. The contribution of pneumolysin to the pathogenicity of *Streptococcus pneumoniae. Trends Microbiology* 4:103–106.

Pinar-Sueiro, S., M. Sota, T. X. Lerchundi, A. Gibelalde, B. Berasategui, B. Vilar and J. L. Hernez. 2012. Dacryocystitis: Systematic approach to diagnosis & therapy. *Current Infectious Disease Reports* 14:137–146.

Pires, D. P., D. V. Boas, S. Sillankorva and J. Azeredo. 2015. Phage therapy: A step forward in the treatment of *Pseudomonas aeruginosa* infections. *Journal of Virology* 89:7449–7456.

Rampioni et, G. et al. 2017. Effect of efflux pump inhibition on *Pseudomonas aeruginosa* transcriptome and virulence. *Scientific Reports* 7:11392–11399.

Rashid, Z., K. Farzana, A. Sattar and G. Murtaza. 2012. Prevalence of nasal *Staphylococcus aureus* & methicillin-resistant *Staphylococcus aureus* in hospital personnel & associated risk factors. *Acta Poloniae Pharmaceutica* 69:985–991.

Rayner, C. and W. J. Munckhof. 2005. Antibiotics currently used in the treatment of infections caused by *Staphylococcus aureus. Journal of Internal Medicine* 35:1–5.

Reid, F. R. and T. O. Wood. 1979. *Pseudomonas* corneal ulcer: The causative role of contaminated eye cosmetics. *Archives of Ophthalmology* 97:1640–1641.

Rhem, M. N. et al. 2000. The collagen-binding adhesin is a virulence factor in *Staphylococcus aureus* keratitis. *Infection and Immunity* 68:3776–3779.

Riegel, P., R. Ruimy, D. De Briel, G. Prevost, F. Jehl, R. Christen and H. Monteil. 1995. Genomic diversity and phylogenetic relationships among lipid-requiring diphtheroids from humans and characterization of *Corynebacterium macginleyi* sp. *International Journal of Systematic and Evolutionary Microbiology* 45:128–133.

Roach, D. R. and D. M. Donovan. 2015. Antimicrobial bacteriophage-derived proteins & therapeutic applications. *Bacteriophage* 5:e1062590.

Rosenbaum, J. T., H. L. Rosenzweig, J. R. Smith, T. M. Martin and S. R. Planck. 2008. Uveitis secondary to bacterial products. *Ophthalmic Research* 40:165–168.

Rubins, J. B. and E. N. Janoff. 1998. Pneumolysin: A multifunctional pneumococcal virulence factor. *Translational Research* 131:21–27.

Rumelt, S. 2016. Overview of common & less common ocular infections. In *Advances in Common Eye Infections*. Rumelt editor. Israel: Bar Ilan University.

Ruoff, K. L., C. M. Toutain-Kidd, M. Srinivasan, P. Lalitha, N. R. Acharya, M. E. Zegans and J. D. Schwartzman. 2010. *Corynebacterium macginleyi* isolated from a corneal ulcer. *Infectious Disease Reports* 2:1568–1572.

Sahin, A., N. Yildirim, S. Gultekin, Y. Akgun, A. Kiremitci, M. Schicht and F. Paulsen, F. 2017. Changes in the conjunctival bacterial flora of patients hospitalized in an intensivecare unit. *Arq Bras Oftalmol* 80:21–24.

Sarkar, I., S. K. Choudhury, M. Bandyopadhyay, K. Sarkar, and J. Biswas. 2015. Clinicobacteriological profile of chronic dacryocystitis in rural india. *Surgery* 5:82–87.

Schaefer, F., O. Bruttin, L. Zografos, and Y. Guex-Crosier. 2001. Bacterial keratitis: A prospective clinical & microbiological study. *British Journal of Ophthalmology* 85:842–847.

Schmelcher, M., D. M. Donovan and M. J. Loessner. 2012. Bacteriophage endolysins as novel antimicrobials. *Future Microbiology* 7:1147–1171.

Schwartz, G. R. and S. W. Wright. 1996. Changing bacteriology of periorbital cellulitis. *Annals of Emergency Medicine* 280:617–620.

Sethupathy, S., K. G. Prasath, S. Ananthi, S. Mahalingam, S. Y. Balan and S. K. Pandian. 2016. Proteomic analysis reveals modulation of iron homeostasis and oxidative stress response in *Pseudomonas aeruginosa* PAO1 by curcumin inhibiting quorum sensing regulated virulence factors and biofilm production. *Journal of Proteome Research* 145:112–126.

Sherwal, B. L. and A. K. Verma. 2008. Epidemiology of ocular infection due to bacteria & fungus–a prospective study. *JK science* 10 (3):127–131.

Shin, H., K. Price, L. Albert, J. Dodick, L. Park and M. G. Dominguez-Bello. 2016. Changes in the eye microbiota associated with contact lens wearing. *M Bio* 7:e00198–e00216.

Shine, W. E., R. Silvany and J. P. McCulley. 1993. Relation of cholesterol-stimulated *Staphylococcus aureus* growth to chronic blepharitis. *Investigative Ophthalmology & Visual Science* 34:2291–2296.

Silva-Costa, C., A. Friaes, M. Ramirez and J. Melo-Cristino. 2015. Macrolide-resistant *Streptococcus pyogenes*: Prevalence and treatment strategies. *Expert Review of Anti-infective Therapy* 13:615–628.

Squissato, V., Y. H. Yucel, S. E. Richardson, A. Alkhotani, D. T. Wong, N. Nijhawan and C. C. Chan. 2015. *Colletotrichum truncatum* species complex: Treatment considerations & review of the literature for an unusual pathogen causing fungal keratitis & endophthalmitis. *Medical Mycology Case Reports* 9:1–6.

Sueke, H. et al. 2013. lukSF-PV in *Staphylococcus aureus* keratitis isolates & association with clinical outcome. *Investigative Ophthalmology & Visual Science* 54:3410–3416.

Suzuki, T. et al. 2007. Suture-related keratitis caused by *Corynebacterium macginleyi*. *Journal of Clinical Microbiology* 45:3833–3836.

Terrada, C., B. Bodaghi, J. Conrath, D. Raoult and M. Drancourt. 2009. Uveitis: An emerging clinical form of Bartonella infection. *Clinical Microbiology and Infection* 15:132–133.

Thielen, T. L., S. S. Castle and J. E. Terry. 2000. Anterior ocular infections: An overview of pathophysiology & treatment. *Annals of Pharmacotherapy* 34:235–246.

Torres-Barcelo, C., F. I. Arias-Sanchez, M. Vasse, J. Ramsayer, O. Kaltz and M. E. Hochberg. 2014. A window of opportunity to control the bacterial pathogen *Pseudomonas aeruginosa* combining antibiotics and phages. *PLoS One* 9:e106628.

Viszwapriya, D., U. Prithika, S. Deebika, K. Balamurugan and S. K. Pandian. S. K. 2016. *In vitro* and *in vivo* antibiofilm potential of 2, 4-Di-tert-butylphenol from seaweed surface associated bacterium *Bacillus subtilis* against Group A Streptococcus. *Microbiological Research* 191:19–31.

Viszwapriya, D., G. A. Subramenium, U. Prithika, K. Balamurugan, and S. K. Pandian. 2016. Betulin inhibits virulence and biofilm of *Streptococcus pyogenes* by suppressing ropB core regulon, sagA and dltA. *Pathogens and Disease* 74:ftw088.

Wald, E. R. 2007. Periorbital & orbital infections. *Infectious Diseases in Clinical* 21:393–308.

Whiston, E. A. et al. 2008. αB-crystallin protects retinal tissue during *Staphylococcus aureus*-induced endophthalmitis. *Infection and Immunity* 76:1781–1790.

Willcox, M. D. 2007. *Pseudomonas aeruginosa* infection and inflammation during contact lens wear: A review. *Optometry and Vision Science* 84 273–278.

Williamson, Y. M. et al. 2008. Adherence of non-typeable *Streptococcus pneumoniae* to human conjunctival epithelial cells. *Microbial Pathogenesis* 44:175–185.

Wilson, L. A. and D. G. Ahearn. 1977. *Pseudomonas*-induced corneal ulcers associated with contaminated eye mascaras. *American Journal of Ophthalmology* 84:112–119.

Wong, E. S., C. W. Chow, W. K. Luk, K. S. Fung and K. K. A. Li. 2017. 10-Year review of ocular methicillin-resistant *Staphylococcus aureus* infections: Epidemiology, clinical features, & treatment. *Cornea* 36:92–97.

Wretlind, B. and O. R. Pavlovskis. 1983. *Pseudomonas aeruginosa* elastase and its role in pseudomonas infections. *The Journal of Infectious Diseases* 5:S998–S1004.

Wu, P. Z. J., H. Zhu, A. Thakur and M. D. P. Willcox. 1999. Comparison of potential pathogenic traits of staphylococci that may contribute to corneal ulceration & inflammation. *Clinical & Experimental Ophthalmology* 27:234–236.

Wu, Y. T., L. S. Zhu, K. C. Tam, D. J. Evans and S. M. Fleiszig. 2015. *Pseudomonas aeruginosa* survival at posterior contact lens surfaces after daily wear. *Optometry and Vision Science* 92:659–664.

Yesilkaya, H., A. Kadioglu, N. Gingles, J. E. Alexander, T. J. Mitchell and P. W. Andrew. 2000. Role of manganese-containing superoxide dismutase in oxidative stress and virulence of *Streptococcus pneumoniae. Infection and Immunity* 68:2819–2826.

Yokoi, N., K. Okada, J. Sugita, and S. Kinoshita. 2000. Acute conjunctivitis associated with biofilm formation on a punctal plug. *Japanese Journal of Ophthalmology* 44:559–560.

Zhu, H. et al. 2017. Antimicrobial blue light therapy for infectious keratitis: Ex vivo & in vivo studies. *Investigative Ophthalmology & Visual Science* 58:586–593 .

Ziha-Zarifi, I., C. Llanes, T. Köhler, J. C. Pechere and P. Plesiat. 1999. In vivo emergence of multidrug-Resistant mutants of *Pseudomonas aeruginosa* overexpressing the active efflux system MexA-MexB-OprM. *Antimicrobial Agents and Chemotherapy* 43:287–291.

12
Role of Bacteria in Urinary Tract Infections

Gnanasekaran JebaMercy, Kannan Balaji, and K. Balamurugan

Contents

12.1 Introduction

Microorganisms are the first forms of life to develop on Earth, approximately 4 billion years ago (Altermann & Kazmierczak, 2003). Microorganisms are beneficial in many aspects to the society, but at the same time, they are also highly harmful in regard to infections and diseases. Severe public health problems like urinary tract infections (UTIs) are caused by a range of pathogens, most importantly *Proteus* spp. and *Staphylococcus* spp. The social costs of these infections were about USD 3.5 billion per year in the United States. It is an important cause of morbidity in infants and older people. UTI can be caused by both gram-positive and -negative bacteria. In this chapter, we are going study the virulence factors of UTI pathogens *Proteus mirabilis* and *Staphylococcus aureus*.

Proteus spp. belongs to the group of opportunistic pathogens and is widely distributed in the natural environment. Under favorable conditions, they can cause severe infections in individuals who are immunocompromised. These bacteria can also be the main agents of nosocomial infections and infections of the urinary tract, respiratory tract, and wound infections (Tiwana et al., 1999; Warren, 1996). Gram-positive bacterium *S. aureus* leads to serious complications in many animal models (Lindsay, 2010). It can colonize in the *Caenorhabdtis elegans* gut and invade into other organs by disrupting the gut epithelial cells further leading to mortality (Garsin et al., 2001; Sifri et al., 2003).

12.2 Urease

Urease is an enzyme used for the hydrolysis of urea. A healthy individual produces about 30 g of urea per day (Newsholme & Leech, 2011). Most of the UTI pathogens have urease activity as a part of its virulence character. The urease-dependent process is associated with bacterial UTI, including those caused by *Proteus* and *Staphylococcal* spp. Urease produced by these bacteria can lead to the formation of stones because of the precipitation of the minerals struvite and carbonate apatite. The stone formation is used to protect the pathogen (Griffith et al., 1976). These stones can block the flow of urine and lead to tissue damage (Schaffer & Pearson, 2015) Ammonia produced during the above process can damage the glycosaminoglycan surface of the urothelium, and it will allow the entry for the other bacterial infections (Rutherford, 2014). It was shown that urease from *Proteus mirabilis* can hydrolyze urea several times faster than urease synthesized by other species of bacteria (Jordan & Nicolle, 2014).

12.3 *Proteus* spp.

The opportunistic pathogens, *Proteus* genus, are commonly encountered in nosocomial infections, and it is comparatively more resistant to antibiotic agents and is, therefore, one of the most difficult organisms to deal clinically. During the introduction of ampicillin and nalidixic acid into the army of antibiotics, it was found that these treatments were satisfactory in the control of *Proteus* spp. infections (Stratford, 1964). It is a gram-negative, motile, urease-positive, lactose-negative, and indole negative organism. It is an important cause for the UTIs in patients who are hospitalized.

12.3.1 Virulence factors

Gram-negative *Proteus* rods are widely distributed in the natural environment, and they can especially be found in polluted water and soil. They are also the inhabitant of human and animal intestines. *Proteus* bacilli are opportunistic pathogens; under favorable conditions, they can cause severe infections in individuals who are immunocompromised. These bacteria can also be the main agents of nosocomial infections, infections of the respiratory tract, wound infections, and rheumatoid arthritis (Tiwana et al., 1999; Warren, 1996; Różalski et al., 1997). *Proteus* have developed several virulence determinants such as fimbriae, flagella, urease, amino acid deaminases, proteases, hemolysins, swarming growth, and LPS which enable them to colonize, survive, and multiply in the host (Coker et al., 2000). One of the mechanisms which support the survival of gram-negative bacteria in the host is resistance to the bactericidal action of normal human serum (Kumar et al., 1997; Grzybek-Hryncewicz et al., 1981).

Proteus mirabilis causes 90% of all *Proteus* infections. It is facultatively anaerobic, rod-shaped bacterium. It shows swarming motility, and urease activity. *P. mirabilis* usually isolated from the urine of patients and individuals with the urinary catheters. Infection can lead to serious complications including acute pyelonephritis, stone formation in the bladder and kidney, encrustation, and obstruction of the catheter, fever, and bacteremia (Griffith et al., 1976; Mobley & Warren, 1987; Rubin et al., 1986). *P. mirabilis* is the second-leading cause of bacteremia (Setia et al., 1984). It has a many virulence factors like urease (Griffith et al., 1976; Mobley & Warren, 1987; Jones et al., 1990; Johnson et al., 1993), fimbriae (Mobley & Belas, 1995), flagella (Belas & Flaherty, 1994; Allison et al., 1994), hemolysin (Mobley et al., 1991), IgA protease (Loomes et al., 1990; Wassif et al., 1995), and amino acid deaminase (Massad et al., 1995). And it has a distinct type of protease ZapA metalloprotease, which is capable of degrading antibacterial immune peptides (Allison et al., 1992; Belas et al., 2004) (Figure 12.1). In different from other pathogens, *P. mirabilis* exhibits a unique kind of character called swarming migration, vegetative rods undergo differentiation at the colony border into long, aseptate filaments that have 50-fold more flagella per unit cell surface area (Allison et al., 1992; Williams et al., 1978). Another member of this

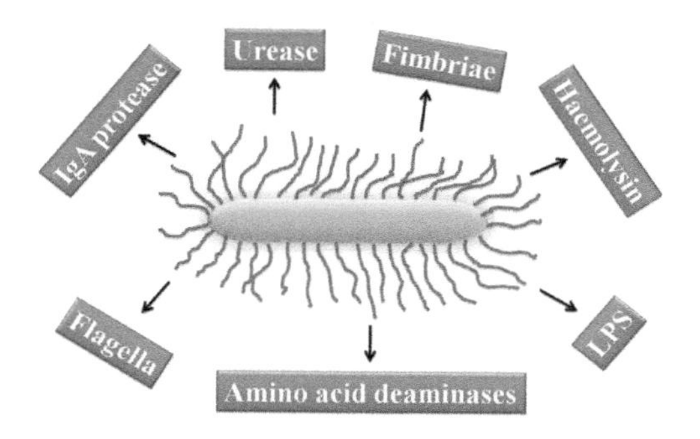

Figure 12.1 Virulence determinants of *P. mirabilis*.

opportunistic group, *Proteus vulgaris* is also known to cause urinary, blood, and wound infections. The *P. vulgaris* strains isolated from patients with UTI are often insensitive to the bactericidal action of normal human serum which poses a severe clinical problem (Kwil et al., 2013).

12.3.2 Interaction with host

Interaction of *Proteus* spp. with *C. elegans* was reported recently and the study established *C. elegans* as a model for studying phenotypic changes and regulation of MAP kinase immune pathway in the host against the bacterial infection. Being in the group of opportunistic pathogens, *Proteus* spp. causes large number of nosocomial infections. The *Proteus* spp. does not cause death in wild type *C. elegans* (JebaMercy & Balamurugan, 2012). *Proteus* spp. can cause mortality in MAP kinase pathway mutant *C. elegans.* And the author showed the involvement of innate immune pathways specific players at the mRNA level against *Proteus* spp. infections. In addition, the lipopolysaccharides (LPS) from *P. mirabilis* treated with mutant *C. elegans* showed structural changes compared with wild type worms exposed, and it clearly indicates *P. mirabilis* changes its internal system according to the specific host for effective infection (JebaMercy et al., 2013).

12.4 *Staphylococcus aureus*

Staphylococci are gram-positive, cluster-forming coccus, non-motile, non-spore-forming facultative anaerobes. Not only UTIs, but *S. aureus* can also cause a variety of illnesses from minor infections to life-threatening diseases. It can cause superficial skin lesions UTIs. *S. aureus* is one of the major causes of nosocomial infection of surgical patients and infections related with indwelling medical procedures. Enterotoxin and super antigens from *S. aureus* cause food poisoning and toxic shock syndrome (Lowry et al., 1998).

The ability of *S. aureus* to cause a wide spectrum of disease has been attributed to its ability to produce a broad array of pathogenicity factors. These factors can be subdivided into three general groups: cell-associated products, secreted exoproteins, and regulatory loci. Cell-associated products, including adhesins of the microbial surface components recognizing adhesive matrix molecules (MSCRAMM) family and capsular polysaccharide, facilitate binding to host tissue and help resist host immune responses. The exoproteins like cytolysins and extracellular proteases are known to fight against host immunity and results in tissue invasion and nutrient absorbance (Lowy et al., 1998; Sifri et al., 2003).

S. aureus produces a large number of virulence factors, and these include extracellular toxins like alpha-toxin and cell-wall-associated proteins, which are important for colonization, immune evasion, and tissue destruction. Treatment of *S. aureus* infections has become complicated by the emergence of widespread antimicrobial resistance (Garvis et al., 2002).

In 1959, Methicillin was introduced to treat penicillin-resistant *S. aureus*–related diseases. *Methicillin-resistant S. aureus* (MRSA) was emerged in the UK around 1961 and spread around European countries and then to Japan, Australia, and the United States. The methicillin-resistance gene (mecA) encodes a methicillin-resistant penicillin-binding protein that is not present in susceptible strains and is believed to have been acquired from a distantly related species. mecA is carried on a mobile genetic element, the staphylococcal cassette chromosome mec (SCCmec), of which four forms have been described that differ in size and genetic composition. Many MRSA isolates are multiresistant and are susceptible only to glycopeptide antibiotics such as vancomycin and investigational drugs. MRSA isolates that have decreased susceptibility to glycopeptides (glycopeptide intermediately susceptible *S. aureus* [GISA]), reported in recent years, are a cause of great public health concern (Martin et al., 2002). Community-related MRSA is associated with increased disease severity, ranging from cutaneous abscesses to deadly necrotizing pneumonia. USA300 and USA400 are the two dominant CA- MRSA strains. The USA400 strain causes deadly infections but in less frequency, whereas USA300 was widespread and mainly related with community infections and life-threatening infections like necrotizing pneumonia (Wu et al., 2010).

12.4.1 Lipoteichoic acid

Lipoteichoic acid (LTA) is a surface-associated adhesion amphiphile from gram-positive bacteria made up of a polymer of repetitive 1,3-phosphodiester-linked glycerol-l-phosphate units with a glycolipid anchor (Leopold & Fischer, 1992). It gets released from the gram-positive bacterial cells during bacteriolysis induced by host factors like lysozyme, cationic peptides from leucocytes, or during the antibiotic treatment. It binds to target cell receptors like CD14 or toll-like receptors. LTA can interact with antibodies and activate the passive immune kill phenomenon. LTA is similar to the endotoxin lipopolysaccharide and shares several of its pathogenetic properties (Dishon et al., 1967). Chemotaxis of human neutrophils and phagocytosis were also inhibited by LTA (Raynor et al., 1981; Card et al., 1994). LTA also induced expression of macrophage inflammatory protein 1 alpha. These findings suggest that LTA may have a role in the regulation, recruitment, and activation of leukocytes in inflammatory sites (Nonogaki et al., 1995). LTA, teichoic acid, and peptidoglycan each inhibit proliferation of fibroblasts by a still undefined mechanism and also act on T cells to activate nuclear factor kappa B. LTA, therefore, seems to be a versatile immunomodulator that can alter cell responses in inflammatory conditions (Elgavish et al., 2000).

12.4.2 Interaction with host

According to Sifiri et al. (2003). *S. aureus* infects *C. elegans*, ultimately leading to worm death, and key aspects of *S. aureus* pathogenesis and interaction with the innate immune system have been mechanistically conserved from nematodes through vertebrates. *C. elegans* will be a great model to study the novel staphylococcal genes required for the mammalian pathogenesis and host innate immune defense systems (Sifri et al., 2003). Death in *C. elegans* is mainly due to the disruption of the gut epithelium and assault of the internal organs by the invading pathogen (Garsin et al., 2003; Sifri et al., 2003; Irazoqui et al., 2010). Deformed anal region (Dar) was identified in *S. aureus* infection and is mainly relying on both β-catenin and MAPK pathways (Irazoqui et al., 2010). *S. aureus* requires protease and alpha-hemolysin for the pathogenesis in *C. elegans* (Sifri et al., 2003). The p38 MAPK and the catenin pathways are required for the host immune system against *S. aureus*. The significance of this pathway in *S. aureus* pathogenesis is comparable in higher vertebrates, where catenin activates NFκB-mediated immune gene expression.

12.5 Treatment and future prospective

Bacterial UTIs are recurrent infections in the normal and the nosocomial environments. Two types of antimicrobial treatment required for all the UTIs: (1) rapid and effective response to therapy and prevention of recurrence (2) prevention of emergence of resistance (Wagenlehner & Naber, 2006). The important disadvantage of antibiotic therapies is the development of antibiotic resistance. There is no major improvement in the UTI treatment for the last decade except antibiotic therapy (Nickel, 2002). Development of vaccines to prevent the UTIs in susceptible persons is an exciting emerging project. Recent papers reporting the oral and vaginal substance preparations used to stimulate the patients' immune system against the UTI pathogens. The suitable formulation with appropriate clinical trials will remove the burden of nosocomial UTI infections.

References

Allison, C., L. Emody, N. Coleman and C. Hughes. 1994. The role of swarm cell differentiation and multicellular migration in the uropathogenicity of *Proteus mirabilis*. *The Journal of Infectious Diseases* 169:1155–1158.

Allison, C., H. C. Lai and C. Hughes. 1992. Co-ordinate expression of virulence genes during swarm-cell differentiation and population migration of *Proteus mirabilis. Molecular Microbiology* 6:1583–1591.

Belas, R. and Flaherty D. 1994. Sequence and genetic analysis of multiple flagellin-encoding genes from *Proteus mirabilis.Gene* 128:3341.

Belas, R., J. Manos and R. Suvanasuthi. 2004. *Proteus mirabilis* ZapA metalloprotease degrades a broad spectrum of substrates, including antimicrobial peptides. *Infection and Immunity* 72:5159–5167.

Card, G. L., Jasuja, R. R. and G. L. Gustafson. 1994. Activation of arachidonic acid metabolism in mouse macrophages by bacterial amphiphiles. *Journal of Leukocyte Biology* 56:723–728.

Coker, C., C. A. Poore, X. Li and H. L. T. Mobley. 2000. Pathogenesis of *Proteus mirabilis* urinary tract infection. *Microbes Infection* 2:1497–1505.

Dishon, T., R. Finkel, Z. Marcus and I. Ginsburg. 1967. Cell sensitising products of Streptococci. *Immunology* 13:555–564.

Elgavish, A. 2000. NF-kappaB activation mediates the response of a subpopulation of basal urothelial cells to a cell-wall component of *Enterococcus faecalis. Journal of Cellular Physiology* 182:232–238.

Garsin, D. A., C. D. Sifri, E. Mylonakis, X. Qin, K. V. Singh, B. E. Murray, S. B. Calderwood and F. M. Ausubel. 2001. A simple model host for identifying Gram-positive virulence factors. *Proceedings of the National Academy of Sciences of the United States of America* 98:10892–10897.

Garsin, D. A., J. M. Villanueva, J. Begun, D. H. Kim, C. D. Sifri, S. B. Calderwood, G. Ruvkun and F. M. Ausubel. 2003. Long-lived *C. elegans* daf-2 mutants are resistant to bacterial pathogens. *Science* 300:1921.

Garvis, S., J. M. Mei, J. Ruiz-Albert and D. H. Holden. 2002. *Staphylococcus aureus svrA*: A gene required for virulence and expression of the *agr* locus. *Microbiology* 148:3235–3243.

Griffith, D. P., D. M. Musher and C. Itin. 1976. Urease: The primary cause of infection-induced urinary stones. *Investigative Urology* 13:346–350.

Grzybek-Hryncewicz, K., S. Jankowski, G. Mokracka-Latajka and E. Mróz. 1981. Sensitivity of gram negative bacilli isolated from patients to bactericidal action of human serum. *Immunology Polymerase* 6:41–47.

Irazoqui, J. E., E. R. Troemel, R. L. Feinbaum, L. G. Luhachack, B. O. Cezairliyan and F. M. Ausubel. 2010. Distinct pathogenesis and host responses during infection of *C. elegans* by *P. aeruginosa* and *S. aureus. PLoS Pathogenesis* 6: e1000982.

JebaMercy, G. and K. Balamurugan. 2012. Effects of sequential infections of *Caenorhabditis elegans* with *Staphylococcus aureus* and *Proteus mirabilis. Microbiology and Immunology* 56:825–835.

JebaMercy, G., L. Vigneshwari and K. Balamurugan. 2013. A MAP Kinase pathway in *Caenorhabditis elegans* is required for defense against infection by opportunistic *Proteus* species. *Microbes and Infection* 15:550–568.

Johnson, D. E., R. G. Russell, C. V. Lockatell, C. Zulty, W. Warren and H. L. T. Mobley. 1993. Contribution of *Proteus mirabilis* urease to persistence, urolithiasis and acute pyelonephritis in a mouse model of ascending urinary tract infection. *Infection and Immunity* 61:2748–2754.

Jones, B. D., C. V. Lockatell, D. E. Johnson, J. W. Warren and H. L. Mobley. 1990. Construction of a urease-negative mutant of *Proteus mirabilis*: Analysis of virulence in a mouse model of ascending urinary tract infection. *Infection and Immunity* 58:1120–1123.

Jordan, R. P., L. E. Nicolle. 2014. Preventing infection associated with urethral catheter biofilms. *InBiofilms in Infection Prevention and Control* 287–309.

Kumar, S. and M. D. Mathur. 1997. Sensitivity to the bactericidal effect of humans serum of Proteus strains from clinical specimens. *Indian Journal of Pathology and Microbiology* 40:335–338.

Kwil, I., D. Kaźmierczaka and A. Różalski. 2013. Swarming growth and resistance of *Proteus penneri* and *Proteus vulgaris* strains to normal human serum. *Advances in Clinical and Experimental Medicine* 22:165–175.

Leopold, K. and W. Fischer. 1992. Hydrophobic interaction chromatography fractionates lipoteichoic acid according to the size of the hydrophilic chain: A comparative study with anion-exchange and affinity chromatography for suitability in species analysis. *Analytical Biochemistry* 201:350–355.

Lindsay, J. A. 2010. Genomic variation and evolution of *Staphylococcus aureus*. *International Journal of Medical Microbiology* 300:98–103.

Loomes, L. M., B. W. Senior and M. A. Kerr. 1990. A proteolytic enzyme secreted by *Proteus mirabilis* degrades immunoglobulins of the immunoglobulin A1 (IgA1), IgA2, and IgG isotypes. *Infection and Immunity* 58:1979–1985.

Lowry, F. D. 1998. *Staphylococcus aureus* infections. *New England Journal of Medicine* 339:520–532.

Massad, G., H. Zhao and H. L. Mobley. 1995. *Proteus mirabilis* amino acid deaminase: Cloning, nucleotide sequence, and characterization of *aad*. *Journal of Bacteriology* 177:5878–5883.

Mobley, H. L. T. and I. W. Warren. 1987. Urease-positive bacteriuria and obstruction of long-term urinary catheters. *Journal of Clinical Microbiology* 25:2216–2217.

Mobley, H. L. T., G. R. Chippendale, K. G. Swihart and R. Welch. 1991. Cytotoxicity of the HpmA hemolysin and urease of *Proteus mirabilis* and *Proteus vulgaris* against cultured human renal proximal tubular epithelial cells. *Infection and Immunity* 59:2036–2042.

Mobley, H. L. T. and R. Belas. 1995. Swarming and pathogenicity of *Proteus mirabilis* in the urinary tract. *Trends Microbiology* 3:280–284.

Newsholme, E. and A. Leech. 2011. *Functional Biochemistry in Health and Disease*. 2nd ed. Chichester, UK: John Wiley and Sons.

Nickel, J. C. 2002. Immunological based therapies for urinary tract infection: The future is almost here!. *Reviews in Urology* 4(4):196–197.

Nonogaki, K., A. H. Moser, X. M. Pan, I. Staprans, C. Grunfeld and K. R. Feingold. 1995. Lipoteichoic acid stimulates lipolysis and hepatic triglyceride secretion in rats *in vivo*. *Journal of Lipid Research* 36:1987–1995.

Raynor, R. H., D. F. N. Scott and G. K. Best. 1981. Lipoteichoic acid inhibition of phagocytosis of *Staphylococcus aureus* by human polymorphonuclear leukocytes. *Clinical Immunology and Immunopathology* 19:181–189.

Różalski, A., Z. Sidorczyk and K. Kotełko. 1997. Potential virulence factors of Proteus bacilli. *Microbiology and Molecular Biology Reviews* 61:65–89.

Rubin, R., N. Tolkoff-Rubin and R. Cotran. 1986. Urinary tract infection, pyelonephritis and reflux nephropathy. In *The Kidney*, 3rd edn, 1085–1141. Edited by G. Brenner and F. Rector, Jr. Philadelphia, PA: W. B. Saunders.

Rutherford, J. C. 2014. The emerging role of urease as a general microbial virulence factor. *PLoS pathogens*. 10(5):e1004062.

Schaffer, J. N. and M. M. Pearson. 2015. Proteus mirabilis and urinary tract infections. *Microbiology Spectrum*. 3(5).

Setia, U., I. Serventi and P. Lorenz. 1984. Bacteremia in a long term care facility: Spectrum and mortality. *Archives of Internal Medicine* 144:1633–1635.

Sifri, C. D., J. Begun, F. M. Ausubel and S. B. Calderwood. 2003. *Caenorhabditis elegans* as a model host for *Staphylococcus aureus* pathogenesis. *Infection and Immunity* 71:2208–2217.

Stratford, B. C. 1964. Some observations on the use of ampicillin in the urinary tract infections. *Postgraduate Medical Journal* 40:68.

Tiwana, H., C. Wilson, A. Alvarez, R. Abuknesha, S. Bansal and A. Ebringer. 1999. Cross-reactive between rheumatoid arthritis-associated motif EQKRAA and structurally related sequence found in *Proteus Mirabilis*. *Infection and Immunity* 67:2769–2775.

Wagenlehner, F. M. and K. G. Naber. 2006. Treatment of bacterial urinary tract infections: Presence and future. *European Urology*. 49(2):235–244.

Warren, J. W. 1996. Clinical presentations and epidemiology of urinary tract infections. In *Urinary Tract Infections, Molecular Pathogenesis And Clinical Management*. Eds H. L. T. Mobley, J. W. Warren. Washington DC: ASM Press, 2–28.

Wassif, C., D. Cheek and R. Belas. 1995. Molecular analysis of a metalloprotease from *P. mirabilis*. *Journal of Bacteriology* 177:5790–5798.

Williams, F. D. and R. H. Schwarzhoff. 1978. Nature of the swarming phenomenon in Proteus. *Annual Review of Microbiology* 32:101–122.

Wu, K., J. Conly, J. A. McClure, S. Elsayed, T. Louie and K. Zhang. 2010. *Caenorhabditis elegans* as a host model for community-associated methicillin-resistant *Staphylococcus aureus*. *Clinical Microbiology and Infection* 16:245–254.

Altermann, W. and J. Kazmierczak. 2003. Archean microfossils: A reappraisal of early life on Earth. *Research in Microbiology* 154:611–617.

13 Role of Bacteria in Blood Infections

Kannan Balaji, Gnanasekaran JebaMercy,
and K. Balamurugan

Contents

13.1 Introduction

Blood is a vital human body fluid that plays several important roles like circulation of oxygen to various parts of the body, cleaning of metabolic wastes from cells, regulation of body, and immunity. Human body harbors millions of microorganisms in various parts of the body like oral region, gut, skin, and respiratory tract from birth to death. These microbes exist in several forms like commensal, normal flora, carriers, and pathogens. When it enters the bloodstream, bacteria confers with the host immune system and is compelled to exhibit virulence factors for its survival. Blood infections caused by bacteria are a global burden because of the increase in virulent and drug-resistant strains. Blood-related infections associated with bacteria are in increasing trend in the rate of mortality because drug-resistant strains remain intractable to treat. This chapter focuses on components of blood, blood infections caused by bacteria, bacterial species involved, and their virulence mechanisms.

13.1.1 Blood

Blood is a specialized fluid which supplies oxygen and nutrients to cells present in various parts of the body and carry away metabolic wastes from the cells. Human blood is composed of red blood cells (RBCs), white blood cells (WBCs), platelets, and plasma.

13.1.2 Red blood cells

RBCs or erythrocytes are the most abundant cells in the blood, which comprise about 40% to 45% in whole blood. They are large microscopic disc-like structured cells without nucleus. The red color of the RBCs is by a special protein called hemoglobin, which carries oxygen from lungs to other cells in the body, and in return, it carries away carbon dioxide from cells to lungs.

13.1.3 White blood cells

WBCs or leukocytes are much less abundant than RBCs and account for about 1% in whole blood volume. They are nucleated, and apart from blood, these WBCs are also present in other body organs like the liver, spleen, and lymph nodes. WBCs are the protecting agent in a human body from infections. They comprise neutrophils, lymphocytes, monocytes, basophils, and eosinophils. Among these cells, the lymphocytes are of two types, T and B.

13.1.4 Platelets

Platelets are not cells like RBCs or WBCs; they are cell fragments without a nucleus that involved in the blood-clotting process known as *coagulation*. Platelets, along with several factors, gather at the site of injury and create a fibrin clog to prevent blood flow. Platelets are also called *thrombocytes*.

13.1.5 Plasma

Plasma is a clear, yellow-tinted liquid that carries RBCs, WBCs, and platelets. Plasma are the major component in blood, which comprises about 55% of the total blood volume. It contains sugars, proteins, vitamins, lipids, enzymes, and salts. Plasma carries away the metabolic wastes from the cells and helps the flow of blood through vessels to various body parts.

13.2 Bacterial blood infections

Blood is a nutrient-rich human body fluid and is more prone to bacterial encounters during blood transfusion, through open wounds, burn wounds, and other bacterial infections.

13.2.1 Bacteremia

Presence of bacteria in the bloodstream is termed *bacteremia*. Despite antibiotic treatment, bloodstream infections by bacteria leads to deleterious effects and is associated with high mortality rates (Salomão et al., 1999). *Staphylococcus aureus*, *Escherichia coli*, and *Streptococcus pneumoniae* are the most common cause of bacteremia (Siegman-Igra et al., 2002; Kollef et al., 2011; Vallés et al., 2013). Based on the case of origin, bacteremia are differentiated into hospital-acquired bacteremia (HAB) and community-acquired bacteremia (CAB).

In HAB, the infection is caused by several ways, such as contaminated blood transfusion, organ or stem cell transplants, medical procedures like dental treatments, and contaminated catheters and needles. CAB is defined as the bacteremia associated with several factors like a particular population age group; geographical location; climatic condition; and along with other disease conditions like diabetes, endocarditis, pneumonia, urinary tract infections, and HIV (Christaki & Giamarellos-Bourboulis, 2014).

13.2.1.1 Epidemiology – Extensive use of medical procedures and the emergence of drug-resistant pathogens and nosocomial infections leads to an increasing incidence of bacteremia. The occurrence rate of CAB per year in 100,000 individuals varies according to the geographic locations and reported as 153 episodes in Olmsted County in the United State, 101.2 cases in Victoria, Canada, 92 episodes in Denmark, and 31.1 episodes in Thailand (Viscoli, 2016). The etiology varies depending on the age group, geographical location, climate, environment, and other associated illness (Friedman et al., 2002; Siegman-Igra et al., 2002). Increasing incidence of multidrug resistant bacteremia in after transplants (Oliveira et al., 2007) will be a challenge to medical treatments.

13.2.2 Sepsis

Sepsis is a severe clinical complication by host inflammatory response triggered against any infection. Sepsis is an irregulated host immune response that leads to organ dysfunction (Singer et al., 2016). It is one of the most common causes of hospitalization, and untreated sepsis cases are potentially fatal. Anyone can have sepsis, but it is prevalent in individuals who are older or immunocompromised. Some of the common causes of sepsis are pneumonia, kidney infection, and bacteremia. It ranges from mild sepsis to septic shock syndrome. Untreated sepsis leads to septic shock syndrome, which results in difficulty in breathing and decrease in platelet counts, urine output, and blood pressure. In addition, septic shock causes blood clots that cause organ failure and ultimately leads to death.

13.2.2.1 Epidemiology – Sepsis is a critical illness and the leading cause of mortality worldwide. The United States spends more than $20 billion (5.2%) of its total medical costs on sepsis treatment (Torio & Andrews, 2011). The reported cases of sepsis are increasing, especially with greater recognition and in aging populations with more comorbidities. In the United States, the reports of severe sepsis are estimated to be 300 cases per 100,000 population (Iwashyna et al., 2012; Gaieski et al., 2013). Studies on patients admitted to the Intensive Care Unit in Europe and China states that 37.4% and 37.3%, respectively, were diagnosed with severe sepsis (Vincent et al., 2006; Zhou et al., 2014). Despite advanced medical treatments, one in four patients who develop severe sepsis will die during their hospitalization. In the case of septic shock, the mortality rate is approaching an alarming rate of 50% (Singer et al., 2016).

13.2.3 Meningitis

Meninges are the membrane layers present in brain and spinal cord. Fluid present in the meninges get infected by bacteria or viruses and cause inflammation in the meninges. This condition is called *meningitis*. A range of pathogenic bacteria can cause bacterial meningitis. *Streptococcus pneumonia* is the most common causative pathogen. Other bacterial pathogens like *Neisseria meningitides*, *Haemophilus influenza*, *Listeria monocytogenes*, and Group B *Streptococcus* also causes meningitis.

13.2.3.1 Epidemiology – Bacterial meningitis is a life-threatening infection in the central nervous system and is a leading cause of mortality (Matthijs et al., 2010); 13,974 cases of bacterial meningitis were reported in 27 states in the United States in late 1970s (Schlech et al., 1985). Until the clinical use of pneumococcal vaccine, every year almost 6000 people reported the development of meningitis (Chávez-Bueno & McCracken, 2005). Bacterial meningitis was even more a serious problem in many developing countries. Sub-Saharan Africa is referred to as the meningitis belt for its meningococcal meningitis prevalence (Campagne et al., 1999; de Gans & van de Beek, 2002). In Dakar, Senegal, 50 cases of bacterial meningitis were reported out of 100,000 individuals, where 1 in 250 children get infected within 1 year from birth (Greenwoods, 1987). With the advent of proper diagnostic methods and effective vaccines against the causative pathogens, significant reductions have been seen in the disease burden. However, differentiation between acute bacterial meningitis and acute viral meningitis has always remained a challenge because of its prognostic significance.

13.2.4 Pericarditis

The pericardium is a thin protective membrane present around the heart, and inflammation in this membrane is called *pericarditis*. There are several factors that cause pericarditis include bacteria, virus, parasite, or fungal infection and trauma from injury or any surgery. In the case of bacterial pericarditis, pathogenic bacteria enters and causes infection in the pericardium. The most common causative pathogens are *Staphylococcus* spp., *Streptococcus* spp., and *Pneumococcus* spp. Bacterial infections from other parts of the body like pneumonia and bloodstream infections and after surgery are some of the routes that cause bacterial pericarditis.

13.2.4.1 Epidemiology – Bacterial pericarditis has become rare with an incidence of 1 of 18,000 individuals by the arrival of antibiotics and effective diagnostic methods. But pericarditis is a rapid-progressive, high-risk infection with a mortality rate of 100% if left untreated. Mostly diagnosed after death during a postmortem examination, the mortality rate remains 40%, even with treatment because of the associated sequelae such as severe sepsis, septic shock, and cardiac constriction and tamponade (Klacsmann et al., 1977; Sauleda et al., 1993).

13.2.5 Endocarditis

Endocarditis is an infection inside the heart lining, heart valves, or blood vessels. Bacteria present in the bloodstream is called *bacterial endocarditis* or *infective endocarditis*. Heart valves are devoid of the immune response system.

13.2.5.1 Epidemiology – Infective endocarditis is associated with long-term hospitalization and requires surgery for treatment (Moreillon & Que, 2004). In the United States, hospitalization of patients rose from 28,195 in 1998 to 43,419 in 2009 (Bor et al., 2013). Increasing rate of transplant surgeries, prosthetic valves, and cardiovascular implantable electronic devices in recent years revised the trend worldwide. In addition, nosocomial infections and the postsurgery scenario makes endocarditis a serious issue: the mortality rate remains unchanged as approximately 25% (Ambrosioni et al., 2017).

13.2.6 Osteomyelitis

Infection in bone regions are termed *osteomyelitis*. Bacteria present in bloodstream infections or open wounds of fractured bones or bone surgery are the common mode of infection. *S. aureus* is the most common causative bacteria. Untreated osteomyelitis causes impaired blood circulation, which ultimately leads to bone death (i.e., osteonecrosis). In some cases, osteomyelitis spreads to nearby bone joints and causes septic arthritis.

13.3 Predominant bacteria in blood infections

13.3.1 *Streptococcus* spp.

The genus *Streptococcus* comprises of gram-positive bacteria, which are found to inhabit a wide range of hosts. In humans, streptococci are often found to colonize the mouth and pharynx. However, in certain circumstances, they may also inhabit the skin, heart, or muscle tissue. *S. pyogenes*, *S. pneumonia*, and Group B *Streptococcus* are the important blood infection-causing pathogens.

13.3.1.1 *Streptococcus pyogenes* – *S. pyogenes* or Group A *Streptococcus* is a gram-positive, nonmotile, non-spore-forming coccus that occurs in chains. *S. pyogenes* colonizes the throat or skin and produces a variety of pyogenic infections in mucous membranes, tonsils, skin, and deeper tissues. *S. pyogenes* causes blood-related infections, including toxic streptococcal syndrome, bacteremia, sepsis, pneumonia, and meningitis (Cunningham, 2000). They are the common cause of puerperal sepsis or childbed fever. *S. pyogenes* is also responsible for streptococcal toxic shock syndrome, and it has gained notoriety as the "flesh-eating bacterium," which invades skin and soft tissues, and in severe cases, leaves infected tissues or limbs destroyed (Stevens, 1999).

S. pyogenes delivers a number of blood infection-related postinfection sequelae, including acute glomerulonephritis and reactive arthritis and is associated with disability and death in children worldwide. *S. pyogenes* have become such a serious pathogen by means of its improved virulence factors and superantigens. As a blood-infective pathogen, it has developed complex virulence mechanisms to avoid host defenses.

13.3.2 *Staphylococcus aureus*

S. aureus is a gram-positive cocci that causes a range of pathogenicity from normal commensal to invasive infection. *S. aureus* is one of the leading causes of blood infections like bacteremia and sepsis. Nosocomial and medical device implant-related infections, especially with multidrug resistant *S. aureus*, are of serious concern. *S. aureus* bacteremia is the most common cause of sepsis in pediatric population, which is the leading cause for hospital visits and admission to intensive care units (Schlapbach et al., 2015; Munro et al., 2018).

13.3.3 *Escherichia coli*

E. coli is a gram-negative bacilli, colonized in the human intestine as commensal, and causes urinary tract infections. *E. coli* is the leading cause of bloodstream infections and a report states about 53,000 senior citizens older than 65 years of age are affected by *E. coli* bacteremia every year. In addition, emergence of extended-spectrum beta-lactamases producing *E. coli* strains showed resistance to antibiotic treatment and cause very high mortality rate in senior adults (Tumbarello et al., 2010).

13.3.4 *Klebsiella pneumonia*

K. pneumonia is a gram-negative, rod-shaped, nonmotile bacteria with a protective polysaccharide capsule and is part of the Enterobacteriaceae family. This capsule helps the bacteria to escape against many host defense mechanisms. *K. pneumonia* is the cause of bloodstream-associated infections like bacteremia, sepsis, and pneumonia.

13.3.5 *Neisseria meningitides*

N. meningitides is a gram-negative diplococci that is colonized in the human respiratory tract as a commensal. *N. meningitides* is the important cause of meningitis globally (Hill et al., 2010). It also causes life-threatening meningococcemia, sepsis, and disease-associated bloodstream infections.

13.3.6 Other bacteria

In addition to the pathogens already discussed, there are several other bacteria that are involved in bloodstream infections and sepsis. Some of them are *Pseudomonas aeruginosa, Listeria monocytogenes, Haemophilus influenza, Salmonella* spp., and *Shigella* spp.

13.3.7 Polymicrobial infections

The ancient paradigm that one microbe causes one infection has become entrenched by our understanding of microbiology since the time of Robert Koch. In addition to single species infections, blood infections will cause serious effects and complicate the treatment process in polymicrobial infections. Two different species in combination with virus or fungi form a kind of microbial team called a *polymicrobial infection*. Complex microbial interactions lead to the emergence of drug resistance, disease manifestations, and quorum sensing. According to Del Pozo et al. (2007), biofilm communities in most circumstances, including human infections, tend to be polymicrobial communities. Combining multiple bacteria with other microbes like fungal species or viruses in a single community provides numerous advantages such as passive resistance (Weimer et al., 2011), metabolic teamwork (Fischbach & Sonnenburg, 2011; Elias & Banin, 2012), by-product influence (Carlsson, 1997), quorum sensing systems (Elias & Banin, 2012), and many more competitive advantages. A review by McCuller (2013) discusses the existence of polymicrobial interactions between the influenza virus and specific bacteria in pneumonia.

13.4 Pathogenesis and virulence mechanisms

13.4.1 Emergence of resistance

The impact of antibiotic-resistant strains is an alarming scenario in the case of bacterial pathogenesis. In the case of bloodstream infection, the encounters of antibiotic-resistant strains is mostly by nosocomial contaminations in patients in the Intensive Care Unit (Tabah et al., 2012). The emergence of methicillin-resistant *S. aureus* (MRSA), extended-spectrum beta-lactamases producing *E. coli*, metallo-beta-lactamase producing *E. coli* and

K. pneumoniae, and carbapenemase-producing *Enterobacteriaceae* are the challenges in treating bloodstream infections (Laupland and Church, 2014).

13.4.2 Membrane proteins

In *S. pyogenes*, the major virulence factor is M protein present in the membrane, which shows antiphagocytic effect by interfering with opsonization via the alternative complementary pathway (Bisno, 1991). M protein is composed of two polypeptide chains complexed in an alpha-helically coiled configuration anchored in the cell membrane and appears as fibrils on the cell surface. The chains comprise four repeat blocks (labeled A–D), each differing in size and amino acid sequence, within which there are seven-residue repeats of nonpolar amino acids. These streptococcal M proteins have been used to divide *S. pyogenes* into different M serotypes. Beall et al. (1996) introduced the polymerase chain reaction (PCR)-based M protein serotyping method where a specific primer pair was employed for amplification and identification of the *emm* allele with the help of Genebank-submitted sequences. In *K. pnemoniae*, outer membrane protein A is one of the major pathogenesis factors involved in suppression of host immune responses (March et al., 2011).

13.4.3 Capsules

The *S. pyogenes* contains an encapsulation called a *hyaluronic acid capsule* (Figure 13.1). It is composed of a high-molecular-weight polymer consisting of alternating residues of N-acetylglucosamine and glucuronic acid. This capsule protects *S. pyogenes* by resisting phagocytosis. An acapsular isogenic mutant *S. pyogenes* lost its ability to resist phagocytic killing and showed decreased virulence up to 100-fold (Moses et al., 1997). Like *K. pneumoniae, N. meningitidis* strains are characteristics of a protective polysaccharide capsule layer. In *N. meningitides*, these capsules are used for serotyping, and a total of 13 serogroups have been described so far. The majority of the invasive or bloodstream infections will be manifested by strains belonging to only five specific serogroups.

13.4.4 Extracellular virulence products

S. pyogenes exerts several extracellular products, of which two distinct hemolysins streptolysin O and streptolysin S have been well characterized (Hackett & Stevens, 1992; Fontaine et al., 2003; Harder et al., 2009). Streptolysin O derives its name from its oxygen lability and is reversibly inhibited by oxygen and irreversibly inhibited by cholesterol. It is a member of a family of highly conserved pore-forming cytolysins (Bhakdi et al., 1985, 1993) and stimulates targeted autophagy (O'Seaghdha & Wessels, 2013).

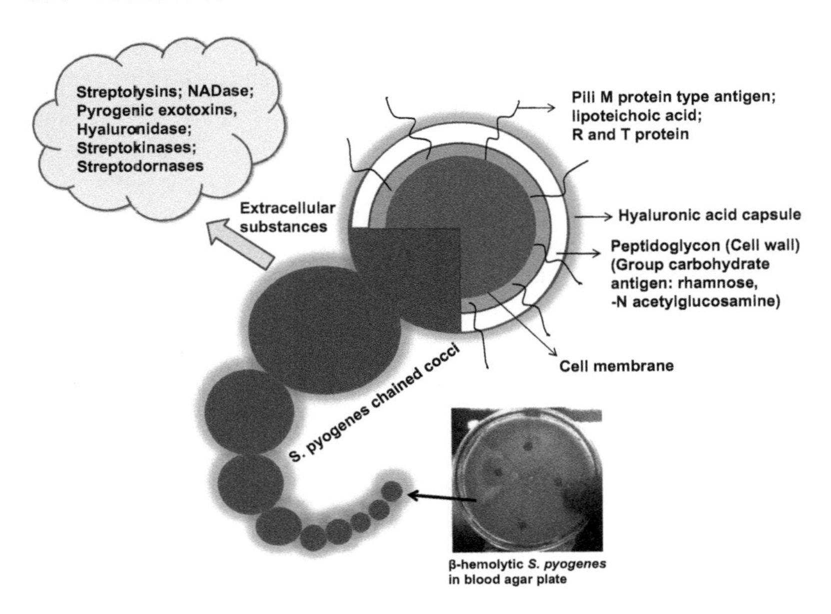

Figure 13.1 Representative image of virulence factors present in a pathogen *Streptococcus pyogenes*.

In addition to its effect on erythrocytes, extracellular products released by such bacteria are toxic to a variety of cells and cell fractions, including polymorphonuclear leukocytes, platelets, tissue-culture cells, lysosomes, and isolated mammalian and amphibian hearts (Duncan & Schlegel, 1975; Bisno et al., 2003). Streptolysin S is a hemolysin produced by *S. pyogenes* in the presence of serum or several other substances such as serum albumin, alpha-lipoprotein, and ribonucleic acid. It exists in intracellular, cell-surface-bound, and extracellular forms, and it is one of the most potent cytotoxins known (Sierig et al., 2003; Datta et al., 2005; Sumitomo et al., 2011). Other extracellular antigenically distinct enzymes are DNases A, B, C, and D, which are involved in degradation of DNA, streptokinase, which causes dissolution of clots by catalyzing the conversion of plasminogen to plasmin, streptococcal pyrogenic exotoxin B (*speB*), a potent protease that cleaves the PMN binding site (Ji et al., 1996; Bisno et al., 2003), streptococcal inhibitor of complement, which inhibits lysis of the bacterium by binding to the insertion site of complement (Fernie-King et al., 2002).

The pathogenesis of *P. aeruginosa* starts from adherence to epithelial cells using adhesins and exoenzyme S. An exotoxin A causes tissue necrosis followed by phospholipase C, which is a hemolysin. The exoenzyme S causes resistance to macrophages by disrupting the normal cytoskeletal organization and destruction of immunoglobulin G and A (Kalifa et al., 2011).

The *N. meningitides* strain is characteristics for its successful immune evasion strategies. Although vaccines have been developed against some serogroups, a universal meningococcal vaccine remains a challenge because of their frequent antigenic varying ability of the organism; it also mimics host structures. *N. meningitides* is established as a successful pathogen with the help of a number of virulence mechanism and includes the capsule layer, lipopolysaccharides (LPS), and a number of surface-expressed adhesive proteins.

13.4.5 Superantigens

Superantigens are extracellular protein toxins released by gram-positive bacteria in the blood that are pyrogenic, increase host susceptibility to endotoxic shock, suppress immunoglobulin production, and have mitogenic activity for specific T-cell subset (Curtis, 1996; Commons et al., 2008). This produces an extensive release of proinflammatory cytokines with massive immune activation. These superantigens are likely to be involved in the pathogenesis of various invasive infections. A total of 11 superantigens have been identified so far, including streptococcal pyrogenic exotoxin (*spe*) A, C, G, H, I, J, K, L, M, streptococcal mitogenic exotoxin (*sme*) Z, and streptococcal superantigen (*ssa*) (Commons et al., 2008). Except *speG*, *speJ*, and *smeZ*, all of the superantigens-encoding genes are associated with bacteriophages (Ferretti et al., 2001; Proft & Fraser, 2003; Proft et al., 2003). *S. aureus* strains secrete around 24 superantigens, which involved in antiphagocytic activity and establishment of toxic shock syndrome Type-1 (Spaulding et al., 2013).

13.5 Conclusion and future prospects

The pathogenesis and the disease virulence caused by the bacteria in bloodstream is still ambiguous. Though blood carries protective immune response machinery, the evolving preventive and resistance mechanisms among bacteria still prevails. Emergence of multidrug resistance and extreme drug resistance shows an alarming need for new treatment strategies. Recently, many researchers focus on the use of bacteriophage therapy with genetically engineered phages to target specific pathogenic bacterial species. In addition, the use of active principles from traditional medicines and anti-quorum sensing molecules against drug-resistant pathogens are recent initiatives in the field of pharmacology.

Despite the recent medical advances, the upcoming challenges in bloodstream infections include global increases in the older population; massive raise in transplantation surgeries; nosocomial infections; and multidrug and extreme drug resistance and persisters, which keep the blood-related infections an increasing trend. To overcome these issues, a proper epidemiological survey on disease prevalence, early diagnosis, and proper dosage of antibiotics needs to be practiced to prevent the global burden of antibiotic resistance and well-being of mankind.

References

Ambrosioni, J., M. Hernandez-Meneses, A. Téllez et al. 2017. The changing epidemiology of infective endocarditis in the twenty-first century. *Current Infectious Disease Reports* 19:21.

Beall, B., R. Facklam, and T. Thompson. 1996. Sequencing *emm*-specific PCR products for routine and accurate typing of group A streptococci. *Journal of Clinical Microbiology* 34:953–958.

Bhakdi, S., J. Tranum-Jensen, and A. Sziegoleit. 1985. Mechanism of membrane damage by streptolysin-O. *Infection and Immunity* 47:52–60.

Bhakdi, S., U. Weller, I. Walev, E. Martin, D. Jonas, and M. Palmer. 1993. A guide to the use of pore-forming toxins for controlled permeabilization of cell membranes. *Medical Microbiology and Immunology* 182:167–175.

Bisno, A. L. 1991. Group A streptococcal infections and acute rheumatic fever. *New England Journal of Medicine* 325:783–793.

Bisno, A. L., M. O. Brito, and C. M. Collins. 2003. Molecular basis of group A streptococcal virulence. *The Lancet Infectious Diseases* 3:191–200.

Bor, D. H., S. Woolhandler, R. Nardin, J. Brusch, and D. U. Himmelstein. 2013. Infective endocarditis in the US, 1998–2009: A nationwide study. *PLOS One* 8 (3):e60033.

Campagne, G., A. Schuchat, S. Djibo, A. Ousseini, L. Cisse, and J. P. Chippaux. 1999. Epidemiology of bacterial meningitis in Niamey, Niger, 1981–1996. *Bulletin of the World Health Organization* 77:499–508.

Carlsson, J. 1997. Bacterial metabolism in dental biofilms. *Advances in Dental Research* 11:75–80.

Chávez-Bueno, S. and G. H. McCracken. 2005. Bacterial meningitis in children. *Pediatric Clinics North America* 52 (3):795–810.

Christaki, E and E. J. Giamarellos-Bourboulis. 2014. The complex pathogenesis of bacteremia: From antimicrobial clearance mechanisms to the genetic background of the host. *Virulence* 5 (1):57–65.

Commons, R., S. Rogers, T. Gooding, M. Danchin, J. Carapetis, R. Robins-Browne, and N. Curtis. 2008. Superantigen genes in group A streptococcal isolates and their relationship with *emm* types. *Journal of Medical Microbiology* 57:1238–1246.

Cunningham, M.W. 2000. Pathogenesis of Group A Streptococcal infections. *Clinical Microbiology Reviews* 13:470–511.

Curtis, N. 1996. Invasive group A streptococcal infection. *Current Opinion Infectious Diseases* 9:191–202.

Datta, V., S. M. Myskowski, L. A. Kwinn, D. N. Chiem, N. Varki, R. G. Kansal, M. Kotb, and V. Nizet. 2005. Mutational analysis of the group A streptococcal operon encoding streptolysin S and its virulence role in invasive infection. *Molecular Microbiology* 56:681–695.

de Gans, J. and D. van de Beek. 2002. Dexamethasone in adults with bacterial meningitis. *New England Journal of Medicine* 347:1549–1556.

Del Pozo, J. L. and R. Patel. 2007. The challenge of treating biofilm-associated bacterial infections. *Clinical Pharmacology & Therapeutics* 82:204–209.

Duncan, J. L. and R. Schlegel. 1975. Effect of streptolysin O on erythrocyte membranes, liposomes, and lipid dispersions. A protein-cholesterol interaction. *The Journal of Cell Biology* 67:160–174.

Elias, S. and E. Banin. 2012. Multi-species biofilms: Living with friendly neighbours. *FEMS Microbiology Reviews* 36:990–1004.

Fernie-King, B. A., D. J. Seilly, A. Davies, and P. J. Lachmann. 2002. Streptococcal inhibitor of complement inhibits two additional components of the mucosal innate immune system: Secretory leukocyte proteinase inhibitor and lysozyme. *Infection and Immunity* 70:4908–4916.

Ferretti, J. J., W. M. McShan, D. Ajdic, D. J. Savic, G. Savic, K. Lyon, C. Primeaux, S. Sezate, A. N. Suvorov, and S. Kenton. 2001. Complete genome sequence of an M1 strain of *Streptococcus pyogenes*. *Proceeding of the National Academy Sciences* 98:4658–4663.

Fischbach, M. A. and J. L. Sonnenburg. 2011. Eating for two: How metabolism establishes interspecies interactions in the gut. *Cell Host & Microbe* 10:336–347.

Fontaine, M. C., J. J. Lee, and M. A. Kehoe. 2003. Combined contributions of streptolysin O and streptolysin S to virulence of serotype M5 *Streptococcus pyogenes* strain Manfredo. *Infection and Immunity* 71:3857–3865.

Friedman, N. D., K. S. Kaye, J. E. Stout, S. A. McGarry, S. L. Trivette, J. P. Briggs, W. Lamm, C. Clark, J. MacFarquhar, and A. L. Walton. 2002. Health care—Associated bloodstream infections in adults: A reason to change the accepted definition of community-acquired infections. *Annals of Internal Medicine* 137:791–797.

Gaieski, D. F., J. M. Edwards, M. J. Kallan, and B. G. Carr. 2013. Benchmarking the incidence and mortality of severe sepsis in the United States. *Critical Care Medicine* 41 (5):1167–1174.

Greenwood, B. M. 1987. The epidemiology of acute bacterial meningitis in tropical Africa. In *Bacterial Meningitis*, pp. 93–113. Edited by J. D. Williams and J. Burnie, London, UK: Academic Press.

Hackett, S. P. and D. L. Stevens. 1992. Streptococcal toxic shock syndrome: Synthesis of tumor necrosis factor and interleukin-1 by monocytes stimulated with pyrogenic exotoxin A and streptolysin O. *Journal of Infectious Diseases* 165:879–885.

Harder, J., L. Franchi, R. Munoz-Planillo, J. H. Park, T. Reimer, and G. Nunez. 2009. Activation of the Nlrp3 inflammasome by *Streptococcus pyogenes* requires streptolysin O and NF-β activation but proceeds independently of TLR signaling and P2 × 7 receptor. *The Journal Immunology* 183:5823–5829.

Hill, D. J., N. J. Griffiths, E. Borodina, and M. Virji. 2010. Cellular and molecular biology of *Neisseria Meningitidis* colonization and invasive disease. *Clinical Science (London, England: 1979)* 118 (9):547–564.

Iwashyna, T. J., C. R. Cooke, H. Wunsch, and J. M. Kahn. 2012. Population burden of long-term survivorship after severe sepsis in older Americans. *Journal of the American Geriatrics Society* 60 (6):1070–1077.

Ji, Y., L. McLandsborough, A. Kondagunta, and P. P. Cleary. 1996. C5a peptidase alters clearance and trafficking of group A streptococci by infected mice. *Infection Immunity* 64:503–510.

Khalifa, A. B. H., D. Moissenet, V. H. Thien, and M. Khedher. 2011. Virulence factors in Pseudomonas aeruginosa: Mechanisms and modes of regulation. *Annal de Biologie Clinique* 69 (4):393–403.

Klacsmann, P. G., B. H. Bulkley, and G. M. Hutchins. 1977. The changed spectrum of purulent pericarditis: An 86 year autopsy experience in 200 patients. *The American Journal Medicine* 63:666–673.

Kollef, M. H., M. D. Zilberberg, A. F. Shorr, L. Vo, J. Schein, S. T. Micek, and M. Kim. 2011. Epidemiology, microbiology and outcomes of healthcare-associated and community acquired bacteremia: A multicenter cohort study. *Journal of Infection* 62:130–135.

Laupland, K. B. and D. L. Church. 2014. Population-based epidemiology and microbiology of community-onset bloodstream infections. *Clinical Microbiology Reviews* 27 (4):647–664.

March, C., D. Moranta, V. Regueiro, E. Llobet, A. Tomás, J. Garmendia, and J. A. Bengoechea. 2011. *Klebsiella pneumoniae* outer membrane protein A is required to prevent the activation of airway epithelial cells. *Journal of Biological Chemistry* 286:9956–9967.

McCullers, J. A. 2013. Do specific virus-bacteria pairings drive clinical outcomes of pneumonia? *Clinical Microbiology and Infection* 19:113–118.

Moreillon, P. and Y. A. Que. 2004. Infective endocarditis. *The Lancet* 363:139–149.

Moses, A. E., M. R. Wessels, K. Zalcman, S. Alberti, S. Natanson-Yaron, T. Menes, and E. Hanski. 1997. Relative contributions of hyaluronic acid capsule and M protein to virulence in a mucoid strain of the group A streptococcus. *Infection Immunity* 65:64–71.

Munro, A. P. S., C. C. Blyth, A. J. Campbell, and A. C. Bowen. 2018. Infection characteristics and treatment of Staphylococcus aureus bacteraemia at a tertiary children's hospital. *BMC Infectious Diseases* 18:387.

Oliveira, A. L., M. Souza, V. M. H. Carvalho-Dias et al. 2007. Epidemiology of bacteremia and factors associated with multi-drug-resistant gram-negative bacteremia in hematopoietic stem cell transplant recipients. *Bone Marrow Transplantation* 39:775–781.

O'Seaghdha, M. and M. R. Wessels. 2013. Streptolysin O and its co-toxin NAD-glycohydrolase protect group A streptococcus from xenophagic killing. *PLoS Pathogens* 9:e1003394.

Proft, T. and J. D. Fraser. 2003. Bacterial superantigens. *Clinical & Experimental Immunology* 133:299–306.

Proft, T., P. D. Webb, V. Handley, and J. D. Fraser. 2003. Two novel superantigens found in both group A and group C streptococcus. *Infection and Immunity* 71:1361–1369.

Salomão, R., O. Rigato, A. C. Pignatari, M. A. Freudenberg, and C. Galanos. 1999. Bloodstream infections: Epidemiology, pathophysiology and therapeutic perspectives. *Infection* 27:1.

Sauleda, J. S., J. A. Barrabés, G. Permanyer-Miralda, and J. Soler-Soler. 1993. Purulent pericarditis: Review of a 20-year experience in a general hospital. *Journal of the American College of Cardiology* 22:1661–1665.

Schlapbach, L. J., L. Straney, J. Alexander, G. MacLaren, M. Festa, A. Schibler, A. Slater, and ANZICS Paediatric Study Group. 2015. Mortality related to invasive infections, sepsis, and septic shock in critically ill children in Australia and New Zealand, 2002–2013: A multicentre retrospective cohort study. *Lancet Infectious Diseases* 15:46–54.

Schlech, W. F., J. I. Ward, J. D. Band, A. Hightower, D. W Fraser, and C. V. Broome. 1985. Bacterial meningitis in the United States, 1978 through 1981. The national bacterial meningitis surveillance study. *JAMA* 253 (12):1749–54.

Siegman-Igra, Y., B. Fourer, R. Orni-Wasserlauf, Y. Golan, A. Noy, D. Schwartz, and M. Giladi. 2002. Reappraisal of community-acquired bacteremia: A proposal of a new classification for the spectrum of acquisition of bacteremia. *Clinical Infectious Diseases* 34:1431–1439.

Sierig, G., C. Cywes, M. R. Wessels, and C. D. Ashbaugh. 2003. Cytotoxic effects of streptolysin O and streptolysin S enhance the virulence of poorly encapsulated group A streptococci. *Infection Immunity* 71:446–455.

Singer, M., C. S Deutschman, C. W. Seymour et al. 2016. The third international consensus definitions for sepsis and septic shock (Sepsis-3). *JAMA* 315 (8):801–810.

Spaulding, A. R., W. Salgado-Pabón, P. L. Kohler, A. R. Horswill, D. Y. M. Leung, and P. M. Schlievert. 2013. Staphylococcal and streptococcal superantigen exotoxins. *Clinical Microbiology Reviews* 26 (3):422–447.

Stevens, D. L. 1999. The flesh-eating bacterium: What's next? *Journal of Infectious Disease* 179:S366–S374.

Sumitomo, T., M. Nakata, M. Higashino, Y. Jin, Y. Terao, Y. Fujinaga, and S. Kawabata. 2011. Streptolysin S contributes to group A streptococcal translocation across an epithelial barrier. *Journal of Biological Chemistry* 286:2750–2761.

Tabah, A., D. Koulenti, K. Laupland et al. 2012. Characteristics and determinants of outcome of hospital-acquired bloodstream infections in intensive care units: The EUROBACT International Cohort Study. *Intensive Care Medicine* 38:1930–1945.

Torio, C. M. and R. M. Andrews. 2013. National inpatient hospital costs: The most expensive conditions by payer. Statistical Brief #160. Healthcare Cost and Utilization Project (HCUP) Statistical Briefs.

Tumbarello, M., T. Spanu, R. D. Bidino et al. 2010. Costs of bloodstream infections caused by *Escherichia coli* and influence of extended-spectrum-β-lactamase production and inadequate initial antibiotic therapy. *Antimicrobial Agents and Chemotherapy* 54 (10):4085–4091.

Vallés, J., M. Palomar, F. Alvárez-Lerma, J. Rello, A. Blanco, J. Garnacho-Montero, and I. Martín-Loeches. 2013. GTEI/SEMICYUC Working Group on Bacteremia. Evolution over a 15-year period of clinical characteristics and outcomes of critically ill patients with community-acquired bacteremia. *Critical Care Medicine* 41:76–83.

Vincent, J. L., Y. Sakr, C. L. Sprung, V. M. Ranieri, K. Reinhart, H. Gerlach, R. Moreno, J. Carlet, J. R. Le Gall, and D. Payen. 2006. Sepsis in European intensive care units: Results of the SOAP study. *Critical Care Medicine* 34 (2):344–353.

Viscoli, C. 2016. Bloodstream infections: The peak of the iceberg. *Virulence* 7:248–51.

Weimer, K. E. D., R. A. Juneau, K. A. Murrah, B. Pang, C. E. Armbruster, S. H. Richardson, and Swords, W. E. 2011. Divergent mechanisms for passive pneumococcal resistance to β-lactam antibiotics in the presence of *Haemophilus influenzae*. *Journal of Infectious Diseases* 203:549–555.

Zhou, J., C. Qian, M. Zhao et al. 2014. Epidemiology and outcome of severe sepsis and septic shock in intensive care units in Mainland China. *PLoS One* 9 (9):e107181.

Index

Note: Page numbers in italic and bold refer to figures and tables respectively.

POCKET GUIDES TO
BIOMEDICAL SCIENCES

https://www.crcpress.com/Pocket-Guides-to-Biomedical-Sciences/bookseries/CRCPOCGUITOB

Series Editor
Lijuan Yuan
Virginia Polytechnic Institute and State University

A Guide to AIDS
Omar Bagasra and Donald Gene Pace

Tumors and Cancers: Brain—Central Nervous System
Dongyou Liu

Tumors and Cancers: Head—Neck—Heart—Lung—Gut
Dongyou Liu

A Guide to Bioethics
Emmanuel A. Kornyo

Tumors and Cancers: Skin—Soft Tissue—Bone—Urogenitals
Dongyou Liu

Tumors and Cancers: Endocrine Glands—Blood—Marrow—Lymph
Dongyou Liu

A Guide to Cancer: Origins and Revelations
Melford John

A Beginner's Guide to Using Open Access Data
SaifAldeen S. AlRyalat and Shaher M. Momani

Pocket Guide to Bacterial Infections
K. Balamurugan

Printed in Canada